P9-AQD-602

DATE DUE

Nerve–Muscle Interaction

Nerve–Muscle Interaction

Second edition

G. Vrbová
Department of Neuroscience, University College London, UK

T. Gordon
Department of Anatomy and Developmental Biology, University of Edmonton, Alberta, Canada

and

R. Jones
Multiple Sclerosis Research Centre, Bristol, UK

CHAPMAN & HALL *1995*

London · Glasgow · Weinheim · New York · Tokyo · Melbourne · Madras

**Published by Chapman & Hall, 2–6 Boundary Row, London
SE1 8HN, UK**

Chapman & Hall, 2–6 Boundary Row, London SE1 8HN, UK

Blackie Academic & Professional, Wester Cleddens Road, Bishopbriggs, Glasgow G64 2NZ, UK

Chapman & Hall GmbH, Pappelallee 3, 69469 Weinheim, Germany

Chapman & Hall USA, One Penn Plaza, 41st Floor, New York, NY 10119, USA

Chapman & Hall Japan, ITP-Japan, Kyowa Building 3F, 2–2–1 Hirakawacho, Chiyoda-ku, Tokyo 102, Japan

Chapman & Hall Australia, Thomas Nelson Australia, 102 Dodds Street, South Melbourne, Victoria 3205, Australia

Chapman & Hall India, R. Seshadri, 32 Second Main Road, CIT East, Madras 600 035, India

First edition 1978

Second edition 1995

© 1978, 1995 G. Vrbová, T. Gordon and R. Jones

Phototypeset in 10/12pt Palatino by Intype, London
Printed in Great Britain at the University Press, Cambridge

ISBN 0 412 40490 7

Contents

Preface to the second edition

The previous edition of the book was completed in December 1977. For the past few years we have felt the need to update the book, since in spite of rapid expansion of the subject no comparable comprehensive account of new developments has been published. Understandably the most novel information was obtained from studies of the components of the neuromuscular system, i.e. muscle cells and nerves, for here the new methods of molecular biology could be used most effectively. Thus the discovery of a particular family of helix-loop-helix proteins and their role in muscle development excited much interest. Although this topic is not directly related to nerve–muscle interactions the new edition of the book gives an account of these important results. Similarly, the use of more sophisticated electrophysiological techniques such as patch clamp and *in vitro* studies of neuronal tissues provided more precise information about the characteristic properties of nerve and muscle cells, as well as information about their control.

Observations on developing neuromuscular systems both *in vivo* and *in vitro* re-examined old questions about the importance of nature vs. nurture for the emergence of the adult neuromuscular system and its organization. Unfortunately, in spite of the fact that development is an event where timing is of utmost importance, this factor was not always considered and the static view of the system has led to some unlikely proposals discussed in various chapters of the book. The timing of events in relation to nerve–muscle interactions and the maturity of the motor nerve and muscle fibre during development when they first meet are of particular importance and this is stressed in the new edition.

Since the last edition, our views on the mechanisms that control muscle properties by its motor nerve have also been modified slightly in that, in addition to neuromuscular activity, other important environmental and inherent factors were seen to influence muscle fibre properties.

The response of the adult neurone to injury and the effect this has on the muscle and nerve cell is updated with detailed information on the molecular changes that are induced by these procedures. Similarly, events associated with recovery of function after nerve injury now include new information of events concerned with regeneration of motor nerves, and responses of the muscle to ingrowing new nerves.

The book reflects, we believe, the present state of knowledge in the field. We tried to be true to Pascal who said: 'I hold it equally impossible to know the parts without knowing the whole and to know the whole without knowing the parts.' However, the book does mirror the present state of our scientific knowledge and shows a much greater expansion of our understanding of the 'parts' than the 'whole'. In this sense it is regrettable that our understanding of the mechanisms that operate during nerve–muscle interactions is still far from complete.

University College
London

G.V.
T.G.
R.J.

Preface to first edition

In the second century, Galen recognized that nerve and muscle were functionally inseparable since contraction of muscle occurred only if the nerves supplying that muscle were intact. He therefore concluded that the shortening of a muscle was controlled by the central nervous system while the extension of a muscle could occur in the absence of innervation. Nerves, he thought, were the means of transport for animal spirits to the muscles; the way in which animal spirits may bring about contraction dominated the study of muscle physiology from that time until the historical discovery of Galvani that muscle could be stimulated electrically and that nerve and muscle were themselves a source of electrical energy.

It is now well known that nerves conduct electrically and that transmission from nerve to striated muscle is mediated by the chemical which is liberated from nerve terminals onto the muscle membrane. In vertebrates this chemical is acetylcholine (ACh). Thus the concept of spirits that are released from nerves and control muscle contraction directly, is no longer tenable. Nevertheless the concept of 'substances' transported down nerves which directly control many aspects of muscle has not been abandoned, and has in fact been frequently reinvoked to account for the long-term regulation of many characteristics of muscle (see review by Gutmann, 1976) and for the maintenance of its structural integrity.

There are many examples of dependence of skeletal muscles and motoneurones on each other. During embryonic development, while motoneurones and skeletal muscle fibres can, to some extent, develop independently, they become critically dependent on each other at a certain stage of their development, and disintegrate if they fail to make contact with each other. The dependence of the muscle on its innervation persists, for even in later life the motor nerve is essential for the maintenance of the structural and functional integrity of the muscle.

It is common clinical experience that following damage to the peripheral nerve many changes take place in the paralysed muscle, all of which can be reversed if the motor nerve reinnervates the muscle in time. This finding supported the idea that the nerve exerts a special 'trophic' influence on the muscle it supplies which is mediated by substances transported down the axon.

All normal activity of a muscle is usually initiated by the motor nerve and following damage of the nerve this activity ceases. It soon became apparent that many of the changes that follow nerve damage may be accounted for by the inactivity of the muscle. Ingenious experimental models were invented to investigate the mechanisms by which the motor nerve maintains the structural integrity of the muscle.

In recent years many new exciting approaches have contributed to a better understanding of the problem. Muscle activity can now be reduced more effectively by the use of compounds that paralyse skeletal muscles irreversibly, or prevent the passage of action potentials along the motor nerve. Moreover it is possible to induce muscle activity by electrical stimulation via chronically implanted electrodes. Experimental evidence obtained by using these techniques will be presented and reinterpreted in an attempt to elucidate the mechanisms that the motor nerve employs in maintaining the structural integrity of the muscle it supplies.

Innervated muscle fibres have the potential to develop a variety of different characteristics. During the development of an individual they differentiate and adjust, so as to match the functional requirements of the motoneurones that supply them. The ability of muscle to express particular facets of its genetic potential is essential for the development of the basic functional unit through which movement is accomplished: the motor unit. Experimental results that show how a muscle is induced to differentiate in a particular way will be described.

While the mechanism by which the nerve exerts its influence on the muscle it supplies has been extensively studied, the reverse – the way in which the muscle can influence the nerve – is poorly understood, although there is ample evidence to show that the motoneurone and its axon is much affected by the muscle. During development, when nerve-muscle contacts are first being established, the influence of the muscle on the nerve cell is probably most obvious, for, if by a certain stage of development, the motor nerve terminals fail to make contact with the muscle, the motoneurone dies. Also in later life, the muscle, which constitutes the most intimate environment for the nerve terminal, must have an important influence on the terminal and its neurone. It is therefore not surprising that a change in the muscle such as injury or partial denervation, has profound effects on the nerve terminals that remain within the muscle, as well as on their motoneurones. The possible nature of this influence of nerve on muscle will be examined in the following chapters. We shall also discuss the possibility that activity of the moto-

neurone and muscle influences the development of particular characteristics of the system.

University College,   G.V.
London   T.G.
December 1977   R.J.

Acknowledgements

This book is dedicated to Ernest Gutmann who contributed much to our understanding of long-term interactions between nerve and muscle. Not only his contribution to research, which is apparent from the many papers quoted here, but his attitudes to work, life and people are of continuing value to those of us interested in revealing nature's secrets. In an atmosphere in which research is becoming a more and more competitive activity, Ernest Gutmann's wisdom to put scientific achievements before recognition by peers is particularly important.

We should like to thank all those who had to live and put up with us during the past year while we were revising this book, especially our families and in particular the new generation (Mabon, Hannah, Jan, Matthew and Zoë) who have arrived since the first edition of this book. We could not explain to them adequately why we thought it to be important, and that is our first failure. As well as to our families, we are grateful to our colleagues, particularly Professor Dirk Pette and Dr Roberto Navarrete for their help and advice.

Permission to reproduce copyright material was kindly granted by Cambridge University Press, Macmillan Press, The American Physiological Society, Springer-Verlag, Birkhäuser Verlag (Basel), Schwabe & Co. (Basel), The Wistar Press and Pergamon Press. Permission to reproduce published material was kindly granted by Professor Hamburger, the late Professor Aitken, Dr Émonet-Dénand and Dr Michael Brown.

We are also indebted to Dr M. Scarli and Dr P. Anderson for their help, to Jim Dick for preparing some of the figures and to Debbie Bartram and Honor Neilson for helping with the preparation of the manuscript and communications between UK and Canada.

Early development of muscle

1

The striated muscle fibre is a highly specialized structure that forms the basic unit of skeletal muscles in vertebrates. During co-ordinated movement, muscle fibres are required to perform many different tasks. They adjust to these functional requirements by further differentiation and specialization. Muscle fibres are unique among mammalian cells in that they are multinucleated. This characteristic is a consequence of a unique ability possessed by myogenic cells, i.e. **fusion**. Myoblasts are derived from mesenchymal cells and fuse with each other to form multinucleated myotubes. How myoblasts become myotubes which differentiate further into muscle fibres is briefly discussed in this chapter.

1.1 EARLY STAGES OF MYOGENESIS

During early embryonic development a group of mesenchymal cells stop their rapid proliferation, withdraw from the cell cycle and are committed to becoming **myoblasts** (Tello, 1917, 1922). These mononucleated cells align themselves into rows, fusing with each other to form multi-nucleated cross-striated **myotubes**. Differentiation of these continues as their centrally placed nuclei migrate to the periphery and the synthesis of muscle-specific proteins increases (Yaffe, 1969; Fischman, 1972).

Although it is not entirely clear how multipotential mesenchymal cells become committed myogenic cells, several important observations have been made recently.

Only those cells that have the ability to synthesize proteins typical of the sarcomere, acetylcholine receptors (**AChR**) and other muscle-specific proteins can be considered to be myogenic (Königsberg, 1963; Holtzer and Sanger, 1972; Fambrough and Rash, 1971).

1.1.1 INVOLVEMENT OF HELIX-LOOP-HELIX PROTEINS IN COMMITMENT OF MESENCHYMAL CELLS TO MYOGENIC LINEAGE

A group of proteins that are members of a large family of helix-loop-helix (**HLH**) proteins such as MyoD, myogenin, myf–5 and MRF–4, when introduced to mesenchymal or non-myogenic cells can convert them into cells that are capable of forming myotubes and can initiate the expression of muscle-specific proteins such as myosin heavy chains. MyoD was the first protein to be identified (Davis *et al.*, 1987; Pinney *et al.*, 1988) by subtractive hybridization of cDNA libraries produced in azacydine-treated and in untreated fibroblasts. Subsequently other members of this family of proteins were identified. They appear at different times during development and indicate the myogenic nature of the developing groups of cells (Buckingham *et al.*, 1992). However, there is evidence to suggest that although MyoD family members are involved with muscle differentiation (commitment), they are not the only factors involved. It appears that in mouse myotomes contractile proteins are present before myogenin or MyoD are seen (Cusella-De-Angelis *et al.*, 1992). Moreover, in targeted disruption experiments where endogenous MyoD has been rendered inactive, another myogenic factor (myf-5) is up-regulated and normal muscle development is finally obtained (Rudnicki *et al.*, 1992). It is interesting that these proteins seem to be specific to skeletal muscle and are not found in other types of muscle cells. More detailed reviews on the role of myogenic factors in early muscle commitment and myogenesis are available (Weintraub *et al.*, 1991; Buckingham *et al.*, 1992).

The discovery of the role of helix-loop-helix molecules in myogenesis is an exciting new development. It poses the question as to how this family of proteins is regulated and what factors influence their appearance. Although little is known about regulation of these proteins, myogenic differentiation is influenced by growth factors, suggesting that they also affect the HLH proteins. Investigations of fibroblast growth factor (**FGF**) and epidermal growth factor (**EGF**) and their receptors in culture and *in vivo*, carried out mainly by Hauschka and colleagues, have shown that the binding capacity of EGF receptors declines during differentiation in a myogenic cell line (Lim and Hauschka, 1984). In contrast, receptor numbers and binding capacity are maintained in a differentiation-defective cell line. It has also been demonstrated that receptors for EGF and FGF are lost during differentiation in culture (Olwin and Hauschka, 1988, 1990) and *in vivo*. One proposed mechanism for their action was the suppression of myogenic factors like myogenin. In the latter case, it has been suggested that FGF suppresses myogenin through phosphorylation of its DNA-binding domain by protein kinase C (Li *et al.*, 1992; Konieczny, 1992).

It has also been shown that other growth factors are involved in myoblast differentiation. For example, TGF-ß suppresses differentiation in myoblasts (Ewinger-Hodges *et al.*, 1982) and its binding capacity declines during normal differentiation (Ewton *et al.*, 1988).

The observations described above have led to the proposal that growth factors are generally inhibitory to differentiation. An alternative explanation could be that events elicited by growth factors are necessary for the cells to prepare and achieve competence for terminal differentiation. This question has not been addressed so far, possibly because specific blockers of growth factor receptors are not yet available. Transgenic 'knock-out' experiments probably cannot provide the answers since these growth factors are expressed throughout development and in several tissues, so that elimination of a gene can lead to general rather than specific defects.

Not all growth factors have been found to suppress differentiation. Work by Florini and colleagues has shown that the insulin-like growth factor IGFI stimulates myogenic differentiation by inducing myogenin expression (Florini and Ewton, 1990; Florini *et al.*, 1991a–c). In addition, expression of IGFI and its receptor is induced during myoblast differentiation (Tollefsen *et al.*, 1989; Florini *et al.*, 1991a–c). It seems, therefore, that there is an intricate regulation of differentiation by different types of growth factor. These topics are reviewed in Florini *et al.* (1991a–c) and Olson (1992). Apart from signals from the external environment, the timing of the appearance of these HLH proteins in relation to the developmental state of the cell is probably of crucial importance. The appearance of regulatory molecules has to be appropriately timed.

1.1.2 FUSION OF MYOBLASTS

Skeletal myoblasts have two main characteristics: they eventually withdraw from the cell cycle and never again enter into another cell division. Instead they fuse with each other to form unique, multinucleated cells – myotubes, which later give rise to muscle fibres (Holtzer *et al.*, 1957; Holtzer, 1959; Königsberg, 1963). This early work on fusion has been reviewed by Wakelam (1985).

Fusion of one myoblast with another is an expression of the ability of these cells to recognize each other (Moscona, 1957; Yaffe and Feldman, 1965). While a particular myoblast will fuse with any other myoblast, even across species (i.e. rat myoblasts with chick myoblasts), they will not fuse with other cell types. They can therefore 'recognize' each other.

The presence of Ca^{2+} ions is essential for fusion to take place (Shainberg *et al.*, 1969; Fambrough and Rash, 1971; Patterson and Prives, 1973). It also appears that Ca^{2+} has to enter the cell and that the intracellular levels of Ca^{2+} have to increase for fusion to proceed (Pryzbylski *et al.*, 1989). Moreover, the requirement of prostaglandin E1 (**PGE1**) receptors in addition to the influx of Ca^{2+} is well documented (David and Higginbotham, 1981; Entwistle *et al.*, 1986). PGE receptors appear in myoblasts about 8 hours before fusion (Hausman and Velleman, 1981) and receptor antagonists inhibit fusion. It is possible that Ca^{2+} entry is associated with PGE receptor occupancy, for inhibition of PGE1 receptors also prevents Ca^{2+} influx (David and Higginbotham, 1981). David *et al.* (1990) have suggested that a possible link between PGE1 receptors and Ca^{2+} influx could be proteinkinase C. Activation of proteinkinase C with phorbol esters or prevention of dyacylglycerol breakdown leads to precocious fusion, while inhibition of proteinkinase C prevents Ca^{2+} entry and fusion. Zalin and her colleagues (Entwistle *et al*, 1988a, b; Farzaneh *et al.*, 1989) also argued that proteinkinase C activation mediates fusion and that this enzyme can be activated by prostanoid-generated Ca^{2+} fluxes (Farzaneh *et al.*, 1989; David *et al.*, 1990), by cyclic nucleotides (Zalin 1977, 1987) or by membrane depolarization under conditions where prostanoid synthesis is inhibited. Blocking of AChR delayed but did not entirely prevent fusion (Entwistle *et al.*, 1988a, b).

It appears that myoblasts have a limited set of options: they can divide to produce more myoblasts, or fuse with each other to generate myotubes. However, once committed they are unlikely to revert to any other cell type. Figure 1.1 illustrates some features of these early stages of myogenesis.

1.1.3 ACETYLCHOLINE RECEPTORS AND OTHER MEMBRANE CHARACTERISTICS

Simultaneously with the synthesis of other muscle-specific proteins, various subunits of the immature acetylcholine receptor molecules are synthesized, assembled and incorporated into the membrane (Fambrough and Rash, 1971; Sytkowski *et al.*, 1973; Dryden *et al.*, 1974).

These original observations were carried out either using physiological techniques such as following the response of cells to ACh, or by binding studies with radioactive α-bungarotoxin (Devreotes and Fambrough, 1975). Recently, with the discovery of the genes that code for the various AChR subunits, more detailed information about the time course and mechanisms that regulate the synthesis of AChR have become available. This more detailed analysis also disclosed that myoblasts and myotubes have a different AChR, which resembles that of extrajunctional receptors on mature muscle fibres and which has different channel properties (for review see Schütze and Role, 1987; Salpeter *et al.*, 1992).

Other membrane properties and ion channels have also been extensively studied in myoblasts and myotubes in tissue culture, and these can be taken to be characteristic of immature myogenic cells. It has been reported that the membrane potential of myotubes is relatively low and

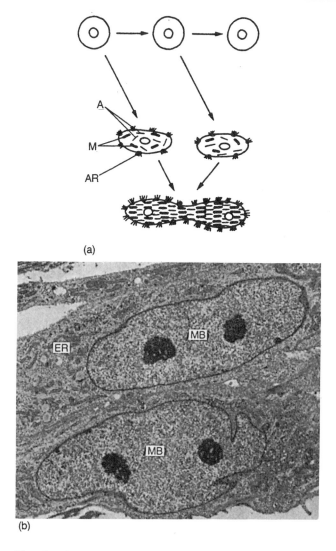

(a)

(b)

Figure 1.1 The development of skeletal muscle fibres: (a) presumptive myoblasts (top), myoblasts (centre) and a myotube after fusion of two myoblasts (bottom); (b) the ultrastructural appearance of two myoblasts (MB) about to fuse. In the myotube, actin (A) and myosin (M) molecules become aligned to form sarcomeres, and AChR clusters (AR) are scattered on the surface.

increases with time (Fischbach *et al.*, 1971; Dryden *et al.*, 1974). Very early on, myotubes are capable of producing active electrogenic responses and conducting them along the entire length of the fibre (Kano and Shimada, 1973; Purves and Vrbová, 1974). The time course of these responses is typically slow but increases with age. Consistent with this is the early appearance of Na$^+$ channels shown in a rat myoblast line (Kidokoro, 1975) and subsequently on myoblasts obtained from rat primary cultures (Gonoi *et al.*, 1985; Weiss and Horn, 1986a, b). However, these Na$^+$ channels have different affinities to tetrodotoxin (**TTX**) in that they are less sensitive to its blocking action than those of adult muscle fibres. Nevertheless, a few TTX-sensitive Na$^+$ channels have been found also on myoblasts (Gonoi *et al.*, 1985; Weiss and Horn, 1986a, b).

Different types of Ca^{2+} channels have been described in immature muscle cells. T-type Ca^{2+} channels are not present in adult muscle fibres, but immature muscle cells have both L and T types (Kano *et al.*, 1989; Gonoi and Hasegawa, 1988). The interesting feature of T-type Ca^{2+} channels is that they open at membrane potentials close to the resting potential, so that even small fluctuations of potential will allow Ca^{2+} entry. This may be important for the induction of a number of developmental changes that may depend on Ca^{2+} entry.

A variety of different types of K$^+$ channels were found in myotubes cultured from chicken breast muscles using patch clamp techniques. There were interesting differences between myoblasts and myotubes. Large conductance K$^+$ channels are rare in myoblasts, where low-conductance K$^+$ channels predominate. The converse applies to myotubes, where the high-conductance K$^+$ channels predominate. Myotubes do not seem to have any ATP-dependent K$^+$ channels. Arresting fusion does not prevent the shift of the low-conductance channels towards high-conductance ones seen during the transition from myoblasts to myotubes, but the ATP-dependent K$^+$ channels develop only if fusion is allowed to proceed (Zemková *et al.*, 1989). It could be that these ATP-dependent channels are more closely related to the contractile activity of the cell and may have a protective influence when energy depletion threatens the muscle fibre.

After fusion the proportion of contractile proteins increases rapidly (Allen and Pepe, 1965; Przybilski and Blumberg, 1966) and spontaneous contractions become more frequent (Yaffe, 1969). Gradually signs of the developing T-systems are seen as the surface membrane of the myotube starts to invaginate (Ezerman and Ishikawa, 1967). Exactly how these developmental events of the myoblast and myotube are orchestrated is not known but it is clear that muscle development can proceed to some extent without innervation, as suggested by early observations on amphibian larvae and chick embryos (Harrison, 1904; Hamburger, 1939).

Even in adults, development of new muscle fibres can be induced. If adult skeletal muscles are destroyed either mechanically (Studitskij, 1974) or pharmacologically using local anaesthetic, the existing satellite cells that are dormant in adult skeletal muscles become activated and myoblast-like. They are then able to recapitulate the developmental sequence of embryonic myoblasts and produce mature muscle fibres (Carlson, 1988).

1.2 EMERGENCE OF PHENOTYPIC DIVERSITY

1.2.1 TONIC AND PHASIC MUSCLE FIBRES

Skeletal muscles of adult vertebrates are composed of a variety of muscle fibres. These can differ from each other according to their membrane properties, contractile characteristics, pattern of innervation, sensitivity to ACh, types of sarcomeric proteins and other biochemical and physiological properties.

In lower vertebrates and birds two functionally distinct types of muscle fibre can be distinguished: **tonic fibres**, which produce a sustained contraction, and *phasic fibres*, which produce a transient contraction. The tonic fibres are usually innervated at several sites, whereas the phasic ones have only one or two nerve–muscle contacts. These two types of fibre differ from each in their biochemical composition, mechanical response (Figure 1.2a, b), passive cable properties of the membrane (Figure 1.2c–d), sensitivity to ACh (Figure 1.2a) and ultrastructural appearance.

In the chick wing, tonic fibres are segregated in the anterior latissimus dorsi (**ALD**) muscle, and phasic ones in the posterior latissimus dorsi (**PLD**). This segregation allows a comparison between the development of the two types of muscle fibre. Studies of the developmental changes show that the slow muscle develops earlier than the fast one, although the first signs of innervation are seen in both muscles at the same time (Gordon *et al.*, 1974). Oxidative enzymes, however, are much more developed in the tonic ALD than in the phasic PLD (Gordon *et al.*, 1977a). The difference in the rate of development of this enzyme is illustrated in Figure 1.3. The ultrastructural appearance of the two types of muscle fibre is also different: the ALD muscle fibres have a higher proportion of contractile proteins and fewer ribosomes than PLD muscle fibres (Gutmann *et al.*, 1969; Gordon *et al.*, 1975). In spite of these differences in the rate of development of these muscles, some of their characteristics at these early stages of development are similar and diverge later. It is particularly interesting that this is the case with the passive cable properties of their membrane, which later are completely different in the two types of muscle fibre, and with their contractile properties, which also differentiate later in development (Gordon *et al.*, 1977b). Thus the different functional characteristics of these two types of muscle are not yet developed early in embryonic life but differentiate with time.

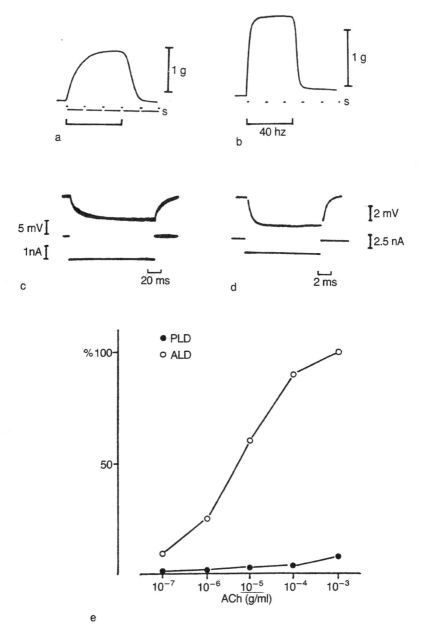

Figure 1.2 Some typical differences between tonic and phasic muscles. Contractions obtained from 2-day-old chick muscles: (a) tonic (ALD); (b) phasic (PLD). Records of electrotonic potentials evoked by (c) ALD muscle; (d) PLD muscle (note the different time scale used). (e) The responses to ALD and PLD muscle to applied ACh: the tension developed by the muscles is expressed as a percentage of maximal tetanic force and plotted against the dose of ACh applied.

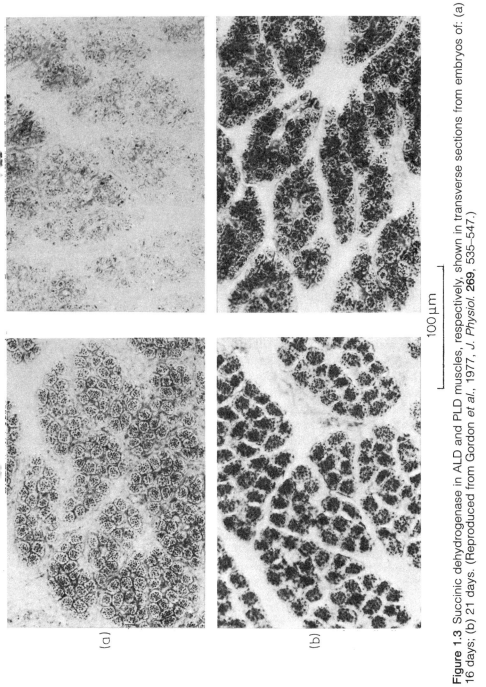

Figure 1.3 Succinic dehydrogenase in ALD and PLD muscles, respectively, shown in transverse sections from embryos of: (a) 16 days; (b) 21 days. (Reproduced from Gordon *et al.*, 1977, *J. Physiol.* **269**, 535–547.)

100 μm

1.2.2 DIFFERENTIATION OF MAMMALIAN MUSCLE FIBRES

The existence of different types of muscle fibre in adult mammalian muscles is well established and has been reviewed recently (Pette and Staron, 1990). The different types can be distinguished using several criteria:

- Histochemical evaluation of the activities of enzymes related to the energy metabolism of the muscle fibre.
- Histochemical identification of the myosin ATPase, and immunohistochemistry of myosin heavy chains (**MHC**).
- The analysis of single fibres and the precise identification of quantitative differences between their various intracellular components.

These criteria allow the distinction of various muscle fibre types based on differences of isozymic forms of myosin and other muscle-specific proteins. These approaches taken together reveal a much greater diversity of muscle fibre types than hitherto expected (Chapter 5).

The demonstration of greater heterogeneity of muscle fibres indicates that the concept of a motor unit composed of completely identical muscle fibre types is unlikely to be true. Indeed, several results have shown that not all muscle fibres belonging to the same motor unit are identical and that their size, as well as the precise activity of oxidative enzymes, varies among different fibres belonging to the same motor unit (Edström and Kugelberg, 1968; Albani *et al.*, 1988; Martin *et al.*, 1988). This is an important point because it illustrates that the motor nerve is not the only factor that determines muscle fibre properties and that the same axon can successfully interact with a varied population of muscle fibres.

Interestingly, not only in adults but also during early development before distinct motor units are formed, there is heterogeneity of muscle fibres. Its existence was first demonstrated by Butler *et al.* (1982) and by Phillips and Bennett (1984) for the embryonic chick-wing musculature. Their results showed that the appearance of muscle fibres containing different isoforms of MHC can proceed even in the absence of innervation. Similar results were obtained on developing rat muscles by Condon *et al.* (1990). These results have led to a number of proposals regarding the regulation of muscle fibre types during early development and the role of this early heterogeneity in the formation of distinct motor units.

The heterogeneity of developing muscles is based predominantly on observations of MHC isoforms in muscle fibres. Early in the development of the rat limb bud, a number of myotubes reacts with an antibody to the adult slow MHC, while other fibres react only to an antibody raised against embryonic MHC. Figure 1.4 shows that, while most fibres of EDL and tibialis anterior (**TA**) from an 18-day-old rat embryo react to antibody to fast myosin (Figure 1.4a), the response to antibodies to slow myosin HC is more selective (Figure 1.4b). Some of the fibres contain both types of myosin and the situation gets more complicated when, during further development, new muscle fibres are formed (Condon *et al.*, 1990; Sheard *et al.*, 1991; Dhoot, 1992). Many investigators imply that the muscle fibres which contain only slow MHC are the ones destined to become the adult slow fibres. Since the formation and distribution of these fibres is independent of innervation they suggest that muscle fibre types are predetermined and different types are derived from separate clones (Miller and Stockdale, 1987; Condon *et al.*, 1990; Miller, 1991). However, results of experiments where the myosin heavy chain isoforms were defined during very early stages of chick development, before distinct muscles are cleaved from the primordial myogenic tissue, show that at this stage two MHC isoforms were present in all cells. These were an MHC isoform almost identical to that found in the heart ventricular muscle, and a fast MHC isoform. Only later in development, when the myogenic cells migrated to their appropriate position, was the expression of the ventricular MHC downregulated and the 'appropriate' myosin HC started to appear (Sweeney *et al.*, 1989). Consistent with these obser- vations are recent results obtained on mouse embryos, where the tem- poral sequence of induction of different MHC isoforms in developing muscles was studied at the mRNA level (Lyons *et al.*, 1990). These studies show that during the initial stages of muscle development the sequence of expression of the myosin gene in mouse somites proceeds in a prepro- grammed fashion, in which the mRNAs coding for cardiac MHC are replaced by those coding for embryonic MHC, subsequently followed by mRNA for perinatal MHC. This sequence is apparently similar in all myogenic cells. Only later, in the mouse, at E15, can differences in the expression of myosin genes between developing muscle fibres be seen (Lyons *et al.*, 1990). From these results it appears that all myogenic cells undergo a similar preprogrammed developmental sequence and differentiate only after its completion.

Different isozymic forms of other components of the myofibril, such as troponin, have also been described. A different isozymic form of troponin T is expressed in rat muscle during embryonic postnatal devel- opment (Sabry and Dhoot, 1991) but the regulation of this protein has not been explored to the same extent as that of myosin.

In spite of the obvious variations in enzymes and protein isoforms of developing muscle fibres, the differentiation into muscle fibre types is difficult to evaluate, since the myogenic cells studied are at the different stages of development and, in addition to their age, many other factors such as myotube length, and therefore mechanical conditions, will vary. The role of these factors will be discussed later.

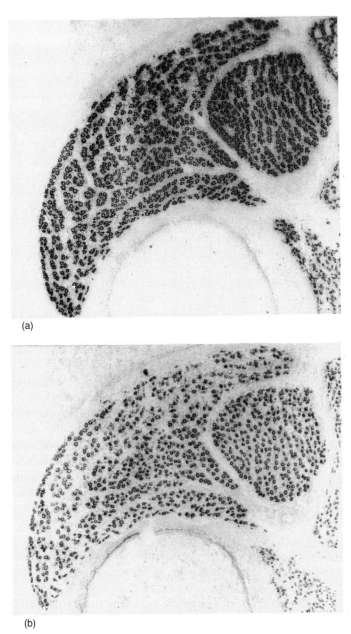

(a)

(b)

Figure 1.4 Cross-sections through a leg from an 18-day-old embryo reacted with antibodies to (a) slow myosin and (b) fast myosin. There is already some segregation into fast and slow muscle fibres. (Reproduced from Dhoot, 1992, *Histochem.* **97**, 479–486.)

The isozymic forms of sarcomeric proteins, i.e. myosin, actin and the troponin complex, are often thought to be associated with physiological contractile properties of the muscle fibre. This association was initially inspired by findings of Bárány (1967) and Bárány and Close (1971), where the speed of force development correlated with the ATPase activity of the myosin molecule. Since the myosin ATPase is associated with the MHC, the different isozymic forms of MHC may confer on to the muscle fibres a particular contractile speed.

In adults, there is a correlation between different sarcomeric proteins and contractile speeds, but such correlations have not been demonstrated in developing muscles. During early embryonic life chick muscles, no matter whether they will become slow- or fast-contracting, are all slow-contracting (Gordon et al., 1975). In kittens and rats, too, future slow and fast muscles are slow-contracting and relaxing (Buller et al., 1960; Close, 1964). In rats, the increase in contraction speed correlates better with the development of the cytosolic Ca^{2+} buffering protein parvalbumin (Leberer and Pette, 1986) than the changes of sarcomeric proteins. Denervation of neonatal muscles or paralysis of embryonic muscles reduces the rate at which the muscle speed increases, but does not altogether prevent this increase (Brown, 1973; Lewis, 1973), indicating that innervation independent processes can bring about some degree of maturation of the muscle cell. However, complete differentiation into slow and fast muscles does not occur in either denervated or inactivated muscles (Brown, 1973; Buller et al., 1960; Gutmann et al., 1974; Lewis, 1973).

1.2.3 MECHANISMS INVOLVED IN GENERATING PHENOTYPIC DIVERSITY

The observations of early diversification of muscle fibres, which was independent of innervation, led to the proposal that myoblasts are intrinsically different and that those populations originating from a particular clone give rise to a particular muscle fibre type. These ideas originated from *in vitro* experiments in which those myotubes derived from a single, cloned myoblast ancestor isolated from chick embryos at a certain stage of development gave rise to myotubes of a similar phenotype. Myotubes derived from ancestors isolated later in development differed from those derived earlier, suggesting a diversity of the original ancestral myoblasts (see Miller, 1991). These experiments were extended to show that it is possible to isolate three distinct 'colonies' of myotubes, each derived from a single myoblast lineage:

- One that expresses only the fast MHC isoforms (70%).
- A second colony that expresses both fast and slow MHC (30%).

- In very few instances (1–2%), a colony where each myotube expresses only slow MHC.

The proposal of the predetermined fate of muscle fibres was largely based on these *in vitro* findings (Miller and Stockdale, 1987). To what extent *in vitro* results are representative of events occurring during *in vivo* development is not certain; and, while these findings are interesting, muscle development in the whole animal may be different.

The fact that diversification of the isozymic forms of myosin in embryonic muscles occurs in the absence of innervation (Butler *et al.*, 1982) may be taken to support the idea that muscle fibre types are predetermined. Nevertheless this initial independence on neuronal activity does not necessarily mean that the development of early myogenic cells is preprogrammed. Mechanical factors related to the stretch imposed on particular sets of muscle fibres may influence their gene expression. Differences in the anatomical position of particular myotubes at a given developmental stage and their involvement in force production are likely to play a role in inducing particular isoforms of MHC. Thus myotubes that span from tendon to tendon would participate in movement in a mechanically different manner from those that are short and do not make a substantial contribution to force production. Indeed, studies on the lumbrical muscles have revealed that primary myotubes, which usually contain slow myosin, span from tendon to tendon, whereas secondary myotubes, which are formed later, are short (Duxson *et al.*, 1989). This finding, taken together with the observation that stretch can influence the development of developing muscles *in vitro* (Vandenburgh *et al.*, 1990), is just one of the possible epigenetic factors that may contribute to the development of a heterogeneous population of myotubes.

Consistent with such a proposal are results that show neonatal muscles to be influenced by mechanical conditions. In the soleus muscle of the rat the number of muscle fibres containing slow MHC increases after birth, as the animal is beginning to use the muscle for support. If the supporting function of this muscle is interfered with, the number of muscle fibres with slow MHC fails to increase (Lowrie *et al.*, 1989). Thus whatever the origin and phenotype of the muscle fibres, other factors have a strong influence on the expression of a particular phenotype.

The importance of environmental factors in the development of a particular phenotype is also clear from results of experiments on reinnervated muscles following nerve crush in the perinatal period. In the rat TA and EDL muscles reinnervated by their own nerve, the distribution of fibres containing slow MHC was completely different from that seen in normal animals, presumably because the few axons of 'slow' motor units colonized muscle fibres that were destined to become fast and converted these into slow fibres (Lowrie *et al.*, 1988). These results on reinnervated neonatal muscles differ from those of Soileau *et al.* (1987) who reported that in the rat soleus, after neonatal nerve injury, there is no rearrangement of muscle fibre types because the reinnervating axons return to the muscle fibres they originally supplied. The discrepancy between these two sets of results is due to the fact that Soileau *et al.* (1987) examined their muscles 2–3 weeks after the nerve injury, while Lowrie *et al.* (1988) studied the muscles at several time intervals following recovery from the nerve injury. Figure 1.5 illustrates that the rearrangement of fibre types was only completed after 6–8 weeks. Intervals of 2–3 weeks, as used by Soileau *et al.* (1987), are not long enough to transform original fast fibres into slow ones.

These results taken together show that:

- in neonatal animals, axons from slow motor units can make connections with muscle fibres that would otherwise be fast; and
- that these muscle fibres have the ability to express different types of myosin isoforms.

Thus muscle fibre type of developing animals can be regulated by external factors, which can be neuronal activity as well as mechanical conditions of the growing muscle cell.

CONCLUSIONS

Muscle fibres are derived from mesenchymal cells. These cells become committed to the myogenic lineage and begin to synthesize muscle-specific proteins. A particular family of nuclear proteins, i.e. helix-loop-helix (HLH) proteins, appears to play a key role in this commitment. Mononucleated myogenic cells (myoblasts) then fuse with each other to form multinucleated myotubes. The ability of myoblasts to fuse with each other is a unique property of these cells. Fusion is controlled and initated by several factors, all of which involve an increase of intracellular Ca^{2+} concentration. Thus Ca^{2+} is a key factor in cell fusion.

Membranes of myotubes contain Na^+, K^+ and Ca^{2+} channels, as well as ACh receptors, but the types of channel expressed and the form of the AChR change with development of the myotube. Myotubes also express a variety of isoforms of sarcomeric proteins and these seem to characterize myotubes of particular generation, so that there is in the muscle of embryos a heterogeneous population of muscle fibres. Whether this early heterogeneity is related to later differentiation into muscle fibre types is being discussed.

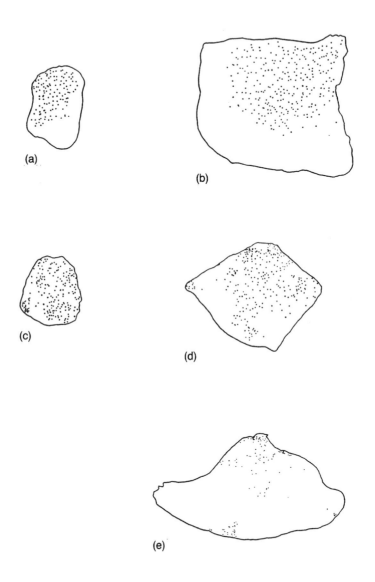

Figure 1.5 Camera lucida drawings of slow muscle fibres in fast muscles of different ages; each dot represents a slow muscle fibre: (a), (b) control TA and EDL muscles; (c), (d) the same muscles 30 days after a nerve crush at 6 days; (e) TA muscle 60 days after nerve crush at 6 days. (Based on results from Lowrie, Dhoot and Vrbová, 1988, *Muscle and Nerve*, **11**, 1043–1050.)

REFERENCES

Albani, M., Lowrie, M. B. and Vrbová, G. (1988) Reorganisation of motor units in reinnervated muscles of the rat. *J. Neurol. Sci.* **88**, 195–206.

Allen, R. R. and Pepe, F. A. (1965) Ultrastructure of developing muscle cells in the chick embryo. *Amer. J. Anat.* **116**, 115.

Bárány, M. (1967) ATPase activity of myosin correlated with speed of shortening. *J. Gen. Physiol.* (Suppl. 2) 197–218.

Bárány, M. and Close, R. I. (1971) The transformation of myosin in cross-innervated muscles. *J. Physiol.* **213**, 455–474.

Brown, M. D. (1973) The role of activity in the differentiation of slow and fast muscles. *Nature* **244**, 178–179.

Buckingham, M., Biben, Ch., Coloba, F. *et al.* (1992) Myogenesis in the mouse. The expression of regulatory and structural genes, in *Neuromuscular Development and Disease* (eds A. M. Kelly and H. M. Blau), Raven Press, NY, pp. 59–72.

Buller, A. J., Eccles, J. C. and Eccles, R. M. (1960) Differentiation of fast and slow muscles in the cat hind limb. *J. Physiol.* **150**, 399–416.

Butler, J., Cosmos, E. and Brierly, J. (1982) Differentiation of muscle fibre types in aneural brachial muscles of the chick embryo. *J. Zool.* **224**, 65–80.

Carlson, B. M. (1988) Nerve muscle inter-relationships in mammalian skeletal muscle regeneration. *Monogr. Dev. Biol.* **21**, 47–56.

Close, R. I. (1964) Dynamic properties of fast and slow skeletal muscles of the rat during development. *J. Physiol.* **173**, 74–95.

Condon, K., Silberstein, L., Blau, H. M. and Thompson, W. J. (1990) Differentiation of fibre types in aneural musculature of the prenatal rat hind limb. *Dev. Biol.* **138**, 275–295.

Cusella-De-Angelis, L., Vivarelli, E., Farmar, K. *et al.* (1992) MyoD Myogenin independent differentiation of primordial myoblasts in mouse somites. *J. Cell Biol.* **116**, 1243–1255.

David, J. D., Fraser, C. R. and Perrot, G. P. (1990) Role of protein kinase C in chick embryo skeletal myoblast fusion. *Dev. Biol.* **139**, 89–99.

David, J. D. and Higgenbotham, C. A. (1981) Fusion of chick embryo skeletal myoblasts. Interactions with PGE1 cAMP and calcium influx. *Dev. Biol.* **82**, 307–316.

Davis, R. I., Weintraub, H. and Lassar, A. B. (1987) Expression of a single transfected cDNA converts fibroblasts to myoblasts. *Cell* **51**, 987–1000.

Devreotes, P. N. and Fambrough, D. M. (1975) Acetylcholine receptor turnover in developing muscle fibres. *J. Cell Biol.* **65**, 335–358.

Dhoot, G. K. (1992) Neural regulation of differentiation of rat skeletal muscle fibre types. *Histochemistry* **97**, 479–486.

Dryden, W. F., Erulkar, S. D. and de la Habba, G. (1974) Properties of the cell membrane of developing skeletal muscle fibres in culture and its sensitivity to acetylcholine. *Clin. Exp. Pharm.* **1**, 369–387.

Duxson, M. J., Usson, Y., and Harris, A. J. (1989) The origin of secondary myotubes in mammalian skeletal muscles: Ultrastructural studies. *Development* **107**, 743–750.

Edström, L. and Kugelberg, E. (1968) Histochemical composition, distribution of fibres and fatiguability of single motor units. Anterior tibialis of the rat. *J. Neurol. Neurosurg. Psychiat.* **31**, 424–433.

Entwistle, A., Curtis, D. H. and Zalin, R. J. (1986) Myoblast fusion is regulated by a prostanoid of the first order independently of the rise in cyclic AMP. *J. Cell Biol.* **103**, 857–866.

Entwistle, A., Zalin, R. J., Bevan, S. and Warner, A. E. (1988a) The control of chick myoblast fusion by ion channels operated by prostaglandins and acetylcholine. *J. Cell Biol.* **106**, 1693–1702.

Entwistle, A., Zalin, R. J., Warner, A. E. and Bevan, S. (1988b) A role for acetylcholine receptors in the fusion of chick myoblasts. *J. Cell Biol.* **106**, 1703–1712.

Ewinger-Hodges, M. J., Ewton, D. Z., Seifert, S. C. and Florini, J. R. (1982) Inhibition of myoblast differentiation in vitro by a protein isolated from liver cell medium. *J. Cell Biol.* **93**, 395–401.

Ewton, D. Z., Spizz, G., Olson, E. N. and Florini, J. R. (1988) Decrease in transforming growth factor-ß binding and action during differentiation in muscle cells. *J. Biol. Chem.* **263**, 4029–4032.

Ezerman, E. B. and Ishikawa, H. (1967) Differentiation of the sarcoplasmic reticulum and T-system in developing chick skeletal muscles *in vitro. J. Cell Biol.* **35**, 405–420.

Fambrough, D. M. and Rash, J. E. (1971) Development of acetylcholine sensitivity during myogenesis. *Dev. Biol.* **26**, 55–68.

Farzaneh, F., Entwistle, A. and Zalin, R. J. (1989) Proteinkinase C mediates the hormonally regulated plasma membrane fusion of avian embryonic skeletal muscle. *Exp. Cell Res.* **18**, 298–304.

Fischbach, G. D., Nameroff, M. and Nelson, P. G. (1971) Electrical properties of chick skeletal muscle fibres developing in cell culture. *J. Cell Physiol.* **78**, 289–300.

Fischman, D. A. (1972) Development of striated muscle, in *The Structure and Function of Muscle*, 2nd edn, Vol. 1 (ed. G. H. Bourne), Academic Press, London, pp. 75–142.

Florini, J. R. and Ewton, D. Z. (1990) Highly specific inhibition of IGF-I-stimulated differentiation by an antisense oligonucleotide to myogenin mRNA. No effects on other actions of IGF-I. *J. Biol. Chem.* **265**, 13435–13437.

Florini, J. R., Ewton, D. Z. and Magri, K. A. (1991a) Hormones, growth factors and myogenic differentiation. *Ann. Rev. Physiol.* **53**, 201–216.

Florini, J. R., Ewton, D. Z. and Roof, S. L. (1991b) Insulin-like growth factor-I stimulates myogenic differentiation by induction of myogenin gene expression. *Mol. Endocrinol.* **5**, 718–724.

Florini, J. R., Magri, K. A., Ewton, D. Z. *et al.* (1991c) 'Spontaneous' differentiation of skeletal myoblasts is dependent on autocrine secretion of insulin-like growth factor II. *J. Biol. Chem.* **266**, 15917–15923.

Gonoi, T., Sherman, S. and Catterall, W. A. (1985) Voltage clamp analysis of tetrodotoxin sensitive and insensitive sodium channels in rat skeletal muscles developing *in vitro. J. Neurosci.* **5**, 2559–2564.

Gonoi, T. and Hasegawa, S. (1988) Post-natal disappearance of transient calcium channels in mouse skeletal muscle: effects of denervation and culture. *J. Physiol. (Lond.)* **401**, 617–637.

Gordon. T., Perry, R., Tuffery, A. R. and Vrbová, G. (1974) Possible mechanisms determining synapse formation in developing skeletal muscles of the chick. *Cell Tiss. Res.* **155**, 13–25.

Gordon, T., Perry, R., Srihari, T. and Vrbová, G. (1977a) Differentiation of slow and fast muscles in chickens. *Cell Tiss. Res.* **180**, 211–222.

Gordon, T., Purves, R. D. and Vrbová, G. (1977b) Differentiation of electrical and contractile properties of slow and fast muscle fibres. *J. Physiol.* **269**, 535–547.

Gordon, T., Perry, R., Spurway, N. C. and Vrbová, G. (1975) Development of slow and fast muscles in the chick embryo. *J. Physiol.* **254**, 24–25.

Gutmann, E., Hanzlíková, V. and Holečková, E. (1969) Development of fast and slow muscles of the chicken *in vivo* and their latent period in tissue culture. *Exp. Cell Res.* **56**, 33–38.

Gutmann, E., Melichna, J. and Syrový, I. (1974) Developmental changes in contraction time myosin properties and fibre patterns of fast and slow skeletal muscles. *Physiol. Bohemoslov.* **23**, 19–27.

Hamburger, V. (1939) Motor and sensory hyperplasia following limb bud transplantations in chick embryos. *Physiol. Zool.* **12**, 268–284.

Harrison, R. G. (1904) An experimental study of the relation of the nervous system to the developing musculature in the embryo. *Am. J. Anat.* **3**, 197–220.

Hausman, R. E. and Velleman, S. G. (1981) Prostaglandin E1 receptors on chick embryo myoblasts. *Biochem. Biophys. Res. Comm.* **103**, 213–218.

Holtzer, H. (1959) Some further uses of antibodies for analysing the structure and development of muscle. *Exp. Cell Res. Suppl.* **7**, 234–243.

Holtzer, H., Marshall, J. and Finck, H. (1957) An analysis of myogenesis by the use of fluorescent antimyosin. *J. Biophys. Biochem. Cytol.* **3**, 705–724.

Holtzer, H. and Sanger, J. W. (1972) Myogenesis: Old views rethought, in *Research in Muscle Development and the Muscle Spindle* (eds D. Barker, R. J. Przybylszki and J. P. van der Meulen), Excerpta Medica Int. Congress Series, pp. 122–134.

Kano, M. and Shimada, Y. (1973) Tetrodotoxin-resistant electric activity in chick skeletal muscle cells differentiated *in vitro*. *J. Cell Physiol.* **81**, 85–90.

Kano, M., Wakuta, K. and Satoh, R. (1989) Two components of calcium channel current in embryonic chick muscle cells developing in culture. *Dev. Brain Res.* **47**, 101–112.

Kidokoro, Y. (1975) Sodium and calcium components of the action potential in a developing muscle cell line. *J. Physiol. (Lond.)* **244**, 145–159.

Konieczny, S. F. (1992) Functional properties associated with MRF4 and other members of the bHLH muscle regulatory factor family, in *Neuromuscular Development and Disease* (eds A. M. Kelly and H. M. Blau), Raven Press, NY, pp. 29–45.

Köningsberg, I. R. (1963) Clonal analysis of myogenesis. *Science* **140**, 1273–1284.

Leberer, E. and Pette, D. (1986) Neural regulation of parvalbumin expression in mammalian skeletal muscles. *Biochem. J.* **235**, 67–73.

Lewis, D. M. (1973) The effect of denervation on the differentiation of twitch muscles in the kitten hind limb. *Nature* **241**, 285–286.

Li, L., Zhou, J., James, G. et al. (1992) FGF inactivates myogenic helix-loop-helix proteins through phosphorylation of a conserved proteinkinase C site in their DNA binding domains. *Cell* **71**, 1181–1194.

Lim, R. W. and Hauschka, S. D. (1984) A rapid decrease in epidermal growth factor-binding capacity accompanies the terminal differentiation of mouse myoblasts in vitro. *J. Cell Biol.* **98**, 739–747.

Lowrie, M. B., Dhoot, G. K. and Vrbová, G. (1988) The distribution of slow myosin in rat muscles after neonatal nerve crush. *Muscle and Nerve* **11**, 1043–1050.

Lowrie, M. B., Moore, A. F. K. and Vrbová, G. (1989) The effect of load on the phenotype of the developing soleus muscle. *Pflügers Archiv. Euro. J. Physiol.* **415**, 204–208.

Lyons, G. E., Ontell, M., Cox, R. *et al.* (1990) The expression of myosin genes in developing skeletal muscle in the mouse embryo. *J. Cell Biol.* **111**, 1465–1476.

Martin, T. P., Bodine-Fowler, S. C., Roy, R. R. *et al.* (1988) Metabolic and fibre size properties of the rat tibialis anterior motor units. *Am. J. Physiol.* **255**, 43–50.

Miller, J. B. (1991) Myoblasts, myosins, myoDs and the diversification of muscle fibres. *Neuromuscular Disorders* **1**, 7–17.

Miller, J. B. and Stockdale, F. E. (1987) What muscle cells know that nerves don't tell them. *Trends Neurosci.* **10**, 325–329.

Moscona, A. A. (1957) The development *in vitro* of chimeric aggregates of dissociated embryonic chick muscle cells. *Proc. Nat. Acad. Sci. USA* **43**, 184–194.

Olson, E. N. (1992) Interplay between proliferation and differentiation within the myogenic lineage. *Dev. Biol.* **154**, 261–272.

Olwin, B. B. and Hauschka, S. D. (1988) Cell surface fibroblast growth factor and epidermal growth factor receptors are permanently lost during skeletal muscle terminal differentiation in culture. *J. Cell Biol.* **107**, 761–769.

Olwin, B. B. and Hauschka, S. D. (1990) Fibroblast growth factor levels decrease during chick embryogenesis. *J. Cell Biol.* **110**, 503–509.

Patterson, B. and Prives, J. (1973) Appearance of acetylcholine receptors in differentiating cultures of chick breast muscle. *J. Cell Biol.* **50**, 241–245.

Pette, D. and Staron (1990) Cellular and molecular diversities of mammalian skeletal muscle fibres. *Rev. Physiol. Biochem. Pharm.* **116**, 1–76.

Pinney, D. F., Pearson-White, S. H., Konieczny, S. F. *et al.* (1988) Myogenic lineage determination and differentiation: evidence for a regulatory gene pathway. *Cell* **53**, 781–793.

Phillips, W. D. and Bennett, M. R. (1984) Differentiation of fibre types in wing muscles during embryonic development: effect of neural tube removal. *Dev. Biol.* **106**, 457–468.

Przybylski, R. J. and Blumberg, J. M. (1966) Ultrastructural aspects of myogenesis in the chick. *Lab. Invest.* **15**, 836–863.

Przybylski, R. J., MacBride, R. G. and Kirby, A. C. (1989) Calcium regulation of cell myogenesis. 1. Cell content critical to myotube formation *in vitro*. *Cell Dev. Biol.* **25**, 830–838.

Purves, R. D. and Vrbová, G. (1974) Some characteristics of myotubes cultured from slow and fast chick muscles. *J. Cell Physiol.* **84**, 94–100.

Rudnicki, M. A., Braun, T., Hinuma, S. and Jaenisch, R. (1992) Inactivation of MyoD in mice leads to up-regulation of the myogenic HLH gene Myf-5 and results in apparently normal muscle development. *Cell* **71**, 383–390.

Sabry, M. A. and Dhoot, G. K. (1991) Identification and patterns of transitions of cardiac, adult slow and slow skeletal muscle-like embryonic isoforms of troponin T in developing rat and human skeletal muscles. *J. Muscle Res. Cell Mot.* **12**, 262–270.

Salpeter, M. M., Buonenno, A., Eftimi, R. *et al.* (1992) Regulation of molecules at the neuromuscular junction, in *Neuromuscular Development and Disease* (eds A. M. Kelly and H. M. Blau), Raven Press, NY, pp. 257–283.

Schütze, S. M. and Role, L. W. (1987) Developmental regulation of nicotinic acetylcholine receptors. *Ann. Rev. Neurosci.* **10**, 403–457.

Shainberg, A., Yagil, C. and Yaffe, D. (1969) Control of myogenesis in vitro by Ca^{2+} concentration in nutritional medium. *Exp. Cell Res.* **58**, 163–167.

Sheard, P. W., Duxson, M. J. and Harris, J. (1991) Neuromuscular transmission to identified primary and secondary myotubes: a re-evaluation of polyneuronal innervation patterns in rat embryos. *Dev. Biol.* **148**, 459–472.

Soileau, L. C., Silberstein, L., Blau, H. M. and Thompson, J. W. (1987) Reinnervation of muscle fibre types in newborn rat soleus. *J. Neurosci.* **7**, 4176–4194.

Studitskij, A. N. (1974) The neural factor in the development of transplanted muscles, in *Exploratory Concepts in Muscular Dystrophy* II, Elsevier, NY, pp. 351–366.

Sweeney, L. J., Kennedy, J. M., Zak, R. *et al.* (1989) Evidence for the expression of a common myosin heavy chain phenotype in future fast and slow muscle during initial stages of avian myogenesis. *Dev. Biol.* **133**, 367–374.

Sytkowski, A. J., Vogel, Z. and Nirenberg, M. W. (1973) Development of acetylcholine receptor clusters on cultured muscle cells. *Proc. Nat. Acad. Sci. USA* **70**, 270–276.

Tello, J. F. (1917) Genesis de las terminaciones nerviosas mortices y sensitas I En el sistema locomotor de los vertebrados superiores histogenesis muscular. *Trabos Lab. Invest. Biol. Univ. Madrid* **15**, 101–199.

Tello, J. F. (1922) Die Entstehung der motorischen und sensiblen Nervenendigungen in dem lokomotorischen System der höheren Wirbeltiere. *Z. Anat. Entw. Gesch. Organ.* **64**, 248–440.

Tollefsen, S. E., Sadow, J. L. and Rotwein, P. (1989) Coordinate expression of insulin-like growth factor II and its receptor during muscle differentiation. *Proc. Nat. Acad. Sci. USA* **86**, 1543–1547.

Vandenburgh, H. H., Hatfaludyi, S., Karlisch, P. and Shansky, J. (1990) Mechanically induced alterations in cultured skeletal myotube growth, in *The Dynamic State of Muscle Fibres* (ed. D. Pette), De Guyter, Berlin, pp. 151–164.

Wakelam, M., (1985) The fusion of myoblasts. *Biochem. J.* **228**, 1–122.

Weintraub, H., Davis, R., Tapscott, S. *et al.* (1991) The MyoD gene family: A nodal point during specification of the muscle cell lineage. *Science* **251**, 761–766.

Weiss, R. E. and Horn, R. (1986a) Functional differences between two classes of sodium channels in developing myoblasts and myotubes of rat skeletal muscle. *Science* **233**, 361–364.

Weiss, R. E. and Horn, R. (1986b) Single channel studies of TTX-sensitive and TTX-resistant sodium channels in developing rat muscle reveal different open channel properties. *Ann. NY Acad. Sci.* **479**, 152–161.

Yaffe, D. (1969) Cellular aspects of differentiation in vitro. *Curr. Topics Dev. Biol.* **4**, 37–75.

Yaffe, D. and Feldman, M. (1965) The formation of hybrid multinucleated muscle fibres from myoblasts of different genetic origin. *Dev. Biol.* **11**, 300–317.

Zalin, R. J. (1977) Prostaglandins and myoblast fusion. *Dev. Biol.* **59**, 241–248.

Zalin, R. J. (1987) The role of hormones and prostanoids in the in vitro proliferation and differentiation of human myoblasts. *Exp. Cell Res.* **172**, 265–281.

Zemková, H., Vyskočil, F., Tolar, M. *et al.* (1989) Single channel potassium currents during differentiation of embryonic muscle in vitro. *Biochemica et Biophysica Acta* **986**, 146–150.

Development of motoneurones

2

2.1 LINEAGE AND EXPRESSION OF THE CHOLINERGIC PHENOTYPE

Motoneurones are derived from neuroepithelial cell precursors in the ventricular zone of the neural tube (Langman and Hadden, 1970; for reviews see Bennett, 1983). After the final mitosis, neuroblasts begin to migrate from the ventricular zone towards their final location in the motor columns of the ventral horn. This migration is believed to occur along pathways provided by radial glial cells (Hendrikson and Vaughn, 1974; Leber *et al.*, 1990). These initial stages of motoneurone development are independent of influences from their target muscles, since both proliferation and subsequent migration are unaffected by early limb bud removal (Hamburger, 1958; Hughes and Tschumi, 1958).

There is little information as to the timing of generation of motoneurone pools. Early studies indicate that those motoneurones that are born first will settle medially, whereas those born later will migrate more laterally (Holliday and Hamburger, 1977). Whether this is related to the emergence of distinct motoneurone pools is not clear.

It has been proposed that motoneurones are clonally related to the muscle fibres they innervate (Moody and Jacobson, 1983). However, results from recent studies on chick embryos with retroviral labelling of motoneurone progenitors do not support this proposal, since multicellular clones derived from single labelled neuroblasts contained different types of neurones, motoneurones and interneurones as well as glial and ependymal cells (Leber *et al.*, 1990). The latter results indicate that motoneurone precursors are multipotential and that epigenetic factors may be involved in determining their final phenotype. Moreover, the formation of distinct motoneurone pools does not appear to be lineage-dependent since clonally related motoneurones were not restricted to a single motor pool (Leber *et al.*, 1990). Similar results were obtained by

Vogel *et al.* (1988) using a chimeric analysis for studying motoneurone lineage in the mouse.

One of the first events that distinguishes motoneurones from other spinal cord neurones is the appearance of neurotransmitter-related enzymes, choline acetyltransferase (**ChAT**) and acetylcholinesterase (**AChE**). These enzymes are present in motoneurones soon after their final mitosis, at the time when their axons have not yet reached their target muscles (Phelps *et al.*, 1984). There are indications that, once cholinergic characteristics are established, interaction with the target muscle is necessary in order to increase and maintain a high level of expression of the enzymes involved in ACh synthesis (Giacobini-Robecchi *et al.*, 1975; Betz *et al.*, 1980; reviewed by Vaca, 1988). During embryonic and early postnatal development the level of ChAT increases (Burt, 1975; O'Brien and Vrbová, 1978; Pilar *et al.*, 1981; Phelps *et al.*, 1984). Electrical activity or membrane depolarization seems to be important for the up-regulation of ChAT since tetrodotoxin (TTX) blockade of action potentials in spinal cord cultures leads to a reduction or delayed development of ChAT activity (Brenneman *et al.*, 1983). Conversely, chronic membrane depolarization induced by high K^+ or activation of NMDA receptors causes an increase in ChAT activity (Ishida and Deguchi, 1983; Brenneman *et al.*, 1990).

2.2 DEVELOPMENT OF ELECTRICAL EXCITABILITY

The number, type and topographical distribution of ion channels and neurotransmitter receptors in the motoneurone membrane undergo developmental changes. These will influence the excitability and firing pattern of the motoneurone. Most of the information available on the development of ion channels and receptor molecules is derived from studies of acutely dissociated motoneurones from chick embryos which continued their development and were examined in tissue culture (O'Brien and Fischbach, 1986a; Fruns *et al.*, 1987; Kreiger and Sears, 1988; McCobb *et al.*, 1989, 1990). Whether results obtained *in vitro* represent an accurate picture of developmental events occurring *in vivo* remains to be seen. The recent methods of recording from identified motoneurones in spinal cord slices (Takahashi, 1990; Manabe *et al.*, 1991) or isolated spinal cords of the chick embryo may be more appropriate.

In motoneurones from acutely dissociated E4 chick embryos, i.e. shortly after axons reach the undifferentiated limb bud, the somatodendritic action potential was generated by inward currents carried by both Na^+ and Ca^{2+} ions (McCobb *et al.*, 1989; 1990). With development, the magnitude of the Na^+ currents increased and, as a consequence of this, the amplitude, overshoot and rate of rise of the action potential increased. Thus, Na^+ channels are present in the soma and axon of the motoneurone from the time of onset of electrical excitability and their numbers increase with age (Ziskind-Conhaim, 1988; McCobb et al., 1990).

Several Ca^{2+} currents with different voltage-dependence, kinetics and pharmacological sensitivity have been described in adult mammalian neurones (Llinás, 1988). With development, the contribution of these various calcium conductances to the action potential changes. Figure 2.1 illustrates this. Motoneurones from E4 chick embryos have predominantly low-threshold (T-type) calcium currents and these decrease during subsequent development in vitro. In contrast, the high-threshold calcium currents (L and N type) increase in magnitude after E4 and reach peak values at E11 (McCobb et al., 1989).

Rat neonatal motoneurones have both high- and low-threshold Ca^{2+} conductances (Harada and Takahashi; 1983; Walton and Fulton, 1986; Berger and Takahashi, 1990). The high-threshold conductance is responsible for the after-depolarization (**ADP**) seen during the falling phase of the action potential (Walton and Fulton, 1986). Low-threshold calcium conductances are particularly prominent during the first few days after birth and are associated with spontaneous oscillations of the membrane potential.

The generation of high-frequency 'doublet' or burst firing observed in neonatal rat motoneurones in vitro may also be accounted for by the presence of the low-threshold calcium conductance (Navarrete and Walton, 1989). The high-frequency 'doublet' or burst firing in young motoneurones is consistent with the predominance of phasic EMG activity pattern observed in neonatal muscles in vivo (Navarrete and Vrbová, 1983) and may be relevant to the functional matching of motoneurone firing patterns to muscle contractile properties in immature animals.

A number of voltage-dependent and neurotransmitter-gated K^+ currents have been described in adult motoneurones. The wide variety of K^+ channels in excitable cells provides for a rich control of excitability and firing patterns (Schwindt and Crill, 1984; Rudy, 1988; Kiehn, 1991). Developmental changes of these outward K^+ currents are likely to contribute to the motoneurone firing patterns. The classical voltage-dependent K^+ current (I_K or delayed rectifier), which is mainly responsible for the repolarization of the action potential, is detectable at relatively high levels in motoneurones obtained from very young (E4) chick embryos and undergoes a moderate increase with development (McCobb et al., 1990). In contrast, the density of transient (I_A) K^+ current is very low at E4 and increases markedly during development (McCobb et al., 1990), accounting for the shortening in the duration of the action potential with age. Figure 2.1 summarizes this and allows the comparison of changes in K^+ and Ca^{2+} conductances with age.

In neonatal rat motoneurones, the I_K and I_A types of K^+ currents are

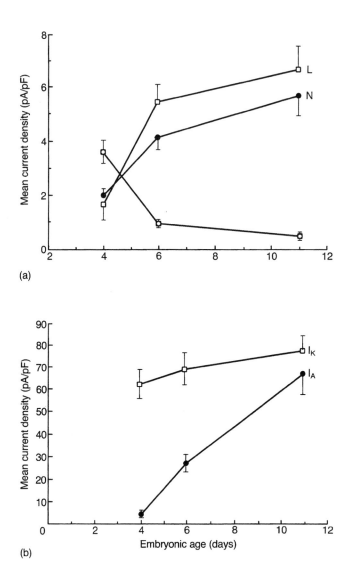

(a)

(b)

Embryonic age (days)

Figure 2.1 Developmental changes of Ca²⁺ and K⁺ currents in embryonic chick motoneurones plotted against age: mean values (SEM) of current densities (pA/pF) measured in motoneurones after 1 day in culture isolated at E4, 6 and 11. (a) Ca²⁺ currents through T, N and L type channels; (b) sustained K⁺ currents (I_K) and transient K⁺ currents (I_A). (Reproduced from McCobb *et al.*, 1989, *Neuron*, **2**, and 1990, *J. Neurosci.*, **10**, with permission from the authors, Cell Press and *Journal of Neuroscience*.)

present. In addition, Ca^{2+} dependent K^+ currents are seen (Takahashi, 1990) and these are important for the magnitude and duration of the after-hyperpolarization (**AHP**) of the action potential, an electrophysiological parameter that in adults distinguishes motoneurones to mammalian slow and fast muscles (Takahashi, 1990; Manabe *et al.*, 1991).

Synaptic activation of motoneurones is mediated by neurotransmitter-gated ion channels located in their somatodendritic membrane. There is little information as to the time course of development of the membrane chemosensitivity to the various neurotransmitters released by afferent inputs to motoneurones. Best studied are the responses to excitatory amino acids, which are known to be released by primary afferent terminals, and interneurones (Ziskin-Conhaim, 1990). The acquisition of chemosensitivity to excitatory amino acid neurotransmitters was studied in tissue cultures composed either exclusively of motoneurones or in mixed spinal cord cell populations from E6 chick embryos (O'Brien and Fischbach, 1986a, b, c). The sensitivity to glutamate was apparently modulated by the presence of spinal interneurones since it was greater in cultures that contained several cell types than in those composed exclusively of motoneurones (O'Brien and Fischbach, 1986d). The sensitivity to glutamate was not homogeneous along the surface of individual neurones where areas of high sensitivity (hot spots) alternated with areas of low sensitivity along dendritic processes. This arrangement is analogous to the distribution of AChR on adult skeletal muscle fibres, where the density of AChR at the synaptic site is much higher than outside (Chapter 3).

2.3 DEVELOPMENT OF FUNCTIONAL SPECIALIZATION

In adults, motoneurones are differentiated with respect to their electrophysiological properties and recruitment order (see Kernell, 1990; and Chapter 5). Motoneurones to slow muscles have a much longer AHP and are recruited more readily (Eccles *et al.*, 1958). Studies of electrophysiological properties of unidentified motoneurones in foetal and newborn kittens (Naka, 1964; Kellerth *et al.*, 1971) and neonatal rats (Fulton and Walton, 1986) indicate that their intrinsic properties undergo changes with age and differentiate into functional sub-classes. Unlike in adults, the duration of AHP in young kittens was short in motoneurones to both the 'slow' and 'fast' muscles (Huizar *et al.*, 1975). With further development, the duration of AHP increased in motoneurones to slow muscles while it remained short in motoneurones to fast muscles.

More recently, the properties of functionally identified motoneurones of the developing rat were studied. Analysis of the input resistance and action potential characteristics indicated that even during early development 'fast' flexor motoneurones (TA/EDL) differed from predominantly 'slow' extensor motoneurones (SOL/LG) (Navarrete *et al.*, 1988). These results therefore suggest that, in rats, phenotypic differences between motoneurones to slow and fast muscles may already be present at birth.

The control of the motoneurone firing pattern depends not only on its intrinsic electrophysiological properties but also on the type and quality of the synaptic input that activates it. There is evidence for continued reorganization of synaptic inputs to motoneurones during the first few weeks of postnatal development in mammals (Conradi and Ronnevi, 1975; Ronnevi, 1979; Gilbert and Stelzner, 1979; Stelzner, 1982). In particular, brainstem descending noradrenergic and serotonergic pathways undergo developmental changes in the distribution and transmitter synthesis of their terminals. This may be important for the modulation of motoneurone excitability (Commissiong, 1983; Rajofetra *et al.*, 1989). The period of maturation of these noradrenergic and serotonergic inputs in the lumbosacral cord corresponds to the onset of antigravity functions and locomotor activity of the hindlimbs when postural tonic EMG activity can first be recorded from the slow extensor soleus muscle (Navarrete and Vrbová, 1983). Thus, changes in synaptic inputs to motoneurones have to be considered in order to account for changes in motoneurone firing patterns during early development. It may indeed be that changes in afferent synaptic activity associated with the maturation of descending and interneuronal inputs to motoneurones may regulate the functional specialization of the motoneurone subpopulation.

2.4 TARGET DEPENDENCE OF MOTONEURONES

Early studies on the embryonic interactions between motoneurones and their target muscles suggested that a quantitative relationship exists between the number of motoneurones in the spinal cord and the amount of muscle tissue available for innervation (Shorey, 1909; Hamburger, 1958). Hamburger (1934) removed a limb bud in 2-day-old chick embryos and subsequently found that the size of the anterior horn on the operated side was much reduced (Figure 2.2). This reduction was due mainly to a decrease in the number of motoneurones in the lateral motor column on the side of the missing limb bud. This result was the first to suggest that there is a retrograde influence of the target on embryonic motoneurones. It was originally thought that the proportion of surviving motoneurones roughly corresponded to the amount of target available (Hamburger, 1977; Holliday and Hamburger, 1976) but not all studies agree with this conclusion. A substantial number of motoneurones still die even in what appear to be favourable conditions of target availability, while it is possible for an excess number of motoneurones to be supported by relatively little muscle tissue (for review, see Oppenheim, 1991). Nevertheless, in spite of these quantitative discrepancies there is

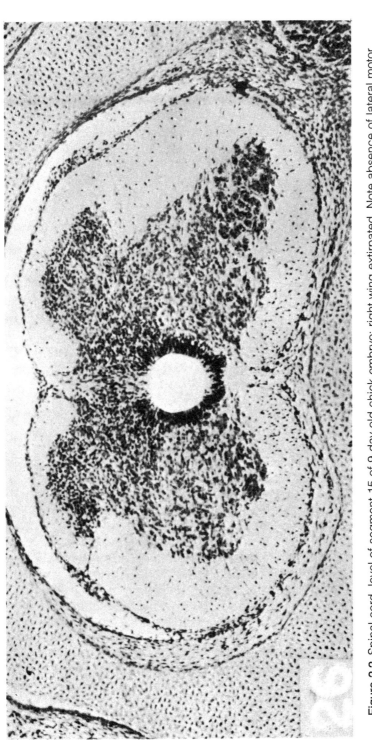

Figure 2.2 Spinal cord, level of segment 15 of 9-day-old chick embryo; right wing extirpated. Note absence of lateral motor column on right (apparent left) side. (Reproduced from Hamburger, 1958, *Am. J. Anat.* **102**, 365.)

substantial evidence to show that some kind of interaction with the target is important for regulating motoneurone survival.

It has been suggested that motoneurone survival is dependent on a trophic factor released from the muscle. Some support for the presence of trophic factors in muscle have come from studies in which muscle extracts in various stages of purification have been shown to improve the survival of embryonic motoneurones both *in vivo* and *in vitro* (Oppenheim, 1991). The idea of a specific substance present in the target that regulates the number of motoneurones which survive is conceptually simple and therefore attractive. However, the mechanism of action of such a substance on motoneurones is not yet known and may well be different from that of nerve growth factor (**NGF**) on sensory and sympathetic neurones (Purves and Lichtmann, 1985).

An alternative to the protective effects of target-derived neurotrophic molecules is the possibility that, as a result of its interaction with the target, the phenotype of the developing motoneurone changes and it becomes less target dependent after having undergone this change.

One part of the motoneurone where radical changes are seen to occur as a result of contact with the target muscle is the nerve terminal, which on reaching the muscle is transformed from a growing into a secreting structure. This transition is accompanied by a rapid increase in transmitter release from the growth cone and a cessation of neurite elongation (Xie and Poo, 1986). Such changes of the terminal could be induced by simple biophysical interactions in the microenvironment of the growth cone which occur as a result of the response of the muscle cells to the released transmitter. The fact that many terminals of a single motoneurone undergo this change may stimulate the cell body to increase the synthesis of the enzyme responsible for the production of ACh, and reduce the synthesis of growth-associated proteins. The cell is thus transformed into a transmitting neurone, and in this way the target induces the motoneurone to modify some features of its phenotype. It could be that associated with this change are other modifications of the cell's phenotype which allow the motoneurone to survive. It is possible that the motoneurone, in order to survive, has to become competent to withstand the ever increasing afferent input that occurs in the developing spinal cord and redistribute or down-regulate its glutamate receptors. The interaction with the target may be required for the motoneurone to develop this ability (Lowrie and Vrbová, 1992). Figure 2.3 illustrates this proposal. Recently evidence in support of this proposal was provided by experiments that show that motoneurones destined to die can be rescued, if their NMDA receptors are blocked (Mentis *et al.*, 1993). Thus it could be that the target regulates the number of surviving motoneurones, and then indirectly the size of the future motor unit.

In summary, our knowledge of developmental changes of ion channels

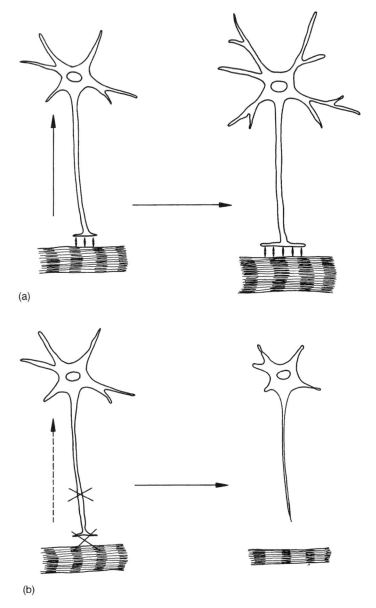

(a)

(b)

Figure 2.3 Possible mechanism for postnatal motoneurone death. As a result of interaction between the motoneurone and muscle, a retrograde signal is sent to the cell body. (a) In the presence of the retrograde signal, the motoneurone develops normally and is able to withstand increased synaptic activation. (b) If interaction with the target is prevented, the retrograde signal is diminished and the target-deprived motoneurone eventually degenerates. (Reproduced from Lowrie and Vrbová, 1992, *TINS*, **15**, 75–80, with permission from Elsevier Science Publishers.)

and neurotransmitter receptors in the somatodendritic membrane of the motoneurone has expanded rapidly in the past few years. Results from these recent studies indicate that, in general, inward Na^+ and Ca^{2+} currents predominate in embryonic motoneurones while outward K^+ currents are relatively low at this stage and increase rapidly during late embryonic and early postnatal development. As a consequence, the duration of the action potential in embryonic motoneurones is much longer than in mature motoneurones and may, in turn, result in greater Ca^{2+} entry via voltage-dependent Ca^{2+} channels. Furthermore, because of the central role of Ca^{2+} as an intracellular second messenger, these activity-related events may influence the growth and differentiation of the motoneurone. Changes in K^+ and Ca^{2+} currents are likely to contribute to the developmental modifications of motoneurone firing patterns and thus affect muscle fibre differentiation.

CONCLUSIONS

Motoneurones are derived from neuroepithelial cell precursors in the ventricular zone of the neural tube. They are among the first neuronal cells to develop, and migrate ventrolaterally to their final destination. An early characteristic feature of motoneurones is the presence of acetyl-cholinesterase and choline acetyltransferase. These enzymes are present in the motoneurones almost as soon as they emerge and their activity increases with age. The electrical excitability, distribution of ion channels and glutamate receptors change with development. The initial development of motoneurones proceeds without contact with the target.

However, at a critical stage of development a proportion of motoneurones die and the remaining motoneurones become dependent on connections with the target. They do not survive if contact and interaction with the muscle fibres is prevented.

REFERENCES

Bennett, M. R. (1983) Development of neuromuscular synapses in striated muscle development. *Physiol. Rev.* **63**, 915–1048.

Berger, A. T. and Takahashi, T. (1990) Serotonin enhances low voltage activated calcium current in rat spinal motoneurones. *J. Neurosci.* **10**, 1922–1928.

Betz, H., Bourgeois, J.-P. and Changeaux, J.-P. (1980) Evolution of cholinergic proteins in developing slow and fast skeletal muscles in chick embryo. *J. Physiol.* **302**, 197–218.

Brenneman, D. E., Neale, A. F., Habig, W. H. *et al.* (1983) Developmental and neurochemical specificity of neuronal deficits produced by electrical impulse blockade in dissociated spinal cultures. *Dev. Brain Res.* **9**, 13–27.

Brenneman, D. E., Forsythe, I. D., Nicol, J. and Nelson, P. G. (1990) N-Methyl-D-

Aspartate receptors influence survival in developing spinal cord cultures. *Dev. Brain Res.* **51**, 63–68.

Burt, A. M. (1975) Choline acetyltransferase and acetylcholinesterase in developing spinal cord. *Exp. Neurol.* **47**, 173–180.

Commissiong, J. W. (1983) Development of catecholaminergic nerves in the spinal cord of the rat. *Brain Res.* **264**, 197–208.

Conradi, S. and Ronnevi, L. O. (1975) Spontaneous elimination of synapses on cat spinal motoneurones after birth. Do half the synapses on the cell disappear? *Brain Res.* **92**, 505–510.

Eccles, J. C., Eccles, R. M. and Lundberg, A. (1958) The action potentials of alpha motoneurones supplying fast and slow muscles. *J. Physiol.* **142**, 272–291.

Fruns, M., Kreiger, C. and Sears, T. A. (1987) Identification and electrophysiological investigation of embryonic mammalian motoneurones in culture. *Neurosci. Lett.* **83**, 82–88.

Fulton, B. P. and Walton, K. D. (1986) Electrophysiological properties of neonatal rat motoneurones studied *in vitro*. *J. Physiol.* **370**, 651–678.

Giacobini-Robecchi, M., Giacobini, E., Filogamo, G. and Changeaux, J.-P. (1975) Effects of type A toxin from *Clostridium botulinum* on the development of skeletal muscles and their innervation in chick embryo. *Brain Res.* **83**, 107–121.

Gilbert, M. and Stelzner, D. J. (1979) The development of descending and dorsal root connections in the lumbrosacral spinal cord of the postnatal rat. *J. Comp. Neurol.* **184**, 821–838.

Hamburger, V. (1934) The effects of wing bud extirpation on the development of the central nervous system. *J. Exp. Zool.* **68**, 449–494.

Hamburger, V. (1958) Regression versus peripheral control of differentiation in motor hyperplasia. *Am. J. Anat.* **102**, 365–410.

Hamburger, V. (1977) The developmental history of the motoneurone. *Neurosci. Prog. Res. Bul.* **15**, 1–37.

Harada, Y. and Takahashi, T. (1983) The calcium component of the action potential in spinal motoneurones. *J. Physiol. (Lond.)* **335**, 89–100.

Hendrikson, C. K. and Vaughn, J. E. (1974) Fine structural relationship between neurites and radial glial processes in developing mouse spinal cord. *J. Neurocytol.* **3**, 659–675.

Holliday, M. and Hamburger, V. (1976) Reduction of the naturally occurring motoneurone loss by enlargement of the periphery. *J. Comp. Neurol.* **170**, 311–320.

Holliday, M. and Hamburger, V. (1977) An autoradiographic study of the formation of the lateral motor column in the chick embryo. *Brain Res.* **132**, 197–208.

Hughes, A. F. and Tschumi, P. A. (1958) The factors controlling the development of the dorsal root ganglia and ventral horn in *Xenopus laevis*. *J. Anat. (Lond.)* **92**, 498–527.

Huizar, P., Kuno, M. and Myata, Y. (1975) Differentiation of motoneurones and skeletal muscles in kittens. *J. Physiol. (Lond.)* **252**, 465–479.

Ishida, I. and Deguchi, T. (1983) Effects of depolarizing agents on choline acetyltransferase activities in primary cell cultures of spinal cord. *J. Neurosci.* **3**, 1818–1823.

Kellerth, J.-O., Mellström, A. and Skoglund, S. (1971) Postnatal excitability changes of kitten motoneurones. *Acta Physiol. Scand.* **83**, 31–41.

Kernell, D. (1990) Spinal motoneurones and their muscle fibres: mechanisms and long-term consequences of common activation patterns, in *The Segmental*

Nervous System (eds M. D. Binder and L. M. Mendell), Oxford University Press, pp. 36–57.

Kiehn, O. (1991) Plateau potentials and active integration in the 'final common pathway' for motor behaviour. *TINS* **14**, 68–73.

Kreiger, C. and Sears, T. A. (1988) Development of voltage-dependent conductances in murine spinal cord neurones in culture. *Can. J. Physiol. Pharmacol.* **66**, 1328–1336.

Langman, J. and Hadden, C. C. (1970) Formation and migration of neuroblasts in the spinal cord of the chick embryo. *J. Comp. Neurol.* **138**, 419–452.

Leber, S. M., Breedlove, S. M. and Sanes, J. R. (1990) Lineage arrangement and death of clonally related motoneurones in chick spinal cord. *J. Neurosci.* **10**, 2451–2462.

Llinás, R. R. (1988) The intrinsic electrophysiological properties of mammalian neurones: insights into central nervous system function. *Science* **242**, 1654–1664.

Lowrie, M. B. and Vrbová, G. (1992) Dependence of postnatal motoneurones on their targets: a review and hypothesis. *TINS* **15**, 75–80.

McCobb, D. P., Best, P. M. and Beam, K. G. (1989) Development alters the expression of calcium currents in chick limb motoneurones. *Neuron* **2**, 1633–1643.

McCobb, D. P., Best, P. M. and Beam, K. G. (1990) The differentiation of excitability in embryonic chick motoneurones. *J. Neurosci.* **10**, 2974–2984.

Manabe, T., Araki, I., Takahashi, J. and Kuno, M. (1991) Membrane currents recorded from sexually dimorphic motoneurones of the bulbocavenous muscle in neonatal rats. *J. Physiol. (Lond.)* **440**, 419–435.

Mentis, G. Z., Greensmith, L. and Vrbová, G. (1993) Motoneurones destined to die are rescued by blocking N-Methyl-d-Aspartate receptors by MK–801. *Neurosci.* **54**, 283–285.

Moody, S. A. and Jacobson, M. (1983) Compartmental relationships between aneuran primary spinal motoneurones and somitic muscle fibres that they first innervate. *J. Neurosci.* **3**, 1670–1682.

Naka, K. I. (1964) Electrophysiology of the fetal spinal cord. 1. Action potentials of the motoneurones. *J. Gen. Physiol.* **47**, 1003–1022.

Navarrete, R. and Vrbová, G. (1983) Changes of activity patterns in fast muscle during postnatal development. *Dev. Brain Res.* **8**, 11–19.

Navarrete, R. and Walton, K. D. (1989) Calcium triggered doublet firing in neonatal rat motoneurones *in vitro*. *J. Physiol. (Lond.)* **415**, 70P.

Navarrete, R., Walton, K. D. and Llinas, R. R. (1988) Postnatal changes in the electrical properties of muscle-identified rat motoneurones: an *in vitro* study. *Soc. Neurosci. Abstr.* **14**, 1060.

O'Brien, R. A. D. and Vrbová, G. (1978) Acetylcholine synthesis in nerve endings to slow and fast muscles of developing chicks: effect of muscle activity. *Neuroscience* **3**, 1227–1230.

O'Brien, R. J. and Fischbach, G. D. (1986a) Isolation of embryonic chick motoneurones and their survival *in vitro* *J. Neurosci.* **6**, 3265–3274.

O'Brien, R. J. and Fischbach, G. D. (1986b) Characterisation of excitatory amino acid receptors expressed by embryonic chick motoneurones *in vitro*. *J. Neurosci.* **6**, 3275–3283.

O'Brien, R. J. and Fischbach, G. D. (1986c) Excitatory synaptic transmission between interneurones and motoneurones in chick spinal cord cell cultures. *J. Neurosci.* **6**, 3284–3289.

O'Brien, R. J. and Fischbach, G. D. (1986d) Modulation of embryonic chick motoneurone glutamate sensitivity by interneurones and agonists. *J. Neurosci.* **6**, 3290–3296.

Oppenheim, R. W. (1991) Cell death during the development of the nervous system. *Ann. Rev. Neurosci.* **14**, 453–501.

Phelps, P. E., Barber, R. P., Houser, C. R. *et al.* (1984) Postnatal development of neurones containing choline acetyltransferase in rat spinal cord: an immunocytochemical study. *J. Comp. Neurol.* **229**, 347–361.

Pilar, G., Tuttle, J. and Vaca, K. (1981) Functional maturation of motor nerve terminals in the avian iris: ultrastructure, transmitter metabolism and synaptic reliability. *J. Physiol. (Lond.)* **321**, 175–193.

Purves, D. and Lichtman, J. W. (1985) *Principles of Neural Development*, Sinauer Associates Inc., Sunderland, Massachusetts, pp. 433.

Rajofetra, N., Sandillon, F., Geffard, M. and Privat, A. (1989) Pre- and postnatal ontogeny of serotonergic projections to the rat spinal cord. *J. Neurosci. Res.* **22**, 305–321.

Ronnevi, L. O. (1979) Spontaneous phagocytosis of C-type synaptic terminals on motoneurones in new-born kittens. An electron microscope study. *Brain Res.* **162**, 189–199.

Rudy, B. (1988) Diversity and ubiquity of K+ channels. *Neurosci.* **25**, 729–749.

Schwindt, P. C. and Crill, W. E. (1984) Membrane properties of cat spinal motoneurones, in *Handbook of the Spinal Cord* (ed. R. A. Davidoff), Marcel Dekker, NY, pp. 199–242.

Shorey, M. L. (1909) The effects of the destruction of the peripheral areas on the differentiation of the neuroblasts. *J. Exp. Zool.* **7**, 25–63.

Stelzner, J. D. (1982) The role of descending systems in maintaining intrinsic and spinal function: a developmental approach, in *Brain Stem Control of Spinal Mechanisms* (eds B. Sjolund and A. Bjorklund), Elsevier, Amsterdam, pp. 297–321.

Takahashi, T. (1990) Membrane currents in visually identified motoneurones of neonatal rat spinal cord. *J. Physiol. (Lond.)* **423**, 27–46.

Vaca, K. (1988) The development of cholinergic neurones. *Brain Res. Rev.* **13**, 261–286.

Vogel, M. W., English, A. W. and Herrup, K. (1988) Chimeric analysis of lineage relationships among mouse lumbar motoneurones. *Soc. Neurosci. Abstr.* **14**, 1130.

Walton, K. D. and Fulton, B. P. (1986) The ionic mechanisms underlying the firing properties of rat neonatal motoneurones. *Neurosci.* **19**, 669–683.

Xie, Z. and Poo, M.-M. (1986) Initial events in the formation of neuromuscular synapse: Rapid induction of acetylcholine release from embryonic neurone. *Proc. Nat. Acad. Sci. USA* **83**, 7069–7073.

Ziskind-Conhaim, L. (1988) Physiological and morphological changes in developing peripheral nerves of rat embryos. *Dev. Brain Res.* **42**, 15–28.

Ziskind-Conhaim, L. (1990) NMDA receptors mediate poly- and monosynaptic potentials in motoneurones of rat embryos. *J. Neurosci.* **10**, 125–135.

Encounter of motor nerves with muscle fibres

3

3.1 REGULATION OF ACETYLCHOLINE RECEPTORS AND CHOLINESTERASE

Apart from intracellular muscle-specific proteins, of which the best explored ones are the components of the myofibril, some membrane-associated proteins are also muscle specific and unique for skeletal muscle fibres. Molecules that are part of the surface membrane could also contribute to the establishment of specific connections between motor nerves and muscles and are therefore of special interest to neuro-biologists. Of these, the acetylcholine receptor (**AChR**) is of particular importance because it is the molecule that enables the muscle to become functionally connected to its innervation and respond to the acetylcholine (**ACh**) released from the motor nerve ending. In view of the unique position of this molecule it is not surprising that much research has been carried out to reveal its structure and function, and the developmental regulation of its synthesis and distribution along skeletal muscle fibres. Excellent reviews have recently been published on the development of the AChR (Schuetze and Role, 1987). The AChR is a pentamer comprised of four distinct subunits with the stochiometry of α (2), β (1) γ or ϵ (1) (see Schuetze and Role, 1987). The receptor complex has two binding sites for ACh, both on the α-subunits. Each of the four subunits of the AChR is translated from a separate mRNA, inserted into the endoplasmic reticulum, glycosylated and assembled (Anderson, 1986; Merlie and Smith, 1986; Salpeter and Loring, 1986). A few hours after completion of their synthesis, the assembled AChR molecules are inserted into the membrane (Fambrough, 1979). AChRs incorporated into the surface membrane can later be either stabilized and become 'junctional AChRs', or, as is the case with extrajunctional AChRs, become internalized and degraded by lysosomal enzymes. AChRs outside the endplate region and at immature endplates have a metabolic half life of about 1 day which increases 10-fold at mature endplates (Fambrough, 1979).

The appearance of the AChR in the membrane of a mononucleated myoblasts is one of the first signs of commitment of these cells to a myogenic lineage, and precedes myoblast fusion (Chapter 1, Figure 1.1). The presence of AChRs in myoblasts and their sharp increase after fusion has been demonstrated *in vitro*, in the absence of innervation (Fambrough and Rash, 1971; Dryden *et al.*, 1974; Sytkowski *et al.*, 1973; Fambrough, 1979). After fusion and further development of the myotube, AChRs form clusters. Cluster formation can be induced by various stimuli applied to the surface of muscle fibres both in adults (Jones and Vrbová, 1974) and in developing muscles (Peng *et al.*, 1981). Although the clusters become associated with the area of contact with the motor nerve, they are also found in aneural areas of the muscle fibre and develop in the absence of innervation. However, following the establishment of neuromuscular contacts the only clusters to remain on the muscle fibre are at the site of these contacts. Those AChRs that are localized outside the region of the neuromuscular contact disappear. During development the disappearance of the extrajunctional AChRs is brought about by muscle activity. If activity in chick embryos is prevented by temporary paralysis of the muscle, then the reduction of ACh sensitivity and the AChR outside the endplate region usually seen during development fails to occur (Gordon *et al.*, 1974; Burden, 1977a, b) (Figure 3.1). Thus nerve-induced activity plays a crucial role in regulating the distribution of AChRs. Even in adult muscle fibres, AChRs at extrajunctional sites can reappear after denervation and following complete paralysis of the muscle. Although both during development and in adults activity is an important factor that controls the distribution of AChRs, there are other factors, such as contact phenomena or putative substances released by cells containing the muscle fibre – e.g. calcitonin gene related peptide (CGRP) – that may play an important regulatory role (Changeaux, 1991).

In developing muscle fibres, AChR clusters at newly formed synapses are easily disrupted by removal of extracellular Ca^{2+}, chronic carbachol exposure and elevated external K^+ (Bloch and Steinbach, 1981). In contrast, mature synaptic contacts are resistant to these treatments. Similarly, AChR clusters at developing synapses disperse within hours or days after denervation (Slater, 1982; Kuromi and Kidokoro, 1984) but clusters at mature endplates remain intact for 2 weeks or more after removal of the nerve (Frank *et al.*, 1975; Ko *et al.*, 1977; Braithwaite and Harris, 1979; Slater, 1982; Labovitz and Robbins, 1983). Thus, continued action of the nerve is required for cluster maintenance at newly formed endplates but not at mature contacts. Presumably other factors at the adult neuromuscular junction maintain the AChR distribution in the absence of the

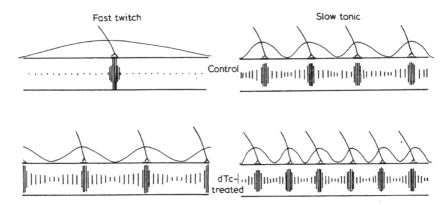

Figure 3.1 Distribution of ACh sensitivity of skeletal muscle fibres in relation to their innervation. The density and size of the vertical bars indicate the degree of chemosensitivity. The curve above each 'muscle fibre' indicates the area along which the depolarization produced on nerve activity may spread and reduce chemosensitivity: on the fast twitch fibre this spread would cover most of the fibre surface; on the slow tonic fibre each nerve ending produces a depolarization that spreads only over a small distance. After curare, since the endplate potential is reduced, the area of spread of the depolarization is reduced, and chemosensitivity will decrease over a smaller area. The fast twitch muscle fibres will become multiply innervated and the distance between successive endplates on the slow tonic muscle will become reduced. (Reproduced from Gordon, Jones and Vrbová, 1976, *Prog. Neurobiol.* 6, 103–136.)

nerve (Slater, 1982). It could be that during maturation an extracellular matrix material such as synapse-specific basal lamina or specific molecules such as 'agrin' are acquired and these play a role in the stabilization of the AChR (reviewed by Sanes, 1989). Alternatively, the AChRs may become anchored to the membrane by cytoskeletal elements (reviewed by Froehner, 1986). Specific antibodies directed against the cytoskeletal elements, vinculin and actinin, stain the neuromuscular junction within a few days after the onset of synaptogenesis in the rat (Bloch and Hall, 1983). Not only does the distribution and metabolic stability of muscle AChRs change during development but also their functional properties become transformed. The existence of different types of AChR was first described on denervated adult muscles, where AChR-associated channels at the endplate region were found to be different from those that appeared extrajunctionally. When activated by ACh, endplate channels conduct larger currents but close more quickly than extrajunctional channels (Neher and Sakmann, 1976; Sakmann, 1978).

During endplate maturation, AChR channel properties change from 'extrajunctional-like' to 'endplate-like'. The mechanisms that control the distribution and also the nature of the AChR channel are shown in Figure 3.2. AChR channels in immature endplates have a longer opening time

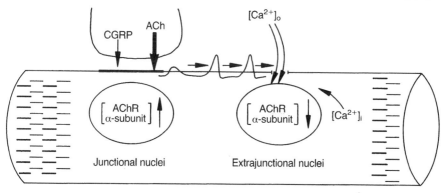

Figure 3.2 Muscle fibre and its place of contact with its motor nerve. The thick line on the muscle fibre illustrates the accumulation of acetylcholine receptors (AChR) at the place of nerve–muscle contact. The action potentials outside this site down-regulate the AChR synthesis of extrajunctional nuclei, possibly via increases of intracellular Ca^{2+}. Calcitonin gene-related peptide (CGRP) released from the nerve may play a role in maintaining the AChRs at the site of nerve–muscle contact.

than at mature endplates. Interestingly, AChRs in extrajunctional regions of developing muscle fibres change in parallel with those at endplates. Therefore, the common practice of describing AChR channels as either 'extrajunctional type' or 'junctional type' is misleading (Sakmann and Brenner, 1978). It is now known that the channel opening time of the AChR is associated with one of its subunits. The AChR with fast channel characteristics have the ε subunit, and those with slow channel properties the γ subunit. (For review, see Schuetze and Role, 1987.)

The fraction of slow embryonic channels at rat endplates decreases during early postnatal life. In the developing soleus endplate, the fraction of slow channels decreases steadily from virtually 100% around birth to <20% three weeks later (Sakmann and Brenner, 1978; Fischbach and Schuetze, 1980; Michler and Sakmann, 1980). There is an increase of mRNA coding for the ε-subunit and a reciprocal loss of that coding for the γ subunit of the AChR receptor (Witzeman et al., 1989; Brenner et al., 1990).

There is considerable evidence demonstrating that AChR channel opening time is also regulated by the nerve and its activity (reviewed by Changeux, 1991). Shortly after birth, when slow channels predominate at the endplate, denervation blocks the normal developmental appearance of fast channels.

The combined results of the denervation studies (Chapters 6 and 8) on neonatal and adult muscles suggest that the appearance of fast AChR channels requires muscle activity, but not necessarily the continued presence of the nerve.

Another important step in endplate development and specialization is the accumulation of the enzyme acetylcholinesterase (**AChE**). This enzyme appears in myogenic cells very early in development, well before ventral root axons reach the myotome. The areas of AChE activity on myogenic cells increase in the vicinity of growth cones that invade the myotome (Tennyson *et al.*, 1971, 1973). Gradually, as innervation proceeds, the AChE accumulates at the place of contact between the motor nerve ending and muscle fibre (Kupfer and Koelle, 1951; Lentz, 1969; Tóth and Karcsú, 1979; Brzin *et al.*, 1981). This accumulation of AChE is one indicator of the morphological differentiation of the subsynaptic sarcolemma and muscle fibre (Kelly and Zacks, 1969; Bennett and Pettigrew, 1974; Brzin *et al.*, 1981). In developing muscle fibres, patches of AChE on the sarcolemma outside the endplate disappear at the time when the enzyme accumulates at the endplate region.

AChE in skeletal muscles can be present in two distinct molecular forms: globular and asymmetric (Massoulié and Bon, 1982). The asymmetric form is present in endplates of mature muscle (Hall, 1973; Sketelj and Brzin, 1985) and is dependent on innervation (Vigny *et al.*, 1976), for it becomes reduced after denervation.

Innervation and activity are known to exert an important influence on the accumulation of AChE at the neuromuscular junction. Following neonatal denervation, AChE fails to accumulate at the neuromuscular junction (Zelená and Szentágothay, 1957). Even with intact innervation AChE does not accumulate at the place of contact between the motor nerve ending and nerve terminal in inactive muscles paralysed with curare (Gordon *et al.*, 1974). However, more sensitive methods have revealed that, even if rat muscles are denervated at birth, small amounts of AChE do accumulate at the site of former contact with the nerve (Sketelj *et al.*, 1991). This continued accumulation of AChE at newly formed endplates is concomitant with the persistence of AChRs (Slater, 1982) and some degree of growth of synaptic folds (Brenner *et al.*, 1983). Moreover, it is regulated by the activity of the muscle fibre.

It has been suggested that the contact with the nerve induces some 'trace', which allows the endplate region of the muscle to continue to develop. There are several proposals as to the nature of the 'trace'. It has been proposed that the presence of synapse organizing factor in the basal lamina under the nerve endings is responsible for the continuous specialization of the endplate region (Nitkin *et al.*, 1987). Such a molecule, called '**agrin**', causes aggregation of AChRs in cultured, aneural myogenic cells (Fallon and Gelfman, 1989; Nastuk and Fallon, 1993). This substance is present in the motoneurone soma (Magill-Solc and McMahan, 1988) but this does not explain its localized presence on a highly selected part of the muscle cell.

In conclusion, profound changes occur during the development of skeletal muscle fibres both in the isoforms and in distribution of the two major proteins that are involved in cholinergic transmission, i.e. the AChR and AChE, and these changes depend on the activity of the muscle fibre.

3.2 DEVELOPMENT OF ION CHANNELS AND PASSIVE ELECTRICAL PROPERTIES

Many of the developmental changes in membrane-associated characteristics such as voltage-gated ion channels and passive membrane properties of the muscle fibre have been described. Most studies have concentrated on ion channels of myoblasts in tissue culture. In addition, developmental transitions in channel properties have been demonstrated in muscles developing *in situ* during postnatal development. Here we shall concentrate on those studies which specifically address the role of innervation, and neuromuscular activity in the regulation of ion channels of the muscle membrane.

The early appearance of Na^+ currents was demonstrated in a rat myoblast cell line L6 (Kidokoro, 1975) and subsequently in myoblasts obtained from rat primary cultures (Gonoi et al., 1985; Weiss and Horn, 1986a, b). The motor nerve appears to play an important role in the developmental induction of TTX-sensitive Na^+ currents. In adult innervated mammalian muscle fibres the majority of Na^+ channels have a high affinity to TTX which also blocks the generation of the action potential (Redfern and Thesleff, 1971; Harris and Thesleff, 1971). However, if adult muscles are denervated or paralysed by botulinum toxin (**BoTx**), a new type of Na^+ channel appears, making the action potential partly resistant to the action of TTX (Mathers and Thesleff, 1978; Bambrick and Gordon, 1987a, b). In newborn rat muscles too, TTX-resistant action potentials are normally found in fully innervated muscle (Harris and Marshall, 1973). The TTX-insensitive Na^+ channels present in denervated muscle fibres and during early development have a different molecular structure and gating properties from those present in adult innervated muscle. TTX-resistant Na^+ channels open at more negative membrane potentials and have longer activation and inactivation decay times, and smaller single channel conductance compared with adult innervated muscle fibres (Gonoi et al., 1989; Kallen et al., 1990). During development, the proportion of TTX-resistant Na^+ channels decreases; and this decrease does not take place, or is slower, if the muscles are denervated or paralysed (Sherman and Catterall, 1984; Gonoi et al., 1989). This developmental transition from TTX-insensitive to TTX-sensitive Na^+ channels is similar to that previously discussed for the AChR.

Experiments using radiolabelled Na^+ channel antagonists such as TTX or saxitoxin (STX), which bind specifically to a receptor site on the extracellular face of the channel (see Chatterall, 1991), show that the number of TTX-sensitive Na[+] channels increases rapidly during the first 3 weeks of postnatal development in the rat (Sherman and Catterall, 1982; Lombert et al., 1983; Bambrick and Gordon, 1988). The most rapid period of incorporation of TTX-sensitive Na[+] channels into the skeletal muscle membrane occurred during the second postnatal week, coinciding with a period of greatly increasing neuromuscular activity (section 4.2). This developmental increase can be inhibited by neonatal denervation (Sherman and Catterall, 1982; Gonoi et al., 1989) or if the muscle is paralysed with BoTx (Bambrick and Gordon, 1988). Since Na[+] channel properties do not mature in inactive muscles, it appears that muscle activity alters the properties of Na[+] channels, inducing a decrease in TTX-sensitive channels and an increase in TTX-sensitive Na[+] channels.

Voltage-gated Ca^{2+} channels are present at high levels in the T-tubular system of adult skeletal muscle. These L-type Ca^{2+} channels are sensitive to the pharmacological action of dihydropyridines and mediate long-lasting inward Ca^{2+} currents in response to strong membrane depolarizations. During prenatal and early postnatal development, both L- and T-type Ca^{2+} channels are present in skeletal muscle fibres (Kano et al., 1987; Gonoi and Hagesawa, 1988). The presence of T-type Ca^{2+} channels in immature muscle fibres is transient and it is replaced by the L-type Ca^{2+} channels during postnatal development (Kano et al., 1989; Gonoi and Hasegawa, 1988). As discussed in section 2.1.2, the presence of T-type Ca^{2+} channels appears to be a general property of developing excitable cells. Unlike L-type channels, T-type channels open at membrane potentials close to the resting potential so that even small fluctuations of the membrane potential of the immature muscle fibre may allow Ca^{2+} entry.

3.3 CHARACTERISTICS OF EARLY NEUROMUSCULAR TRANSMISSION

There are several recent reviews which describe in detail the development of the neuromuscular transmission both in situ and in tissue culture (Salpeter, 1987; Peng, 1987; Sanes, 1989). Here we will specifically review those developmental aspects of neuromuscular transmission that depend on pre- and/or postsynaptic activity and may contribute to the establishment of the adult motor unit topography.

During development, the morphological as well as physiological characteristics of the vertebrate neuromuscular junction undergo profound changes. At the immature mammalian neuromuscular junction several axon profiles are seen to contact the postsynaptic muscle membrane, which at this stage displays few synaptic clefts and presents a rather smooth surface (Figure 3.3a). This changes with development; on twitch fibres, a large terminal contacts a postsynaptic membrane that is folded and has pronounced postsynaptic specialization (Figure 3.3b).

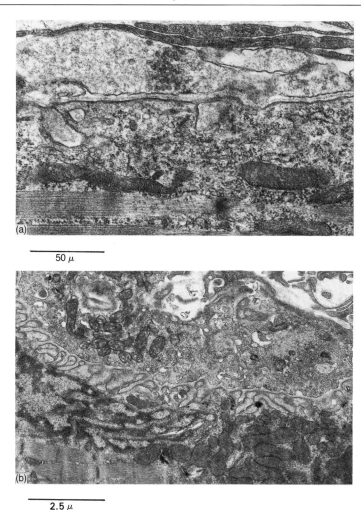

50 μ

2.5 μ

Figure 3.3 Ultrastructural appearance of endplates taken from a soleus muscle of a rat at (a) 9 days old and (b) 1 month old.

Actual chemical transmission starts almost as soon as the axon reaches the muscle mass.

In the chick embryo, electrical stimulation of motor axons first elicits muscle contraction at stage 27–28 (Landmesser and Morris, 1975), at a time when most presynaptic axon profiles are not yet closely associated with AChR clusters in the muscle fibre (Dahm and Landmesser, 1991). The fact that initially neuromuscular transmission does not require synaptic specializations has also been demonstrated in tissue culture studies which show that synaptic potentials can be evoked within a few minutes after a growth cone contacts a receptive myotube (Frank and Fischbach, 1979; Xie and Poo, 1986). It appears that the growth cones can produce spontaneous and evoked ACh release even before the terminals form a physical contact with a myotube (Young and Poo, 1983; Hume *et al.*, 1983). Without contact with a myogenic cell, growth cones release very little ACh; but within 20 minutes of contact with the muscle membrane, both the amplitude and frequency of the spontaneous ACh release increase (Xie and Poo, 1986), suggesting that contact with the target may specifically induce the release of neurotransmitter. The initial ACh-induced depolarizations of the muscle membrane resemble miniature endplate potentials (**MEPPs**) observed in the mature neuromuscular junction, in that their frequency is reduced by low Ca^{2+} and they can be blocked by curare or BTX (Xie and Poo, 1986). However, in contrast to MEPPs in the mature neuromuscular junction, their amplitude can be very large and in some cases trigger muscle contractions (Jaramillo *et al.*, 1988).

During the initial stages of development of the neuromuscular junction, it is possible that the presence of sub-threshold spontaneous synaptic depolarizations constitute the first form of interaction between motoneurones and muscle fibres. This initial form of interaction may be largely independent of action potentials in either nerve or muscle but may nevertheless induce muscle contraction and differentiation.

It is interesting that similar 'giant' spontaneous MEPPs are observed in adult muscles following experimentally induced inactivity of the neuromuscular junction by chronic TTX blockade of action potentials in the nerve (Gundersen, 1990) or after botulinum toxin paralysis (BoTx) (see Thesleff and Molgo, 1983).

Transmission in newly formed neuromuscular junctions is different from that in the adult. Most endplate potentials (**EPPs**) recorded from embryonic muscles have a slow rise time and small amplitude which decreases rapidly in response to repetitive stimulation (Dennis *et al.*, 1981; Pilar *et al.*, 1981; Sheard *et al.*, 1991). Both the frequency of spontaneous MEPPs and the quantal content of the EPP increase markedly during embryonic and early postnatal development (Diamond and Miledi, 1962; Letinsky, 1974; Kelly, 1978; Dennis *et al.*, 1981). This is associated with an increase in the amount of enzymes involved in ACh synthesis (O'Brien and Vrbová, 1978; Pilar *et al.*, 1981). With development, the efficiency of neuromuscular transmission improves gradually as pre- and postsynaptic components of the neuromuscular junction become integrated. The developmental stage of the muscle fibre itself may influence its ability to respond to the transmitter. The size of the response may depend on the type and density of AChRs and the size of the muscle fibre, since smaller cells have higher input resistance and are therefore more excitable.

CONCLUSIONS

Innervation influences the characteristic properties of the two major molecules of the postsynaptic membrane involved in cholinergic transmission, i.e. the acetylcholine receptor and acetylcholinesterase. It also has profound effects on the distribution of the AChR along the muscle fibre. Other membrane-associated molecules of the muscle fibres, particularly ion channels, are also dependent on innervation and change under the influence of the nerve. Most of these effects of innervation on muscle membrane properties are retarded in inactive muscles showing them to be induced by the activity transmitted to the muscle by its motor nerve.

The muscle fibre also has a retrograde influence on the nerve terminal that makes contact with it. The transmitter release, which is initially low, increases after neuromuscular contacts are established, and evidence is presented to suggest that this increase is induced by interaction with the muscle fibre.

During their first encounter, muscle fibres and their nerve terminals are immature and both these structures undergo further development as a result of their interaction with each other.

REFERENCES

Anderson, D. J. (1986) Molecular biology of the acetylcholine receptor: structure and regulation of biogenesis, in *The Vertebrate Neuromuscular Junction* (ed. M. Salpeter, Alan Liss, NY, pp. 285–316.

Bambrick, L. L. and Gordon, T. (1987a) Acetylcholine receptors and sodium channels in denervated and botulinum-toxin treated adult rat muscle. *J. Physiol.* **382**, 69–86.

Bambrick, L. L. and Gordon, T. (1987b) Neuronal regulation of acetylcholine receptors and sodium channels in developing rat muscle. *Neurosci. Abstr.* **13**, 929.

Bambrick, L. L. and Gordon, T. (1988) Neural regulation of (3H) saxitoxin binding site numbers in rat neonatal muscle. *J. Physiol. (Lond.)* **407**, 263–274.

Bennett, M. R. and Pettigrew, A. G. (1974) The formation of synapses in striated muscle during development. *J. Physiol.* **241**, 515–545.

Bloch, R. J. and Hall, Z. W. (1983) Cytoskeletal components of the vertebrate neuromuscular junction: vinculin, a actin and filamin. *J. Cell Biol.* **97**, 217–223.

Bloch, R. J. and Steinbach, J. H. (1981) Reversible loss of acetylcholine receptor clusters at the developing rat neuromuscular junction. *Dev. Biol.* **81**, 386–391.

Braithwaite, A. W. and Harris, A. J. (1979) Neural influence on acetylcholine receptor clusters in embryonic development of skeletal muscles. *Nature* **279**, 549–551.

Brenner, H. R., Meier, T. and Widmer, B. (1983) Early action of nerve determines motor endplate differentiation in rat muscle. *Nature* **305**, 536–537.

Brenner, H. R., Witzemann, V. and Sackman, B. (1990) Imprinting of acetylcholine receptor messenger RNA accumulation in neuromuscular synapses. *Nature* **344**, 544–547.

Brzin, M., Skeletlj, J., Tennyson, V. M. *et al.* (1981) Activity, molecular forms and cytochemistry of cholinesterase in developing rat diaphragm. *Muscle & Nerve* **4**, 505–513.

Burden, S. (1977a) Development of the neuromuscular junction in the chick embryo: The number, distribution and stability of acetylcholine receptors. *Dev. Biol.* **57**, 317–329.

Burden, S. (1977b) Acetylcholine receptors at the neuromuscular junction: Developmental changes in receptor turnover. *Dev. Biol.* **61**, 79–85.

Caterall, W. A. (1991) Structure and function of voltage-gated sodium and calcium channels. *Curr. Top. Neurobiol.* **1**, 5–13.

Changeaux, J.-P. (1991) Compartmentalised transcription of acetylcholine receptor genes during motor endplate epigenesis. *New Biol.* **3**, 413–429.

Dahm, L. M. and Landmesser, L. T. (1991) The regulation of synaptogenesis during normal development and following activity blockade. *J. Neurosci.* **11**, 238–255.

Dennis, M. J., Ziskind-Conhaim, L. and Harris, A. J. (1981) Development of neuromuscular junctions in rat embryos. *Dev. Biol.* **81**, 266–279.

Diamond, J. and Miledi, R. (1962) A study of foetal and newborn rat muscle fibres. *J. Physiol.* **162**, 393–408.

Dryden, W. F., Erulkar, S. D. and de la Habba, G. (1974) Properties of the cell membrane of developing skeletal muscle fibres in culture and its sensitivity to acetylcholine. *Clin. Exp. Pharm.* **1**, 369–387.

Fallon, J. R. and Gelfman, C. E. (1989) Agrin-related molecules are concentrated at acetylcholine receptors in normal and aneural developing muscle. *J. Cell Biol.* **108**, 1527–1535.

Fambrough, D. M. (1979) Control of acetylcholine receptors in skeletal muscle. *Physiol. Rev.* **59**, 165–227.

Fambrough, D. M. and Rash, J. E. (1971) Development of acetylcholine sensitivity during myogenesis. *Dev. Biol.* **26**, 55–68.

Fischbach, G. D. and Schuetze, S. M. (1980) A postnatal decrease in acetylcholine channel open time at rat endplates. *J. Physiol. (Lond.)* **303**, 125–137.

Frank, E. and Fischbach, G. D. (1979) Early events in neuromuscular junction formation *in vitro*: Induction of acetylcholine receptor clusters in the post synaptic membrane and morphology of newly formed synapses. *J. Cell Biol.* **83**, 143–158.

Frank, E., Jansen, J. K. S., Lømo, T. and Westergaard, R. H. (1975) The interaction between foreign and original motor nerves innervating the soleus muscle of rats. *J. Physiol. (Lond.)* **247**, 725–743.

Froehner, S. C. (1986) The role of postsynpatic cytoskeleton in AChR organisation. *TINS* **9**, 37–41.

Gonoi, T. and Hasegawa, S. (1988) Postnatal disappearance of transient calcium channels in mouse skeletal muscle: effects of denervation and culture. *J. Physiol.* **155**, 13–25.

Gonoi, J., Hagihara, Y., Kobayshi, J. *et al.* (1989) Geographutoxin-sensitive and insensitive sodium currents in mouse skeletal muscle developing *in situ*. *J. Physiol. (Lond.)* **414**, 159–177.

Gonoi, T., Sherman, S. and Catterall, W. A. (1985) Voltage clamp analysis of tetrodotoxin-sensitive and -insensitive sodium channels in rat soleus muscles cells developing *in vitro*. *J. Neurosci.* **5**, 2559–2564.

Gordon, T., Perry, R., Tuffery, A. R. and Vrbová, G. (1974) Possible mechanisms determining synapse formation in developing skeletal muscle of the chick. *Cell Tiss. Res.* **155**, 13–25.

Gordon, T., Jones, R. and Vrbová, G. (1976) Changes in chemosensitivity of skeletal muscle as related to endplate formation. *Prog. Neurobiol.* **6**, 103–136.

Gundersen, K. (1990) Spontaneous activity at long-term silenced synapses in rat muscle. *J. Physiol. (Lond.)* **430**, 399–418.

Hall, Z. W. (1973) Multiple forms of acetylcholinesterase and their distribution in endplate and non-endplate regions of rat diaphragm muscle. *J. Neurobiol.* **4**, 343–361.

Harris, J. B. and Marshall, M. W. (1973) Tetrodotoxin-resistant action potentials in newborn rat muscle. *Nature* **243**, 191–192.

Harris, J. B. and Thesleff, S. (1971) Studies on tetrodotoxin-resistant action potentials in denervated skeletal muscle. *Acta Physiol. Scand.* **83**, 382–388.

Hume, R. I., Role, L. W. and Fischbach, G. D. (1983) Acetylcholine release from growth cones detected with patches of acetylcholine rich membranes. *Nature* **305**, 632–634.

Jaramillo, F., Vicini, S. and Scheutze, S. M. (1988) Embryonic acetylcholine receptors guarantee spontaneous contractions in rat developing muscle. *Nature* **335**, 66–68.

Jones, R. and Vrbová, G. (1974) Two factors responsible for the development of denervation hypersensitivity. *J. Physiol.* **236**, 517–538.

Kallen, R. G., Sheng, Z. H., Yang, J. *et al.* (1990) Primary structure and expression of a sodium channel characteristic of denervated and immature rat skeletal muscle. *Neuron* **4**, 232–242.

Kano, M., Wakuta, K. and Satoh, R. (1987) Calcium channel components of action potential in chick skeletal muscle cells developing in culture. *Dev. Brain Res.* **32**, 232–240.

Kano, M., Wakuta, K. and Satoh, R. (1989) Two components of calcium channel current in embryonic chick skeletal muscle cells developing in culture. *Dev. Brain Res.* **47**, 101–112.

Kelly, A. M. and Zacks, S. I. (1969) The fine structure of motor endplate morphogenesis. *J. Cell Biol.* **42**, 154–169.

Kelly, S. S. (1978) The effect of age on neuromuscular transmission. *J. Physiol. (Lond.)* **274**, 51–62.

Kidokoro, Y. (1975) Sodium and calcium components of the action potential in a developing muscle cell line. *J. Physiol. (Lond.)* **244**, 145–159.

Ko, P. K., Anderson, M. J. and Cohen, M. W. (1977) Denervated skeletal muscle fibres develop discrete patches of high acetylcholine receptor density. *Science* **196**, 540–542.

Kupfer, C. and Koelle, G. B. (1951) A histochemical study of cholinesterase during the formation of the motor endplate of the albino rat. *J. Exp. Zool.* **116**, 399–413.

Kuromi, H. and Kidokoro, Y. (1984) Nerve disperses pre-existing acetylcholine receptor accumulation in *Xenopus* cultures. *Dev. Biol.* **103**, 53–61.

Labovitz, S. S. and Robbins, N. (1983) A maturational increase in rat neuromuscular junctional acetylcholine receptors despite disuse or denervation. *Brain Res.* **266**, 155–158.

Landmesser, L. and Morris, D. G. (1975) The development of functional innervation in the hind limb of the chick embryo. *J. Physiol.* **249**, 301–326.

Lentz, T. L. (1969) Development of the neuromuscular junction. 1. Cytological and cytochemical studies on the junction of differentiating muscle in the newt *Triturus. J. Cell Biol.* **42**, 431–443.

Letinsky, M. S. (1974) Physiological properties of developing frog tadpole nerve–muscle junctions during repetitive stimulation. *Dev. Biol.* **40**, 154–161.

Lombert, A., Kazazoglon, T., Delpont, E. *et al.* (1983). Ontogenic appearance of Na⁺ channels characterised as high affinity binding sites for tetrodotoxin during development of the rat nervous and skeletal muscle systems. *Biochem. and Biophys. Res. Comm.* **110**, 894–901.

Magill-Solc, C. and McMahan, U. J. (1988) Motoneurones contain agrin-like molecules. *J. Cell Biol.* **107**, 1825–1833.

Massoulié, J. and Bon, S. (1982) The molecular forms of cholinesterases in vertebrates. *Ann. Rev. Neurosci.* **5**, 57–106.

Mathers, D. A. and Thesleff, S. (1978) Studies on neurotrophic regulation of murine skeletal muscle. *J. Physiol.* **282**, 105–114.

Merlie, J. P. and Smith, M. M. (1986) Synthesis and assembly of acetylcholine receptor, a multisubunit membrane glycoprotein. *J. Memb. Biol.* **91**, 1–10.

Michler, A. and Sakmann, B. (1980) Receptor stability and channel conversion in the subsynaptic membrane of developing mammalian neuromuscular junction. *Dev. Biol.* **80**, 1–17.

Nastuk, M. A. and Fallon, J. R. (1993) Agrin and the molecular choreography of synapse formation. *TINS* **16**, 76–81.

Neher, E. and Sakmann, B. (1976) Noise analysis of drug induced voltage clamp currents in denervated frog muscle fibres. *J. Physiol. (Lond.)* **258**, 705–729.

Nitkin, R. M., Smith, M. A., Magill, C. *et al.* (1987) Identification of agrin, a synaptic organisation protein in *Torpedo* electric organ. *J. Cell Biol.* **105**, 2471–2478.

O'Brien, R. A. D. and Vrbová, G. (1978) Acetylcholine synthesis in nerve endings to slow and fast muscles in developing chicks: effects of muscle activity. *Neurosci.* **3**, 1227–1230.

Peng, P. H. (1987) Development of the neuromuscular function in tissue culture. *CRC Crit. Rev. Anat. Sci.* **1**, 91–131.

Peng, H. B., Cheng, P. C. and Luthert, P. W. (1981) Formation of acetylcholine receptor clusters induced by positively charged latex beads. *Nature* **292**, 831–834.

Pilar, G., Tuttle, J. and Vaca, K. (1981) Functional maturation of motor nerve terminals in the avian iris: Ultrastructure transmitter metabolism and synaptic reliability. *J. Physiol. (Lond.)* **321**, 175–193.

Redfern, P. and Thesleff, S. (1971) Action potential generation in denervated rat skeletal muscle. II. The action of tetrodotoxin. *Acta Physiol. Scand.* **82**, 70–78.

Sakmann, B. (1978) Acetylcholine-induced ionic channels in rat skeletal muscle. *Fed. Proc.* **37**, 2654–2659.

Sakmann, B. and Brenner, H. R. (1978) Changes in synaptic channel gating during neuromuscular development. *Nature* **276**, 401–402.

Salpeter, M. M. (1987) Development and neural control of neuromuscular function and of the junctional acetylcholine receptor, in *The Vertebrate Neuromuscular Junction* (ed. M. M. Salpeter), Alan Liss, pp. 55–115.

Salpeter, M. M. and Loring. R. H. (1986) Nicotinic acetylcholine receptors in vertebrate muscle: Properties, distribution and neural control. *Prog. Neurobiol.* **25**, 297–352.

Sanes, J. R. (1989) Extracellular matrix molecules that influence neural development. *Ann. Rev. Neurosci.* **B12**, 491–516.

Schuetze, S. M. and Role, L. W. (1987) Developmental regulation of nicotinic acetylcholine receptors. *Ann. Rev. Neurosci.* **10**, 403–457.

Sheard, P. W., Duxson, M. J. and Harris, A. J. (1991) Neuromuscular transmission to identified primary and secondary myotubes: a re-evaluation of polyneuronal innervation patterns in rat embryos. *Dev. Biol.* **148**, 459–472.

Sherman, S. J. and Catterall, W. A. (1982) Biphasic regulation of development of the high affinity saxitoxin receptor by innervation in rat skeletal muscle. *J. Gen. Physiol.* **80**, 753–768.

Sherman, S. J. and Catterall, W. A. (1984) Electrical activity and cytosolic calcium regulate levels of tetrodotoxin-sensitive sodium channels in cultured rat muscle cells. *Proc. Nat. Acad. Sci. USA* **81**, 262–266.

Sketelj, J. and Brzin, M. (1985) Asymmetric molecular forms of acetylcholinesterase in mammalian skeletal muscle. *J. Neurosci. Res.* **14**, 95–103.

Sketelj, J., Crne-Finderle, N., Ribarič, S. and Brzin, M. (1991) Interactions between intrinsic regulation and neural modulation of acetylcholinesterase in fast and slow skeletal muscles. *Cell Mol. Neurobiol.* **11**, 35–54.

Slater, C. R. (1982) Neural influences on the postnatal changes in acetylcholine receptor distribution at nerve–muscle junctions in the mouse. *Dev. Biol.* **94**, 23–30.

Sytkowski, A. J., Vogel, Z. and Nirenberg, M. W. (1973) Development of acetylcholine receptor clusters on cultured muscle cells. *Proc. Nat. Acad. Sci. USA* **70**, 270–276.

Tennyson, V. M., Brzin, M. and Slotwiner, P. (1971) The appearance of acetylcholinesterase in the myotome of the embryonic rabbit. An electron microscope, cytochemical and biochemical study. *J. Cell Biol.* **51**, 703–721.

Tennyson, V. M., Brzin, M. and Kemzner, L. T. (1973) Acetylcholinesterase activity in myotube and muscle satellite cells of the foetal rabbit: An electron microscope, cytochemical and biochemical study. *J. Histochem. Cytochem.* **21**, 634–652.

Thesleff, S. and Molgo, J. (1983) A new type of transmitter relaxant at the neuromuscular junction. *Neurosci.* **9**, 1–8.

Tóth, V. L. and Karcsú, S. (1979) Ultrastructural localisation of the acetylcholine activity in the diaphragm of the rat embryo. *Acta Histochem. (Jena)* **64**, 148–156.

Vigny, M., Koenig, J. and Rieger, F. (1976) The motor endplate specific form of acetylcholinesterase: appearance during embryogenesis and reinnervation of rat muscle. *J. Neurochem.* **72**, 1347–1353.

Weiss, R. E. and Horn, R. (1986a) Functional differences between two classes of sodium channels in developing myoblasts and myotubes of rat skeletal muscle. *Science* **233**, 361–364.

Weiss, R. E. and Horn, R. (1986b) Single channel studies of TTX-sensitive and TTX-resistant sodium channels in developing rat muscles reveal different open channel properties. *Ann. NY Acad. Sci.* **479**, 152–161.

Witzemann, V., Barg, B., Criado, M. *et al.* (1989) Developmental regulation of five subunits specific mRNAs encoding acetylcholine receptor subtypes in rat muscle. *FEBS Lett.* **242**, 419–424.

Xie, Z. and Poo, M.-M. (1986) Initial events in the formation of neuromuscular synapse: Rapid induction of acetylcholine release from embryonic neurones. *Proc. Nat. Acad. Sci. USA* **83**, 7069–7073.

Young, S. H. and Poo, M.-M. (1983) Spontaneous release of transmitter from growth cones of embryonic neurones. *Nature* **305**, 634.

Zelená, J. and Szentágothay, J. (1957) Verlagerung der Lokalisation spezifischer Cholinesterase während der Entwicklung der Muskel Innervation. *Acta Histochem.* **3**, 248–296.

Tonic and phasic muscle fibres of lower vertebrates and birds
4

In amphibians and birds, two types of skeletal muscle fibres are well recognized: the **slow tonic** and **fast twitch** fibres. The classification was first based on the different contractile properties of the two types of muscle and their response to ACh (Figure 4.2).

The iliofibularis muscle of the frog contains a bundle of muscle fibres which respond to small doses of applied ACh with a sustained slow contracture. Sommerkamp (1928) called this bundle the *tonus bündel* or **tonic bundle** of the muscle and distinguished it from the rest of the muscle, which responded with a small transient shortening to much higher concentration of ACh. The observed behaviour of the iliofibularis muscle to ACh was the first indication of two fundamental differences between slow tonic and twitch muscle fibres: their degree of sensitivity to ACh and their speed of contraction. In addition to these differences, the muscle fibres differ from each other in their pattern of innervation, characteristic membrane properties and biochemical characteristics (Figures 4.1 and 1.2). Moreover, the morphological and physiological characteristics of nerve fibres to these two types of muscle fibre are different.

4.1 SENSITIVITY TO ACh

The sensitivity to ACh of endplates of slow tonic and fast twitch muscles is rarely compared directly yet the available information suggests that the endplates of fast muscles might be more sensitive to ACh than those of slow tonic muscles. In the chick, the distribution of sensitivity has been mapped in muscle fibres of the fast posterior (**PLD**) and slow anterior (**ALD**) part of the latissimus dorsi muscle, by numerous investigators using iontophoretic application of ACh and intracellular recording (Fedde, 1969; Bennett *et al.*, 1973; Lebeda *et al.*, 1974; Vyskočil and Vyk-

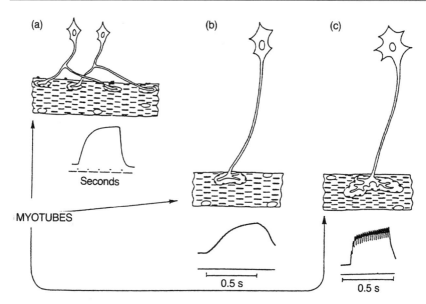

Figure 4.1 Different types of muscle fibre and their innervation patterns: (a) multiply innervated tonic muscle fibre; each muscle fibre is supplied by more than one motoneurone, endplates are distributed on several sites along the muscle; (b) slow twitch muscle fibre, innervated by a relatively small motoneurone at a single endplate; (c) fast twitch fibre, innervated by a large motoneurone at a single endplate. The traces under each muscle fibre illustrate typical isometric contractions. Note that the time marker on (a) indicates 1 s intervals; whereas the bars in (b) and (c) indicate 0.5 s.

lický, 1974). Although the authors do not always comment, differences in sensitivity to ACh were found between the endplates of muscles from ALD and PLD muscles. For example, the peak sensitivity in the PLD was always greater than that of ALD (Vyskočil and Vyklický, 1974).

In the frog iliofibularis and pyriformis muscles, the slow tonic muscle fibres which are multiply innervated show areas of high ACh sensitivity corresponding to endplates, situated about 9 mm from each other. Their maximum sensitivity is also lower than that of the phasic fast fibres (Nasledov, 1969; Nasledov and Thesleff, 1974). Thus, while slow tonic muscle fibres have several endplates, each individual endplate is supplied by a smaller nerve terminal that releases relatively little transmitter per impulse on to a post-synaptic membrane that is less sensitive to ACh than that of a fast muscle. The single nerve terminal of fast twitch muscle fibres releases relatively large amounts of transmitter per impulse on to a post-synaptic membrane that is very sensitive to the transmitter. There may be a correlation between the size of the nerve terminal, the amount of transmitter released, the structural differentiation of the post-synaptic membrane and the sensitivity of the endplates to ACh.

Tonic muscles respond to small doses of diffusely applied ACh by a sustained contraction, whereas twitch muscles do not (Sommerkamp, 1928; Rückert, 1930; Wachholder, 1930). As mentioned above, it was this observation that first indicated the existence of two distinct types of muscle fibre. At this time there was no explanation as to why these muscles had different sensitivities to ACh. Experiments of Langley (1907), however, suggested a possible explanation of this observation. He noticed that the sensitivity of the tonic rectus abdominis muscle to nicotine was diffuse, in contrast to the localized sensitivity of the sartorius muscle: nicotine evoked a contraction of the tonic muscle wherever it was applied, while it was only effective in eliciting a twitch contraction of the sartorius muscle when applied to the region of nerve–muscle contact. Langley proposed that the response of the muscles to particular drugs was due to the presence of specific receptor substances.

Buchthal and Lindhard (1939) demonstrated that application of ACh to the surface of a twitch muscle fibre of the lizard was effective only at the nerve–muscle junction. In the frog, sartorius muscle fibres responded to nicotine (Langley, 1907) or ACh (Kuffler, 1943) only when these drugs were applied to the endplate. These experiments showed that the chemo-sensitive area of the twitch muscle fibres is confined to the endplate region of the muscle, on to which ACh is normally released. This being so, the higher sensitivity of the tonic muscles to ACh (Figure 4.2) is probably due to the large size of the sensitive area on the fibre membrane. The chemosensitive area is greater because tonic muscle fibres have several sites of innervation, and the sensitivity to ACh of the membrane outside the endplate region is high in comparison with that of the twitch fibres, where the sensitivity to ACh outside the endplate region is negli-gible (Figure 4.2). Moreover, the passive cable properties of the slow muscle fibre may enhance the degree and spread of the depolarization in response to ACh.

4.2 MEMBRANE PROPERTIES

The electrical properties of the membrane appear to be related to the amount of transmitter released from the nerve terminal per impulse. If enough transmitter is released to elicit an action potential, the membrane develops the ability to conduct action potentials, and this indeed takes place in the fast twitch muscle fibres. These fibres also have a relatively low input resistance and a short space constant. When too little transmit-ter is released to reach threshold for the initiation of the action potential, the muscle fibre becomes excited by the endplate potential and its dec-remental spread along the fibre. This is the case on slow tonic fibres. To enable the potential to excite a large area of the fibre, the muscle mem-brane develops a high input resistance, so that excitatory inputs produce

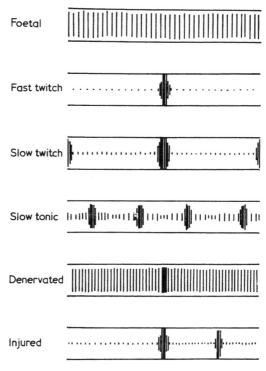

Figure 4.2 Distribution of acetylcholine (ACh) sensitivity on different types of skeletal muscles. The density and size of the vertical lines indicates the degree of chemosensitivity: the closer and longer the lines, the greater the sensitivity.

a greater voltage change, and a long space constant, so that this excitation is then conducted for a relatively long distance along the muscle fibre.

In the frog there is a great difference between the electrical characteristics of membranes of twitch and tonic muscles (Burke and Ginsborg, 1956; Adrian and Peachey, 1965; Stefani and Steinbach, 1969). Similarly, in birds, although the difference is not as great as in frogs, the membrane characteristics of twitch and tonic muscles are very different. In the chicken, for example, the time constant and the membrane resistance per unit area of the slow tonic anterior latissimus dorsi (ALD) muscle are, respectively, ten times and eight times larger than the corresponding values for the membrane of the fast twitch posterior latissimus dorsi (PLD) muscle (Fedde, 1969).

The time constant of twitch muscle fibres in the frog (Adrian and Peachey, 1965) and the chicken is short (Fedde, 1969; Gordon *et al.*, 1977) as compared with long time constants of tonic fibres in the same animals (Stefani and Steinbach, 1969). Thus in all twitch fibres local potentials will decay rapidly within a short distance along the membrane. Therefore, in these muscles, neuromuscular transmission is such that every nerve impulse will cause a muscle impulse to be propagated along the entire muscle fibre, excite the entire muscle fibre and bring about its mechanical activation in response to each nerve impulse.

Multiply innervated muscle fibres of the frog and birds differ in that some of the tonic muscle fibres of birds are sometimes able to generate and propagate action potentials *in vitro* (Ginsborg, 1960b; Harris *et al.*, 1973), while frog muscles normally do not do so (Kuffler and Vaughan-Williams, 1953a; Burke and Ginsborg, 1956; Orkand, 1963; Stefani and Steinbach, 1969). While this is so, action potentials are not recorded in chick tonic muscles following nerve stimulation *in vivo* except at very high rates of stimulation (Hník *et al.*, 1967). Therefore in frogs and birds local potentials are normally conducted between synapses decrementally, the area over which they spread being determined by the size of the potential and the characteristics of the membrane.

Although the tonic muscle fibres of the frog do not normally generate action potentials, the membrane does acquire to some extent the ability to conduct action potentials following denervation of the muscles (Miledi *et al.*, 1971). This is so in spite of the finding that the other electrical characteristics of the membrane (i.e. membrane resistance and time and space constants) are not appreciably altered by interfering with the inner-vation of the muscle. When slow tonic muscles are treated with botulinum toxin, which induces changes in chemosensitivity that have been compared with those following denervation of the muscle, the membrane did not acquire the ability to conduct action potentials (Miledi and Spitzer, 1974). In this case, transmitter release and local potentials were dramatically reduced but not totally abolished. These findings suggest that the ability to conduct action potentials may be a rudimentary property of all muscle membranes.

4.3 CONTRACTILE PROPERTIES

The nature of the response to ACh also reveals another fundamental difference between the two types of muscle, namely their characteristic speed of contraction. The tonic fibres respond to ACh as well as to electrical stimulation or to immersion in solutions containing high concentrations of potassium with a slow, maintained contraction which shows little if any fatigue (Wachholder and von Ledebur, 1931; Wachholder and Nothmann, 1932). Non-tonic twitch muscles respond to these stimuli with a fast contraction.

Normally the characteristic contractile responses are elicited by two distinct groups of nerve fibres in the frog ventral roots which differ in size. Tasaki and Tsukagoshi (1944) found that repetitive stimulation of the small nerve fibres caused slow-graded contractions, and stimulation of the large nerve fibres elicited fast twitch responses in the muscles.

Kuffler and Vaughan-Williams (1953a, b) showed conclusively that there are two distinct types of muscle fibre and that no single muscle fibre ever gave both responses. Each type of muscle fibre is innervated by only one nerve type: the small nerve fibres supplied slow tonic fibres and the large nerve fibres the fast twitch fibres. The muscles together with their nerves are so different in structure and function that they may be considered as two separate and distinct functional nerve–muscle systems, the slow system with small nerves being quite separate from the twitch system and responsible for the slow contractions in a mixed muscle such as the frog iliofibularis muscle.

In these experiments tension was recorded under isometric conditions. Such recordings do not allow us to assess reliably the exact mechanical characteristics of the contractile component. A more adequate way of studying the behaviour of the contractile components is achieved by establishing the force–velocity relations of a muscle.

The force–velocity relation of twitch fibres is well known from many studies on the frog sartorius muscle (Hill, 1950) but relatively few studies were performed on slow tonic muscles. Lännergren (1975) isolated single slow tonic fibres from *Xenopus* and found that their speed of shortening was much less than that of fast twitch muscle fibres. The force–velocity relation is also different in slow tonic muscles from that in fast twitch muscles (Woledge, 1968). It was suggested that this may reflect a more efficient contractile process of the slow muscle fibres.

4.4 MORPHOLOGY IN RELATION TO FUNCTION

Fast, focally innervated twitch muscle fibres have a distinct fibrillar (*'Fibrillenstruktur'*) appearance in the light microscope (Krüger, 1929, 1952; Hess, 1960, 1961) owing to the regular arrangement and uniform width of the myofibrils, separated by sarcoplasm (Peachey, 1961; Page, 1965, 1969; Hess, 1967). Slow, multiply innervated fibres have a more granular and indefinite (*'Felderstruktur'*) appearance, owing to the irregular width of the myofibrils and their less discrete separation. At the ultrastructural level, the slow fibres of the frog lack a distinct M-line at the level of the A-bands and have wavy thick Z-lines with irregular packing of the I-filaments. Furthermore, the slow fibres lack the regularly arranged transverse tubular system and sarcoplasmic reticulum of the fast fibres; the slow fibres have irregularly distributed transverse tubules and a relatively smaller number of triads which represent the sites of contact between the T-tubules and the sarcoplasmic reticulum (Page, 1965, 1969; Hník *et al.*, 1967; Hess, 1967, 1970). In frog slow tonic muscle fibres, triads are rarely found (Hess, 1960). The different appearance of the two types of muscle fibre as seen in birds is illustrated in Figure 4.3.

These structural specializations are paralleled by differences in the way the excitation–contraction coupling is brought about. In fast fibres, coupling of excitation of the surface membrane to the myofilaments occurs only at the level of the I-band (Peachey and Huxley, 1960). Excitation of the surface membrane spreads into the T-tubules of the sarcotubular system and causes a release of calcium from the lateral sacs of the sarcoplasmic reticulum, which in turn activates the neighbouring myofibrils (Sandow, 1965, 1970; Huxley, 1974). Slow tonic fibres respond to local electrical activation at both A- and I-bands and activation spreads as far longitudinally as radially into the muscles (Peachey and Huxley, 1960). Such an arrangement may be effective in the slow muscles where excitation at local sites on the muscle membrane spreads decrementally between endplates.

In fast twitch muscles, in which rapid activation of the contractile machinery is required, diffusion of calcium from the surface membrane would be too slow and in this system the T-tubular system is well elaborated so that calcium is released close to the myofibrils and excitation–contraction coupling is rapid. Thus in twitch muscles the excitation of the nerve propagates rapidly from the endplate along the entire muscle membrane and this in turn is transmitted to the contractile machinery by a system of internal tubes. Evidence for the correlation between structure and function was obtained by Peachey and Huxley (1962) in experiments in which they studied both the physiological response and the ultrastructure of single fibres dissected from the iliofibularis muscle of the frog. They found that the fibres which responded to direct electrical shocks with local slow contractions had large ribbon-like fibrils fused together, which gave the fibres the 'Felderstruktur' appearance described by Krüger. Twitch fibres had well-defined myofibrils delineated by sarcoplasmic elements.

In the slow muscles of the chicken, unlike those of amphibians, association of T-tubules and the sarcoplasmic reticulum does occur at triads or dyads (Hess, 1967; Hník et al., 1967) but these are much less abundant than in the fast twitch muscle fibres. The slow tonic muscle fibres of the chicken can use a local mechanism of excitation–contraction coupling typical of the multiply innervated muscle as well as a more rapid form of coupling when excitation is propagated over the entire membrane.

4.5 METABOLIC DIFFERENTIATION

The fast twitch and slow tonic muscle fibres are specialized for two different types of contraction: fast transient or slow maintained, respectively. The metabolic specialization of the muscle fibres for these functions

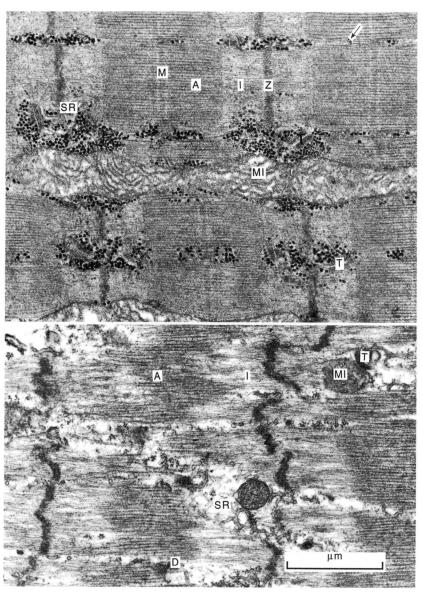

Figure 4.3 Longitudinal sections showing differences between fast twitch and slow tonic muscle fibres in 5-week-old chick (x 27 000). (a) Part of a fibre from the posterior latissimus dorsi muscle. The myofibrils are regularly organized into (I) and (A) bands; the latter band is transected by an H-zone with M-line (M). The Z-lines are straight; the membrane systems are well developed; the sarcoplasmic reticulum (SR) is abundant around the I-bands At this age, triads (T) with both longitudinal and transverse orientation occur towards the periphery of the I-bands; the sarcoplasmic reticular tubules have dense contents. Glycogen (G) is abundant. The mitochondria (MI) occur in rows between the myofibrils. (b) Part of a fibre from the anterior latissimus dorsi muscle. The myofibrils are irregular, so that there is no discrete junctions between the A-bands and I-bands; the Z-lines are also irregular; H-zones and M-lines are absent. The sarcoplasmic reticulum forms an irregular network around the myofibrils. Both dyads (D) and triads occur; the sarcoplasmic reticular tubules have dense contents; the mitochondria are small and scattered.

is apparent in twitch fibres of many vertebrates in which muscles can be regarded functionally as slow or fast (see Gutmann, 1970; Lännergren, 1975; Pette and Staron, 1990). It is known that muscles that are predominantly fast or slow contracting show differences in the activities of enzymes involved in oxidative metabolism (higher in the slow twitch muscles than fast muscles) and those of anaerobic glycolysis, which are higher in the fast muscle fibres (Nachmias and Padykula, 1958; Dubowitz and Pearse, 1960a, b; Pette and Staron, 1990).

The slow twitch muscle fibres and slow tonic fibres of birds and mammals are also characterized by a higher oxidative enzyme capacity which goes together with a high mitochondrial density (Page, 1965, 1969; Hník et al., 1967; Barnard et al., 1982) and a lower capacity for anaerobic glycolysis. Fast fibres of the chick, on the other hand, have a low oxidative enzyme capacity and rely on anaerobic pathways for supply of energy (Koenig and Fardeau, 1973; Hudlická et al., 1973; Melichna et al., 1974; Gordon et al., 1977). This metabolic specialization of muscles with different functions is well matched to their pattern of activation and maintenance of tension. Fast muscle fibres are capable of rapid tension development. Their high capacity for anaerobic glycolysis is essential for a rapid supply of energy to the contracting muscle. Slow muscle fibres, on the other hand, can maintain tension for long periods at the expense of being unable to develop as large a tension as fast muscles per surface area. They often have a high storage capacity for oxygen in their myoglobin which gives them their characteristic red colour, noticed first in mammalian red slow twitch muscles by Kühne (1863).

Exceptions to this are the tonic slow muscle fibres of amphibians. These have very few mitochondria and small energy stores (Lännergren and Smith, 1966; Engel and Irwin, 1967; Smith and Ovalle, 1973). This makes their high resistance to fatigue seem surprising, a situation that may partly be explained by their very low shortening speed which implies a low turnover of crossbridges, so that maintenance of tension may be more economic in these fibres (Lännergren, 1975). However, the difference between the slow tonic fibres of amphibians and those of birds and mammals, which do contain high levels of oxidative enzymes and many mitochondria, is striking and may be related to the low temperature at which these amphibian muscles operate.

Basic differences in proteins, directly related to the rate of development of tension, exist between fast twitch and slow tonic muscles (Melichna et al., 1974). The delay in excitation–contraction coupling introduced by calcium from the membrane to the myofibrils does not in itself explain the difference in the time course of development of tension, since calcium activation of skinned fibres from slow tonic muscles is also slower than that of skinned fast twitch fibres (Page, 1969; Constantin, 1974; Lännergren, 1975). It has been known that the velocity of shortening is related to the rate of splitting of ATP by actomyosin since Prosser (1961) first pointed out that the speed of contraction in fast and slow twitch muscles of invertebrates and vertebrates appeared to be directly related to the actomyosin ATPase activity; the exact relationship between speed and ATPase activity was studied in a variety of fast and slow muscles (see Close, 1972). Bárány (1967) showed clearly that for a large number of invertebrate and vertebrate muscles the intrinsic speed of contraction is proportional to the specific activity of myosin ATPase and suggested that the rate of ATP splitting is one of the major rate-limiting steps in the velocity of muscle shortening.

Two slow myosin HC isoenzymes have been identified in the tonic chick ALD muscle SM1 and SM2 (Matsuda *et al.*, 1982). The ratio between these two isoforms depends on the age of the animal (Hoh, 1979; Reiser *et al.*, 1988) and on the load imposed upon the ALD muscle. The more loaded muscles express higher amounts of the SM2 isoforms (Reid *et al.*, 1989). Thus, as in twitch muscles (see Chapter 5), the particular myosin HC in tonic muscle fibres seems to be associated with the function of the muscle.

4.6 PATTERN OF INNERVATION

It is now well established that while fast twitch muscle fibres are focally innervated, i.e. each muscle fibre is contacted by a nerve fibre at a single site, each slow tonic muscle fibre is contacted by several nerve endings along its surface.

In the frog iliofibularis muscle, Kuffler and Vaughan-Williams (1953a) presented the first evidence for multiple innervation in the slow tonic fibres by showing that when the nerves that supply the slow fibres were stimulated selectively, after blocking the nerve to the fast fibres by a nodal block, local potentials could be recorded from almost any point on the tonic muscle fibres. The pattern of innervation of tonic muscle fibres was later verified by the histochemical demonstration of cholinesterase activity at multiple sites on the muscle in frogs (Couteaux, 1955, 1963) and in chicken (Ginsborg and McKay, 1961). Not only do the muscle fibres have several sites of innervation but also individual muscle fibres are supplied by several axons. When the nerve was stimulated and the current increased in a step-wise fashion, the junctional potentials became increasingly more complex as more axons were recruited.

Focally innervated muscle fibres, on the other hand, have a single endplate which stains for cholinesterase, and local endplate potentials can be recorded only at a single nerve–muscle junction in a partially curarized nerve–muscle preparation (Kuffler, 1943; Fatt and Katz, 1951); each muscle fibre is supplied by only one axon. The pattern of innervation of the two types of muscle fibres is different and also the functional and structural specialization of individual endplates differs in the two types of muscle fibre.

Fast twitch muscles are usually contacted by a large nerve terminal abundantly filled with clear vesicles, while slow tonic muscle fibres are supplied by small terminals (Zelená and Sobotková, 1973). In slow tonic fibres the specialization of the post-synaptic membrane is poorly developed and junctional folds are almost completely absent (Hess, 1960; Page, 1965): nerve terminals to slow muscles release less transmitter per nerve impulse than those to fast muscles. In the chick, for example, the quantal content of endplates of slow muscles is lower than that of fast muscles. The endplate potentials evoked by stimulating the 'slow' nerve fibres are characterized by large variation in amplitude and frequent failures (Vyskočil et al., 1971) and are readily distinguished from those evoked by 'fast' nerves by their failure to initiate action potentials except at high rates of stimulation (Ginsborg, 1960b). How does this different pattern of innervation develop?

Why some muscle fibres develop several endplates on their surface is not known. It is interesting that, although both focally and multiply innervated chick muscle fibres are equally sensitive to ACh during early embryonic development, the sensitivity of the focally innervated muscle rapidly decreases while that of the multiply innervated muscle does not change much (Gordon and Vrbová, 1975). This correlates well with the findings that nerve terminals to focally innervated avian muscle release more transmitter than those to multiply innervated muscle fibres (Fedde, 1969; Vyskočil et al., 1971). In experiments in which the amount of transmitter released was reduced by chronic administration of hemicholinium during embryonic development, the sensitivity to ACh of focally innervated muscles failed to decrease (Gordon and Vrbová, 1975; Burden, 1977a, b). It was found that in these muscles the twitch fibres had become multiply innervated by the ingrowing nerves. It could be that if the nerve releases such a small amount of transmitter, which is insufficient to generate an action potential, only a small area of the muscle fibre will be activated by the decremental spread of the endplate potential and the fibres outside this area will remain sensitive to the transmitter (Figure 3.1). Thus, in a muscle where the nerves release small amounts of ACh, a response is produced which spreads over only a small area of the membrane and further nerve–muscle contacts can be made outside this area. If the large response elicited normally on the fast muscle is lowered – either by reducing the amount of transmitter released so that the situation becomes more akin to the tonic nerve–muscle system, or by reducing the postsynaptic response with post-synaptic d-tubocurarine – the nerves that normally innervate the muscle fibres at a single site will contact the fibres at several sites along each muscle fibre. When chick embryos were treated chronically with botulinum toxin, which also reduces the release of ACh from the nerves, an abundance of endplates was obvious in the leg muscles (Giacobini-Robecchi *et al.*, 1976).

The distance between successive endplates on developing muscle fibres appears to depend on the size of the endplate potential. In experiments where the endplate potential was reduced by administration of curare during embryonic development, the distance between successive endplates on the multiply-innervated anterior latissimus dorsi muscle was reduced and several endplates developed on PLD muscle fibres (Gordon *et al.*, 1974; Dahm and Landmesser, 1991). In embryos, unlike mature animals, the cable properties of slow and fast muscle fibres are similar (Purves and Vrbová, 1974; Gordon *et al.*, 1977) so that the size of the endplate potential depends on the amount of ACh released. It is proposed that the amount of ACh per impulse that is released from the nerve terminals will determine the amplitude and the area of decremental spread of the local depolarization, and hence the size of the area that will become desensitized on nerve–muscle contact.

It is interesting that electrical stimulation of the spinal cord during early embryonic development also leads to an increase in neuromuscular contacts on PLD muscle fibres (Renaud *et al.*, 1978; Fournier le Ray and Fontaine-Perus, 1991). This could be due to the fact that the immature nerve endings are activated at a time when they would otherwise be silent and allowed to mature. Thus, here too, the transmitter released from each ending would be less than if the nerve was activated normally, and so multiple contacts can be made. When the endplate potential is smaller, the area of desensitization is reduced so that the distance between successive endplates is reduced. Since the size of the endplate potential depends on the amount of transmitter released, each nerve fibre specifies the pattern of innervation of the muscle fibre it supplies. Thereafter the contractile characteristics and membrane properties of multiply and focally innervated muscle fibres differentiate.

4.7 HOW ARE TONIC AND PHASIC MUSCLE FIBRES USED?

The word 'tonus' has referred to postural activity since Galen first coined the term. This activity was defined more clearly, by the work of Sherrington, as reflex contraction of muscles that are involved in postural adjustments of the animal (Sherrington, 1904). The tonic muscles of lower vertebrates and birds have been found to be involved in long-lasting contractions that maintain parts of the body in a particular position. For example, in chickens the anterior part of the latissimus dorsi muscle (ALD) stops the wing of the chicken from drooping and holds it in its normal position. This muscle is known to be composed almost entirely of slow tonic muscle fibres (Ginsborg, 1960a, b; Page and Slater, 1965).

In the male frog, tonic long-lasting contractions of the forearm muscles occur during clasping (amplexus). No EMG activity can be recorded from the working muscles while these contractions take place, presumably because only the slow tonic fibres contract and these are activated by slow junctional potentials (Kuffler and Vaughan-Williams, 1953a). Only when the frog was disturbed during clasping could EMG activity from the forearm muscles by recorded, since only then were fast twitch muscles recruited (Kahn, 1919). The wide distribution of slow tonic muscle fibres throughout the body of amphibians led Kuffler and his colleagues to suggest that they play an important role in the maintenance of posture in general, and not only during clasping (Kuffler *et al.*, 1947; Kuffler and Gerard, 1947). While investigating the function of the tonic muscles these authors found that motoneurones to tonic muscle fibres were more readily excited by afferent stimulation. In the lightly anaesthetized and decapitated frog, submaximal stimulation of the proximal end of a cut sensory nerve excited the axons to the slow muscle fibres at low rates of stimulation. The large axons to fast muscle fibres were excited only when either the stimulus strength or the frequency was increased. The finding that increasing the rate of stimulation of an afferent nerve without altering the strength of the stimulus can excite the motoneurone to fast muscle fibres leads to the conclusion that the same afferent fibres excite both types of motoneurones. Although motoneurones to slow muscle fibres are recruited first, they maintain their discharge even when the frequency of stimulation increases so that motoneurones to fast muscles are also recruited. This finding indicates that the slow tonic and fast twitch muscle fibres which are intermingled within the same muscle (iliofibularis) are functional synergists where the slow fibres are extremely economical in the use of energy and can maintain the posture developed by the twitch fibres.

The early investigations, particularly of Kuffler and Vaughan-Williams (1953a, b), also revealed differences between the activity patterns transmitted by the nerves to slow and fast muscle fibres that seemed to match the characteristic properties of the muscles they supplied, and suggested the existence of two separate nerve–muscle systems. The nerves to slow muscle fibres discharge at slow frequencies and this leads to the development of a graded and well-maintained contraction which is ideally suited for the maintenance of posture over long periods. The nerves to fast muscles discharge at high frequencies and this leads to a rapid development of tension, which usually is not maintained. In a heterogeneous muscle the slow fibres act to maintain the tension produced by the fast fibres. In this way the slow fibres, even though they are not numerous, can maintain considerable tension and play an important role in postural reflexes.

CONCLUSIONS

The slow tonic and fast twitch muscle fibre of amphibians and birds have fundamentally different properties and these are matched to their innervation pattern. Fast twitch muscle fibres receive innervation at a single site. Stimulation of the motor axon initiates a muscle action potential which is propagated along the entire membrane and initiates a rapid contractile response. The structural, membrane and biochemical properties of the muscle fibres are such that they enable the occurrence of this rapid contractile response.

Slow tonic muscle fibres, on the other hand, are multiply innervated by nerve endings which release small amounts of transmitter. The muscle fibres do not readily generate action potentials, but their passive membrane properties are specialized so as to conduct the local junctional potentials passively and depolarize the muscle membrane between the sites of nerve–muscle contacts.

The internal membrane system is either absent or poorly developed and excitation–contraction coupling is initiated locally by the depolarization of the surface membrane. Contraction is graded accordingly by the extent and area of membrane depolarization. Contractions are slow, but muscle fibres are able to maintain force for long periods.

Nerves to slow tonic muscle fibres discharge at slow frequencies and are readily recruited. Nerves to twitch muscles fire at higher rates, and these muscle fibres are adapted for rapid movements where the force does not necessarily have to be maintained.

The mechanism by which the different pattern of innervation develops in these two distinct populations of muscle fibres is discussed.

REFERENCES

Adrian, R. H. and Peachey, L. P. (1965) The membrane capacity of frog twitch and slow muscle fibres. *J. Physiol. (Lond.)* **181**, 324–336.

Bárány, M. (1967) ATP-ase activity of myosin correlated with speed of shortening. *J. Gen. Physiol.* **50**, Suppl. 2, 197–218.

Barnard, E. A., Lyles, J. M. and Pizey, J. A. (1982) Fibre types in chicken skeletal muscle and their changes in muscular dystrophy. *J. Physiol.* **331**, 333–355.

Bennett, M. R., McLachlan, E. M. and Taylor, R. (1973) The formation of synapses in reinnervated mammalian striated muscle. *J. Physiol.* **233**, 501–517.

Buchthal, F. and Lindhard, J. (1939) *The Physiology of the Striated Muscle Fibre*, Ejnar Munksgaard, Copenhagen.

Burden, S. (1977a) Development of the neuromuscular junction in the chick embryo. The number, distribution and stability of acetylcholine receptors. *Dev. Biol.* **57**, 317–329.

Burden, S. (1977b) Acetylcholine receptors at the neuromuscular junction: Developmental changes in receptor turnover. *Dev. Biol.* **61**, 79–85.

Burke, W. and Ginsborg, B. L. (1956) The electrical properties of the slow muscle fibre membrane. *J. Physiol.* **132**, 586–598.

Close, R. (1972) Dynamic properties of mammalian skeletal muscles. *Phys. Revs.* **52**, 129–197.

Constantin, L. L. (1974) Contractile activation in frog skeletal muscle. *J. Gen. Physiol.* **63**, 657–674.

Couteaux, R. (1955) Localisation of cholinesterases at neuromuscular junctions. *Int. Rev. Cytol.* **4**, 335–375.

Couteaux, R. (1963) The differentiation of synaptic areas. *Proc. Roy. Soc. B* **158**, 457–480.

Dahm, L. M. and Landmesser, L. T. (1991) The regulation of synaptogenesis during normal development and following activity blockade. *J. Neurosci.* **11**, 238–255.

Dubowitz, V. and Pearse, A. G. E. (1960a) A comparative histochemical study of oxidative enzyme and phosphorylase activity in skeletal muscle. *Histochemie* **2**, 105–117.

Dubowitz, V. and Pearse, A. G. E. (1960b) Reciprocal relationship of phosphorylase and oxidative enzyme in skeletal muscle. *Nature*, **185**, 701–702.

Engel, W. K. and Irwin, R. I. (1967) A histochemical physiological correlation of frog skeletal muscle fibres. *Am. J. Physiol.* **213**, 617–631.

Fatt, P. and Katz, B. (1951) An analysis of the endplate potential recorded with an intracellular electrode. *J. Physiol.* **115**, 320–370.

Fedde, M. R. (1969) Electrical properties of singly and multiply innervated avian muscle fibres. *J. Gen. Physiol.* **53**, 624–637.

Fournier le Ray, C. and Fontaine-Perus, J. (1991) Influence of spinal cord stimulation on the innervation pattern muscle fibres in vitro. *J. Neurosci.* **11**, 3840–3850.

Giacobini-Robecchi, M., Giacobini, G., Filagamo, G. and Changeaux, J.-P. (1976) Effets comparés de l'injection chronique de toxine de *Naja nigrocollis* et de toxine Botulinique A sur le development des racines dorsales et ventrales de la moelle épinière d'embryons de poulet. *C. R. Acad. Sci. Paris* **283**, 271–274.

Ginsborg, B. L. (1960a) Spontaneous activity in muscle fibres of the chick. *J. Physiol. (Lond.)* **150**, 707–717.

Ginsborg, B. L. (1960b) Some properties of avian skeletal muscle fibres with multiple neuromuscular junctions. *J. Physiol. (Lond.)* **154**, 581–598.

Ginsborg, B. L. and Mackay, B. (1961) A histochemical demonstration of two types of motor innervation in avian skeletal muscle. *Bibl. Anat.* **2**, 174–181.

Gordon, T., Jones, R. and Vrbová, G. (1976) Changes in chemosensitivity of skeletal muscle as related to endplate formation. *Prog. Neurobiol.* **6**, 103–136.

Gordon, T., Perry, R., Srihari, T. and Vrbová, G. (1977) Differentiation of slow and fast muscles in chickens. *Cell Tiss. Res.* **180**, 211–222.

Gordon, T., Perry, R., Tuffery, A. R. and Vrbová, G. (1974) Possible mechanisms determining synapse formation in developing skeletal muscles of the chick. *Cell Tiss. Res.* **155**, 13–25.

Gordon, T., Purves, R. D. and Vrbová, G. (1977) Differentiation of electrical and contractile properties of slow and fast muscle fibres. *J. Physiol.* **269**, 535–547.

Gordon, T. and Vrbová, G. (1975) Changes in chemosensitivity of developing chick muscle fibres in relation to endplate formation. *Pflügers Archiv* **360**, 349–364.

Gutmann, E. (1970) Open questions in the study of 'trophic' functions of the nerve cell. *Topical Problems in Psychiat. Neurol.* **10**, 54–61.

Harris, J. B., Marshall, N. W. and Ward, M. R. (1973) Action potential generation in singly and multiply innervated avian muscle fibres. *J. Physiol.* **232**, 51–52P.

Hess, A. (1960) The structure of extrafusal muscle fibres in the frog and their innervation studied by the cholinesterase technique. *Am. J. Anat.* **107**, 129–154.

Hess, A. (1961) Structural differences of fast and slow extrafusal muscle fibres and their nerve endings in chickens. *J. Physiol. (Lond.)* **157**, 221–231.

Hess, A. (1967) The structure of vertebrate slow and twitch fibres. *Invest. Ophthalmol.* **6**, 217.

Hess, A. (1970) Vertebrate slow muscle fibres. *Physiol. Rev.* **50**, 40–62.

Hill, A. V. (1950) A discussion on muscular contraction and relaxation: their physical and chemical basis. *Proc. Roy. Soc. B* **137**, 40–87.

Hník, P., Jirmanova, I., Vyklický, L. and Zelena, J. (1967) Fast and slow muscles of the chick after nerve cross union. *J. Physiol.* **193**, 309–325.

Hoh, J. F. Y. (1979) Developmental changes in chicken skeletal myosin isoenzymes. *Febs. Lett.* **98**, 267–270.

Hudlická, O., Pette, D. and Staudte, H. (1973) The relation between blood flow and enzyme activities in slow and fast muscles during development. *Pflügers Archiv. ges. Physiol.* **343**, 341–356.

Huxley, A. F. (1974) Review lecture: Muscular contraction. *J. Physiol.* **243**, 1–43.

Kahn, R. H. (1919) Beiträge zur Lehre vom Muskeltonus 1. Über den Zustand der Muskeln der vorderen Extremitäten des Frosches Während der Umklammerung. *Pflügers Archiv. ges. Physiol.* **177**, 294–303.

Koenig, J. and Fardeau, M. (1973) Étude histochemique de la morphologie des plaques motrices des grands dorseaux antérieur et postérieur du poulet et des modifications observées après dénervation et réinnervation homologue ou croisée. *Arch. Anat. Mic. Morph. Exp.* **62**, 249–267.

Krüger, P. (1929) Über einen möglichen Zusammenhang zwischen Struktur, Funktion und chemische Beschaffenheit. *Biol. Zentra.* **49**, 616.

Krüger, P. (1952) *Tetanus und Tonus der quergestreiften Skeletmuskeln der Wirbeltiere und des Menschen*, Akademieverlag, Leipzig.

Kuffler, S. W. (1943) Specific excitability of the endplate region in normal and denervated muscle. *J. Neurophysiol.* **6**, 99–110.

Kuffler, S. W. and Gerard, R. W. (1947) The small nerve motor system to skeletal muscle. *J. Neurophysiol.* **10**, 383–394.

Kuffler, S. W., Laporte, Y. and Ransmeier, R. E. (1947) The function of the frog's small-nerve motor system. *J. Physiol.* **10**, 395–408.

Kuffler, S. W. and Vaughan-Williams, E. M. (1953a) Properties of the 'slow' skeletal muscle fibres of the frog. *J. Physiol.* **121**, 318–340.

Kuffler, S. W. and Vaughan-Williams, E. M. (1953b) Small nerve junction potentials. The distribution of small motor nerves to frog skeletal muscles, and the membrane characteristics of the fibres they innervate. *J. Physiol.* **121**, 298–317.

Kühne, W. (1863) Über die Endignung der Nerven in den Muskeln. *Vischow Arch.* **27**, 508–533.

Langley, J. N. (1907) On the contraction of muscle, chiefly in relation to the presence of 'receptive' substances. Part 1. *J. Physiol. (Lond.)* **36**, 347–384.

Lännergren, J. (1975) Structure and function of twitch and slow fibres in amphibian skeletal muscle, in *Basic Mechanisms of Ocular Motility and their Clinical*

Implications (eds G. Lennerstrand and P. Bach-y-Rita), Pergamon Press, Oxford, pp. 63–85.

Lännergren, J. and Smith, R. S. (1966) Types of muscle fibres in toad skeletal muscle. *Arch. Physiol. Scand.* **68**, 263–274.

Lebeda, F. J., Warnick, J. E. and Albuquerque, E. X. (1974) Electrical and chemosensitive properties of normal and dystrophic chicken muscles. *Exp. Neurol.* **43**, 21–37.

Matsuda, R., Bardman, E. and Strohman, R. C. (1982) The two myosin isoenzymes of chicken anterior latissimus dorsi muscle contain different myosin heavy chains encoded by separate mRNAs. *Differentiation* **23**, 36–42.

Melichna, J., Gutmann, E. and Syrový, I. (1974) Developmental changes in contraction properties, adenosine triphosphate activity and muscle fibre pattern of fast and slow chicken muscles. *Physiol. Bohemoslov.* **23**, 511–520.

Miledi, R., Stefani, E. and Steinbach, A. B. (1971) Induction of the action potential mechanism in slow muscle fibres of the frog. *J. Physiol. (Lond.)* **217**, 737–754.

Miledi, R. and Spitzer, N. C. (1974) Absence of action potentials in frog slow muscle fibres paralysed by botulinum toxin. *J. Physiol. (Lond.)* **241**, 183–199.

Nachmias, V. T. and Padykula, H. A. (1958) A histochemical study of normal and denervated red and white muscles of the rat. *J. Biophys. Biochem. Cytol.* **4**, 47–54.

Nasledov, G. A. (1969) On the cholinoreceptor of a tonic muscle fibre. *J. Evol. Biochem. Physiol. (USSR)* **5**, 398–404.

Nasledov, G. A. and Thesleff, S. (1974) Denervation changes in frog skeletal muscle. *Acta Physiol. Scand.* **90**, 370–380.

Orkand, R. K. (1963) A further study of electrical responses in slow and twitch muscle fibres of the frog. *J. Physiol.* **167**, 181–191.

Page, S. G. (1965) A comparison of the fine structures of frog slow and fast twitch fibres. *J. Cell Biol.* **26**, 477–497.

Page, S. G. (1969) Structure and some contractile properties of fast and slow muscles of the chicken. *J. Physiol.* **205**, 131–145.

Page, S. G. and Slater, C. R. (1965) Observations on fine structure and rate of contraction of some muscles from the chicken. *J. Physiol.* **179**, 58–59P.

Peachey, L. D. (1961) Structure and function of slow striated muscle, in *Biophysics, Physiological and Pharmacological Actions*, American Association of Advanced Science, Washington, pp. 391–411.

Peachey, L. D. and Huxley, A. F. (1960) Local activation and structure of slow striated muscle fibres of the frog. *Fed. Proc.* **19**, 257.

Peachey, L. D. and Huxley, A. F. (1962) Structural identification of twitch and striated muscle fibres of the frog. *J. Cell Biol.* **13**, 117–180.

Pette, D. and Staron, R. S. (1990) Cellular and molecular diversities of mammalian skeletal muscle fibres. *Rev. Physiol. Biochem. Pharmacol.* **116**, 1–76.

Prosser, C. L. (1961) Muscle and electric organs, in *Comparative Animal Physiology* (ed. C. L. Prosser), Brown Saunders, Philadelphia, pp. 417–467.

Purves, R. D. and Vrbová, G. (1974) Some characteristics of myotubes cultured from slow and fast chick muscles. *J. Cell Physiol.* **84**, 97–100.

Reid, S. K., Kennedy, J. M., Shimizu, N. *et al.* (1989) Regulation of expression of avian slow myosin heavy chain isoforms. *Biochem. J.* **260**, 449–454.

Reiser, P. J., Greaser, M. L. and Moss, R. L. (1988) Myosin heavy chain composition of single cells from avian slow skeletal muscle is strongly correlated with velocity of shortening during development. *Dev. Biol.* **129**, 400–407.

Renaud, D., Le Douarin, G. H. and Khoskije, A. (1978) Spinal cord stimulation in chick embryo: effect on development of the posterior latissimus dorsi muscle and neuromuscular junctions. *Exp. Neurol.* **60**, 189–200.

Rückert, W. (1930). Die phylogenetische Bedingdtheit tonischer Eigenschaften der quergestreiften Wirbeltiermuskulatur. *Pflügers Archiv. ges. Physiol.* **226**, 323–346.

Sandow, A. (1965) Excitation contraction coupling in skeletal muscle. *Pharm. Revs.* **17**, 265–320.

Sandow, A. (1970) Skeletal muscle. *Ann. Rev. Physiol.* **32**, 87–138.

Sherrington, C. (1904) The correlation of reflexes and the principle of the final common path. *Brit. Assn.* **74**, 728–741.

Smith, R. S. and Ovalle, W. K. (1973) Varieties of fast and slow extrafusal muscle fibres in amphibian hind limb muscles. *J. Anat.* **116**, 1–24.

Sommerkamp, H. (1928) Das Substrat der Dauerverkurzung am Froschmuskel. *Arch. Exper. Pathol. Pharmakol.* **128**, 99–115.

Stefani, E. and Steinbach, A. B. (1969) Resting potential and electrical properties of frog slow muscle fibres. Effect of different external solutions. *J. Physiol.* **203**, 383–401.

Tasaki, I. and Tsukagoshi, M. (1944) Comparative studies on the activities of the muscle evoked by two kinds of motor nerve fibres. Part II: The electromyogram. *Jp. J. Med. Sci. Biophys.* **10**, 2.

Vyskočil, F. and Vyklický, L. (1974) Acetylcholine sensitivity of the chick fast muscle after cross union with the slow muscle nerve. *Brain Res.* **25**, 158–161.

Vyskočil, F., Vyklický, L. and Houston, R. (1971) Quantum content at the neuromuscular junction of fast muscle after cross-union with the nerve of slow muscle in the chick. *Brain Res.* **26**, 443–445.

Wachholder, K. (1930) Untersuchungen über 'Tonische' und 'nicht tonische' Wirbeltiermuskeln. II Mit acetylcholin- und tiegelischer Kontraktur in ihren Beziehungen zueinander und zur tetanischen Kontraktionsform. Die Plastizität der beiden Muskelarten. *Pflügers Archiv. ges. Physiol.* **226**, 255–273.

Wachholder, K. and Nothmann, E. (1932) Jahreszeitliches Schwanken zwischen 'Tonischen' und 'Nicht-tonischen' Verhalten von Wirbeltiermuskeln. *Pflügers Archiv. ges. Physiol.* **229**, 120–132.

Wachholder, K. and von Ledebur, F. (1931) Die Erregbarkeit der 'tonischen' und 'nicht-tonischen' Fasern eines Muskels bei direkter und indirekter Reizung. Ein kritischer Beitrag zur Frage des Isochronismus von Nerve und Muskel. *Pflügers Archiv. ges. Physiol.* **229**, 183–197.

Woledge, R. C. (1968) The energetics of tortoise muscle. *J. Physiol.* **197**, 685–707.

Zelená, J. and Sobotková J. (1973) Ultrastructure of endplates in slow and fast chicken muscles during development. *Folio. Morph.* **21**, 144–145.

Mammalian muscles and motor units

5

5.1 SLOW AND FAST TWITCH MUSCLES

Structural and functional differences between slow tonic and fast twitch muscles in lower vertebrates (Chapter 4) and between slow and fast twitch mammalian muscles were identified more than a hundred years ago (Ranvier, 1874; Grützner, 1884). It is now recognized that most mammalian muscles contain fast and slow twitch muscle fibres and that there is considerable heterogeneity of muscle and motor unit properties. This heterogeneity is functionally important for grading and controlling muscle force during normal movement.

In this chapter, the evidence for muscle and motor unit heterogeneity in mammals and its functional importance is briefly described.

5.1.1 DEVELOPMENT OF FORCE IN MAMMALIAN MUSCLES

In adult mammalian muscles, each muscle fibre is supplied by only one motoneurone but each motoneurone supplies many muscle fibres. The motoneurone and its muscle fibres were referred to as the **motor unit** by Sherrington (1939) because he recognized that the activation of a motoneurone in the spinal cord resulted in the contraction of all the muscle fibres supplied by that motoneurone. The motor unit was also referred to as 'the final common path' because it is through this route that the results of all the processing in the central nervous system is finally transformed into movement. To quote Sechenov (1863): 'The infinite diversity of external manifestations of cerebral activity can be reduced ultimately into a single phenomenon – muscular movement.' The smallest unit of control is the motor unit.

The high safety factor of neuromuscular transmission at skeletal muscle junctions ensures that action potentials in the nerve terminals are transmitted in a one-to-one fashion to the muscle membrane (Katz, 1966).

Muscle action potentials in turn lead to a release of Ca^{2+} from sarco-plasmic reticulum at the triadic junctions between the t-tubules and the terminal cisternae of the sarcoplasmic reticulum (Campbell, 1986). Binding of Ca^{2+} to one of the subunits of the regulatory protein troponin releases the inhibition of the troponin–tropomyosin complex on the thin actin filament, so that the myosin molecules in the thick filament can bind to actin and initiate cross-bridge cycling and muscle contraction. When muscles contract, the thin actin filaments slide past thick myosin filaments (Huxley, 1974; Squire, 1986). The head regions on the myosin molecule form the cross-bridges that power the movement. The details of this process have been elucidated recently by descriptions of the three-dimensional structure of the myosin and actin molecules. These indicate that the myosin head undergoes a conformational change which can account for the movement of the actin filament relative to the myosin thick filament (Rayment et al., 1993). Muscles relax when Ca^{2+} uptake into the sarcoplasmic reticulum exceeds Ca^{2+} release, when excitation ceases (Huxley, 1974).

5.1.2 COLOUR, ENDURANCE AND CONTRACTILE SPEED

Slow twitch muscles were first distinguished from fast twitch muscles by their red colour, visibly slower contraction (Ranvier, 1874; Grützner, 1884) and higher fatigue resistance (Denny-Brown, 1929). The red colour is due to high concentrations of myoglobin and dense capillary networks. Differences in contractile speed result from differences in constituent contractile, regulatory and Ca^{2+} binding proteins which exist in several slow and fast isoforms (see below; reviewed by Pette and Vrbová, 1985, 1992; Pette and Staron, 1990). The fatigue resistance of slow twitch muscles is associated with high resting and active blood flow and high oxidative/glycolytic ratios (Hilton et al., 1970; Hudlická, 1984; Hudlická et al., 1987). These in turn are correlated with dense capillary networks around the fibres and high densities of mitochondria within the fibres (Eisenberg, 1983).

Two main types of mammalian twitch muscle can be distinguished – those that produce force rapidly, and those that contract slowly (Figure 5.1).

Differences in the time to peak of the twitch contraction are due to differing rates of force development. The maximum rate of shortening (V_{max}) of whole muscles and single muscle fibres correlates with the ATPase activity of the myosin heavy chain (Bárány, 1967; reviewed by Green, 1991). To some extent, speed of whole muscle varies as a function of the relative content of fast and slow myosin heavy chains (Lowey et al., 1969; Reiser et al., 1985) and, to a lesser extent, the relative content of slow and fast myosin light chains (Julian et al., 1979; Sweeney et al.,

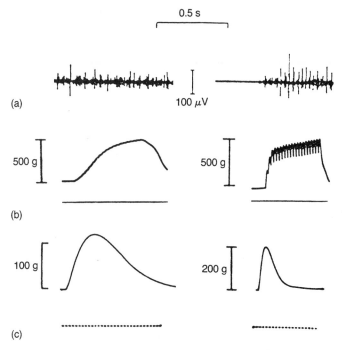

Figure 5.1 Recordings from (left) the rabbit soleus and (right) tibialis anterior. (a) EMG activity recorded from these muscles: continuous activity can be obtained from the soleus muscle, whereas activity in the TA was elicited by pinching the animal's paw. (b) Records of isometric contractions from the same muscles in response to repetitive stimulation at 40 Hz. (c) Records of single isometric contractions. The time-marker under the contractions represents 10 ms between successive dots.

1988; Greaser *et al.*, 1988). Different isoforms of the regulatory proteins, troponin and tropomyosin, may also contribute to differences in contractile speed by conferring different Ca^{2+} sensitivities to the contractile proteins (reviewed by Pette and Staron, 1990; Green, 1991). Finally, there is evidence to suggest that a higher density of dihydropyridine (**DHP**) receptors in the t-tubular membrane (Kandarian *et al.*, 1992) and of junctional 'feet' (Franzini-Armstrong *et al.*, 1988) may contribute to the faster contractile speed of fast twitch muscles as compared with slow twitch muscles. The t-tubular L-type Ca^{2+} channels or DHP receptors are believed to act as voltage sensors in the triadic junction between t-tubules and the terminal cisternae of the sarcoplasmic reticulum: charge movements in the L-type Ca^{2+} channels may be responsible for a transformational change in the Ca^{2+} release channels in sarcoplasmic reticulum via the cytoplasmic 'foot' domain of the Ca^{2+} release channel (Rios and Brum, 1987; Adams and Beam, 1990).

The rate of muscle relaxation correlates with the rate of Ca^{2+} uptake (Stein *et al.*, 1982). Ca^{2+} uptake is significantly higher in fast twitch compared with slow twitch muscles (Sreter and Gergely, 1964; van Winkle and Schwartz, 1978). Higher rates of Ca^{2+} uptake in fast muscle fibres are due to more extensive sarcoplasmic reticular systems (Luff and Atwood, 1971; Eisenberg, 1983; Franzini-Armstrong *et al.*, 1988), fast as opposed to slow isoforms of calcium-ATPase (Damiani *et al.*, 1981; Brandl *et al.*, 1986), higher densities of calcium-ATPase in the longitudinal tubules (Jorgensen *et al.*, 1982, 1983; Leberer and Pette, 1986a) and greater amounts of Ca^{2+} binding proteins in the cytoplasm (Leberer and Pette, 1986a, b). Phospholamban, a phosphorylatable regulatory protein of Ca-ATPase which is expressed only in slow twitch muscle (Jorgensen and Jones, 1986), may also affect contractile speed.

Fast and slow twitch muscles can also be distinguished by their tetanic response to repetitive stimulation and post-tetanic twitch responses (Cooper and Eccles, 1930; Close, 1972). Slow twitch muscles achieve maximum force production at lower rates of stimulation while fast twitch muscles require higher frequencies of activation. Single motor units show similar patterns of response (Figure 5.2).

5.1.3 FUNCTIONAL SIGNIFICANCE OF MUSCLE PROPERTIES

Denny-Brown (1929) first drew attention to the functional matching between the contractile properties of slow and fast twitch muscles and the patterns of recruitment of motor units in the muscles during movement. In the soleus (a slow twitch muscle which extends the ankle), motor units fire at low frequencies and for long periods during posture or in response to reflex activation in the decerebrate animal (Denny-Brown, 1929; Granit *et al.*, 1957). The muscle, in turn, develops fused maximal tetanic forces at these frequencies. In contrast, the motor units in fast twitch gastrocnemius muscle (a synergist in the triceps surae extensor group which extends the ankle) tend to fire in high frequency bursts which elicit brief, large tetanic forces (Denny-Brown, 1929; Granit *et al.*, 1956). If the muscles are stimulated at lower frequencies, tetani are unfused, and do not reach maximal force (Figure 5.2). Thus, contractile speed and endurance appear to be well matched to the pattern of motor unit activity (Kernell, 1992).

5.1.4 MOTONEURONE–MUSCLE MATCHING

The electrophysiological characteristics of the motoneurones that supply slow and fast muscle fibres differ, and are in part responsible for the specific firing patterns in these cells. In addition, the firing patterns, determined largely by the duration of the after-hyperpolarization (**AHP**)

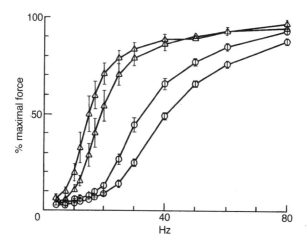

Figure 5.2 The percentage of maximal tetanic force developed by two slow (△) and two fast (○) motor units is plotted against the frequency of stimulation. The motor units were isolated from the rat medial gastrocnemius muscles.

of the motoneuronal action potential, correlated well with the duration of the twitch contraction of the muscles that the motoneurone supplied. It was therefore suggested that the duration of AHP determined the firing rate of the motor units (Eccles *et al.*, 1958). Thus when AHP is long, the firing rate is slow, and *vice versa*. The possibility that the time course of muscle contraction may be related to the firing rate of motoneurones is supported by the observation that the duration of AHP of motoneurones to hindlimb muscles of the cat is longer than in the rat (Eccles *et al.*, 1958; Gardiner and Kernell, 1990) and tetanic contractions of the muscles of the lower limb of the rat fuse at very much higher rates of repetitive stimulation than those in the cat (Granit *et al.*, 1963; Gardiner and Kernell, 1990).

The correlation between duration of AHP and the time course of contraction is illustrated in Figure 5.3. The time course of contraction, in turn, accounts for the shift of the force–frequency curves to lower frequencies in the slow twitch muscles (Cooper and Eccles, 1930). Thus the time course of the twitch contraction appears to be appropriately matched to the motoneurone with respect to membrane properties which govern its firing rates. The rate of discharge of motoneurones in response to current injected into the cell varies according to neuronal cell size and increases more in larger cells. Although small cells have larger postsynaptic excitatory responses to a given input than large cells, the increase in the frequency of firing produced by an increment of current is greater in large motoneurones (Kernell, 1966). Most alpha-motoneurones can be made to sustain repetitive firing given sufficient input

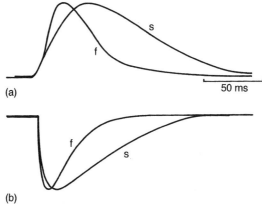

(a)

50 ms

(b)

Figure 5.3 (a) Time course of a single twitch elicited from fast (f) and slow (s) motor units from the gastrocnemius muscle of a cat; (b) hyperpolarization of the motoneurone supplying the corresponding fast (f) and slow (s) units.

(Kernell, 1965b, 1966; Kernell and Monster, 1982); however, motor units in fast twitch muscles usually do not maintain their discharge. This phasic pattern of discharge can in part be accounted for by the tendency of large motoneurones to accommodate to sustained input (Burke and Nelson, 1971). Thus, for large motor units under physiological conditions both the nerve cell and the muscle fibres supplied by it are unable to maintain activity for long periods.

A trend for reflex discharge to outlast the sensory stimulus that initiates it has recently been attributed to the development of plateau potentials in the excited motoneurones. These potentials require serotonergic input from higher centres which is believed to increase a Ca^{2+} current. This leads to maintained motoneurone firing which outlasts the stimulation via afferent inputs (Crone *et al.*, 1988). These findings provide further evidence for the importance of Ca^{2+} currents in controlling firing rates in motoneurones: Ca^{2+} currents underlie the plateau potentials, and Ca^{2+}, entering during the spike, modulates K^+ channels which control the duration of the AHP.

5.2 MUSCLE FIBRE AND MOTOR UNIT HETEROGENEITY

5.2.1 MUSCLE FIBRE AND MOTOR UNIT TYPES

It became clear that most mammalian skeletal muscles are not homogeneous in their muscle fibre composition. Fibre heterogeneity within a muscle was first recognized by the histochemical detection of different staining intensities of myosin ATPase and of oxidative and glycolytic enzymes in human and animal studies (Dubowitz and Pearce, 1960;

Engel, 1962). Even though slow and fast twitch muscles may be composed predominantly of slow or fast muscle fibres, most mammalian muscles contain different muscle fibre types which can be distinguished using histochemical and/or immunocytochemical methods (Stein and Padykula, 1962; Ariano et al., 1973; Dubowitz and Brooke, 1973; Burke and Edgerton, 1975; Gauthier and Lowey, 1977; Pette and Staron, 1990).

Engel (1962) first classified muscle fibres into two types: type I and type II, to correspond with slow and fast twitch muscle fibres, respectively. Nevertheless, histochemical reactions for oxidative enzymes (Stein and Padykula, 1962) or ATPase (Brooke and Kaiser, 1970a; Engel, 1962; Yellin and Guth, 1970) permitted at least three and as many as eight muscle fibre types to be discerned (Brooke and Kaiser, 1970b; Romanul, 1964). For simplicity and for a closer correspondence with known differences in the properties of slow and fast twitch muscles, histochemical procedures were modified to discriminate only three fibre types; fibre typing based on pH sensitivity of myosin ATP-ase and metabolic enzyme profiles has become standard practice (Pette and Staron, 1990) and the muscle fibre types correspond well with motor unit types as described below.

By stimulating single motoneurones or their motor nerves it is possible to elicit tetanic contractions of muscle fibres belonging to the same motor units. The muscle fibres of that motor unit can then be depleted of glycogen with repeated contractions and distinguished from the rest of the muscle fibres by the absence of glycogen in stained cross-sections of the muscle (Edström and Kugelberg, 1968). Using this technique, Burke and his colleagues (1973) were able to show directly that three main types of motor unit can be identified. Initially, a division was made into two groups: those that contain slow type I (**SO**) muscle fibres and those that contain fast contracting motor units having type II muscle fibres. The fast contracting motor units were subdivided into two categories on the basis of their susceptibility to fatigue: fast fatigue resistant (**FR**) and fast fatiguable (**FF**) motor units, whose muscle fibres were found to be type IIa (**FOG**) and type IIb (**FG**) respectively. Figure 5.4 illustrates the main features of these units and their muscle fibres.

More recently immunohistochemical analysis of muscle fibres using antibodies raised against myosin of fast or slow twitch muscles have complemented the histochemical fibre classifications and have identified a fourth fibre type, type IIx (Schiaffino et al., 1985; Gorza, 1990), which may correspond to a fourth motor unit type, which is also fast and is found in many fast twitch muscles, particularly in the physiological flexor group (Kugelberg and Lindegren, 1979; Totosy de Zepetnek et al., 1992a, b; Dum and Kennedy, 1980).

The distinction of three or more muscle fibre types on the basis of the pH sensitivity of myosin ATPase is due to differences in myosin heavy

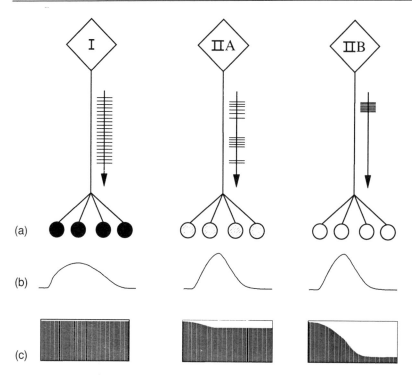

Figure 5.4 Diagrammatic illustration of the three main types of motor unit present in mixed mammalian muscles: (a) the motoneurone and the muscle fibres; (b) time course of muscle contraction; (c) examples of fatigue tests. The horizontal lines in (a) illustrate the activity transmitted to the muscle by the motoneurones, labelled I, IIA and IIB according to the types of muscle fibre that the motoneurones supply (type I =slow; types IIA and IIB =fast).

chain isoforms. Four isoforms have been separated and muscle fibres containing a particular isoform have been classified as types I, IIa, IIb and IIx (IId). Each of these isoforms is coded by a separate gene (reviewed by Pette and Staron, 1990). However, it is now recognized that there are several different isoforms of myosin heavy and light chains; currently, nine different myosin heavy chains and six myosin light chain isoforms have been identified (Pette and Staron, 1990). Since myosin is a hexameric structure containing two heavy chains in combination with two phosphorylatable (regulatory) light chains and two alkali light chains, the number of possible isomyosins far exceeds the three to four fibre types which are commonly described. The possibility that myosin expression is non-uniform along the length of muscle fibres further increases the complexity of isomyosin composition.

Generally, greater maximum shortening velocities are associated with greater amounts of fast heavy chains, whereas slower shortening velocities correlate with increasing amounts of the slow heavy chains (Reiser *et al.*, 1985; Sweeney *et al.*, 1988). Characteristically, motor unit isometric contraction speed varies widely even within predominantly slow or fast twitch muscles (Henneman and Mendell, 1981; Gordon and Patullo, 1993). A continuum of contractile speeds is consistent with a continuum of myosin heavy chain compositions (Pette and Staron, 1990). Several different isoforms of the regulatory proteins, troponin and tropomyosin, may contribute to the range in contractile speeds of different motor units. These are described in detail by Pette and Staron (1990).

5.2.2 METABOLIC ENZYME ACTIVITIES, FIBRE HETEROGENEITY AND MOTOR UNIT FATIGUABILITY

Although three main muscle fibre histochemical types can be discerned on the basis of reciprocal or coexpression of oxidative and glycolytic enzymes (Romanul, 1964; Guth and Yellin, 1971), biochemical analyses of enzyme activities in single muscle fibres have revealed a continuum of enzyme activities within and between muscle fibre types (see Pette and Staron, 1990) rather than discrete groups. The continuous distribution of enzyme activities in muscles is consistent with a continuous range of motor unit susceptibility to fatigue in many fast twitch muscles, for example the rat TA muscle. In this muscle, the physiological measures of fatigue resistance correlate well with oxidative enzyme content (e.g. Kugelberg and Lindegren, 1979) and fibre classifications based on ATPase and metabolic enzyme activities correspond to physiological classifications of motor units (Totosy de Zepetnek *et al.*, 1992b).

Muscle fibres within a motor unit are homogeneous with respect to their histochemical fibre type; the fibres contain the same isoforms of contractile and regulatory proteins and microassays of metabolic enzyme activities show that there is relatively little variance between unit fibres as compared with fibres of the same histochemical type (Kugelberg *et al.*, 1970; Burke *et al.*, 1973; Gauthier *et al.*, 1983; Németh *et al.*, 1981; Martin *et al.*, 1988).

Muscle fibres belonging to a single motor unit are normally distributed in a characteristic mosaic pattern amongst fibres belonging to several different motor units (Kugelberg *et al.*, 1970) giving the characteristic mosaic or checkerboard distribution of fibre types recognized by histochemical staining in muscle (Dubowitz and Brooke, 1973).

5.3 MOTOR UNIT PROPERTIES AND FUNCTION

5.2.1 RANGE IN CONTRACTILE FORCE, SPEED AND ENDURANCE

Physiological isolation of single motor units in slow and fast twitch muscles has revealed a wide spectrum of motor unit properties and

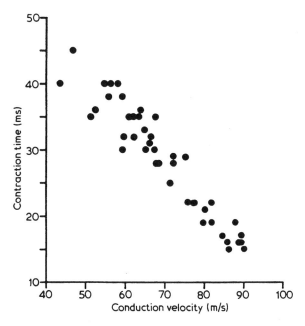

Figure 5.5 Relation between contraction times of 43 motor units and conduction velocities of their motor fibres (first superficial lumbrical muscles, results from six experiments). (From Appelberg and Émonet-Dénand, 1967, *J. Neurophysiol.* **30**, 154–160, with permission of the authors and *Journal of Neurophysiology.*)

extended the observations of Denny-Brown (1929) and Sherrington (1939) that motoneurone and muscle properties are functionally matched.

In 1963, Bessou and colleagues provided the first systematic description of a wide range of contractile properties of motor units in a deep lumbrical muscle of the cat (Bessou *et al.*, 1963). During the same decade, Henneman and his colleagues demonstrated that slow and fast twitch muscles in the hindlimb contain motor unit populations with differences in contractile force and speed and that these were correlated with motor nerve conduction velocity (McPhredran *et al.*, 1965; Henneman and Olson, 1965; Wuerker *et al.*, 1965). Figure 5.5 illustrates this point. The relationship between the electrophysiological properties of axons and contractile responses of the muscle fibres has been demonstrated in several muscles throughout the body and for different species, including humans (reviewed by Burke, 1981; Henneman and Mendell, 1981; Dengler *et al.*, 1988; Cope and Clark, 1991; Clark *et al.*, 1993; Gordon and Pattullo, 1993). Motor unit force, for example, varies over a 10- to 100-fold range, respectively, even in small hand, foot and eye muscles (Appelberg and Émonet-Dénand, 1967; Jami and Petit, 1975; Thomas *et al.*, 1987, 1990) and in the muscles of the rat and mouse (Brown and

Ironton, 1978; Gillespie *et al.*, 1987; Kanda and Hashizume, 1992; Totosy de Zepetnek *et al.*, 1992a).

Three factors contribute to the wide range in motor unit force: number of muscle fibres (innervation ratio, IR); their size; and specific force (force per unit area) (Burke, 1981; Stein *et al.*, 1990). Fibre number and size are the primary determinants of motor unit force while specific force contributes relatively little (Bodine *et al.*, 1987; Chamberlain and Lewis, 1989; Totosy de Zepetnek *et al.*, 1992a). The study of the relationships between motor unit force and conduction velocity of the axon show that the progressively stronger motor units are supplied by axons with faster conduction velocities (Gordon *et al.*, 1991; Totosy de Zepetnek *et al.*, 1992a) consistent with the suggestions of Ramón y Cajal (1928) and Henneman and Olson (1965) that larger axons supply larger peripheral fields. The smallest and most active motor units contain the smallest muscle fibres, which are also the most fatigue resistant.

5.3.2 FORCE GRADATION BY ORDERLY RECRUITMENT

A rank order of motor unit recruitment during movement was first described by Denny-Brown in 1929. Under conditions of reflex activation of ankle extensor muscles in the decerebrate cat, he extrapolated that the progressive activation (recruitment) of motor units with progressively larger-amplitude EMG signals reflected the progressive recruitment of stronger motor units. Henneman and his colleagues extended these findings and demonstrated that motoneurones were recruited into action in order of their size. Using electrophysiological parameters to assess axon size, they demonstrated that the motoneurones with the smallest axons were excited before the largest and that the motoneurones with the smallest axons were also the most readily inhibited during the flexor reflex (Henneman, 1985). Henneman and Olson (1965) noted that the skewed distribution in which small motor units were in the majority is ideal for precise control of force and that a size-dependent susceptibility of motor units to discharge would result in smooth gradation of force during movement. Simply by selectively recruiting varying numbers of the many small motor units, contractile force would be finely graded at low force levels. With the demands for larger force output, larger increments of force may be added through contributions of progressively stronger units (for review, see Henneman, 1985).

Ordered recruitment of progressively stronger units has been demonstrated with few exceptions during voluntary and reflex activation in animals and man (Adrian and Bronk, 1929; Milner-Brown *et al.*, 1973; reviewed by Henneman and Mendell, 1981; see also Clarke *et al.*, 1993). The more generalized size principle of Henneman that recruitment of motoneurones in order of their size leads to progressive recruitment of stronger motor units (Henneman and Mendell, 1981) is more controversial, although recruitment according to conduction velocity has been directly demonstrated (Dengler *et al.*, 1988; Cope and Clark, 1991).

Thus the wide range of motor unit force in skeletal muscles is functionally important for force gradation during muscle contraction. The extensive debate on the relevance of 'size' in orderly recruitment of motor units is beyond the scope of this book and is reviewed in Binder and Mendell (1990).

In addition to its size, the recruitment threshold of a motoneurone depends on the magnitude of its synaptic inputs and the processing of these inputs by the neurone's intrinsic electrophysiological properties (Burke, 1981; Henneman and Mendell, 1981; Pinter, 1990).

The 'intrinsic excitability' of motoneurones generally refers to the responsiveness of motoneurones to injected excitatory current via synaptic mechanisms or via an intracellular microelectrode (Gustafsson and Jankowska, 1976). Rheobase is the minimal current required to generate an action potential at the axon initial segment (Kernell, 1966; Fleshman *et al.*, 1981; Gustafsson and Pinter, 1984a, b). The motoneurones with slow conducting axons, and which supply the low force motor units, are the most excitable and require the lowest current injection to generate action potentials. Rheobase current in turn depends on a systematic variation in steady-state intrinsic membrane properties, in particular the effective membrane conductivity which requires that progressively larger rheobase current is needed to excite cells with progressively higher input conductance (Kernell and Zwaagstra, 1981; Gustafsson and Pinter, 1984a, b; Pinter, 1990).

Generally, electrophysiological and morphological measures of motoneuronal size, including soma-dendritic size and axon size (Cullheim, 1978; Kernell and Zwaagstra, 1981), show comparably small variation between motoneurones and are relatively poorer indicators of motoneuronal excitability than rheobase. Rheobase depends on both intrinsic membrane conductance and motoneurone size (Fleshman *et al.*, 1981; Gustafsson and Pinter, 1984b). As stated by Heckman and Binder (1990): 'In the simplest case in which the synaptic input is uniform (i.e. homogeneously distributed within the motoneurone pool), one would find a 10-fold range of recruitment thresholds attributable solely to rheobase.' Slow (S), fast fatiguable (FF) and fast fatigue resistant (FR) motor units can be subdivided on the basis of their excitability using rheobase values normalized by input resistance (Zengel *et al.*, 1985) which predicts their order of recruitment (Sypert and Munson, 1981).

Spinal cord motoneurone pools which are arranged in columns provide a good target for uniform afferent inputs that are important for a size-related recruitment order (Burke, 1981; Henneman and Mendell, 1981; Henneman, 1985). All 1a afferents from muscle spindles make homonymous connections on to motoneurones supplying the muscle of origin (Henneman and Mendell, 1981) but the individual EPSP amplitudes recorded intracellularly in motoneurones were not the same. EPSPs are largest in the most excitable motoneurones and are inversely related to the axon conduction velocity of the motoneurone (Henneman and Mendell, 1981).

As described above, the rank order in size of EPSPs in motoneurones can be attributed, at least in part, to differences in intrinsic properties of the motoneurones. Recent voltage clamp studies of effective synaptic currents in cat medial gastrocnemius motoneurones demonstrate that differences in synaptic input are also a significant factor in determining the rank order of EPSP size in motoneurones (reviewed by Heckman and Binder, 1990). Effective synaptic currents were highest in the small motoneurones and varied systematically with input resistance. It is therefore apparent that both the magnitude of the synaptic inputs and intrinsic properties of motoneurones account for the order of excitability of motoneurones which underlies orderly recruitment.

Sherrington (1939) was aware of, and emphasized, the importance of the different sensitivities of motor units to the same afferent input, i.e. that motoneurones have different thresholds. He noted that the motor units may in fact be grouped according to their excitability. According to Sherrington:

> The weakest postural reflex is that adapted to standing, where all that is wanted of the great extensors is to support their load to the extent of preventing the limb from sinking under it. This grade of reflex is postural. The motor units it employs in the great extensors are, comfortably with its mild contraction strength, a small proportion only of the total aggregate composing these muscles. But this small percentage is not a scattered population, it is a segregated set which forms at the ankle the soleus part of the great calf muscle, trespassing a little into gastrocnemius.

And he concludes further:

> In these great extensor muscles the 'upstanding' reflex as it grows in strength and changes its functional scope passes over from the restricted field of slow red motor units to large reserves of quick pale motor units which are in fact the main masses of these large extensor muscles. These latter motor units are, to judge from gastrocnemius, individually more powerful than those of the red kernel muscles soleus, thrice as powerful (Eccles and Sherrington, 1930) and nearly thrice as rapid (Denny-Brown, 1929; Cooper and Eccles, 1930).

Studies of motoneuronal properties and the muscle fibres they innervate show that they are well adapted for function: the most excitable motor units are recruited first and develop low forces which can be maintained for long periods; progressively stronger units are recuited when synaptic input is increased as required.

5.3.3 FORCE GRADATION BY RATE CODING

The minimal force of a motor unit is the twitch contraction which is evoked in response to a single action potential of its axon. Depending on the interval between action potentials and the contractile speed of the muscle fibres, repeated stimuli may in turn lead to greater force development (Huxley, 1974; Squire, 1986). Generally there is a steep relationship between the frequency of firing and the tetanic force developed by the muscle, and this relationship is shifted to higher frequencies as the contractile speed of the motor unit increases (Figure 5.2). Adrian and Bronk (1929) recognized that force gradation during reflex or voluntary contractions depends on recruitment of motor units and on modulating the frequency of discharge of the already recruited motor units. Using EMG recordings, Denny-Brown (1929) noted that motor units in slow and fast twitch muscles in the cat tended to discharge at different frequencies. Motor units in the slow twitch soleus tended to fire at low frequencies of 7–15 HZ which were maintained for some time, in contrast to the motor units in the fast twitch gastrocnemius muscles whose minimal discharge frequencies were higher and were not generally maintained (see also Granit et al., 1956, 1957). These findings were later confirmed using intracellular recording from motoneurones by Eccles et al. (1958) and by Kernell (1965a, b, c). Similar observations have been made in human muscles. For example, the mean discharge rate of motor units in the fast biceps muscles during maximal voluntary contractions was 30 Hz, compared with 11 Hz in the slow soleus (Bellamare et al., 1983).

Motoneurones discharging at their minimum rate of maintained discharge will typically elicit a mean force of only 10–25% of maximum which corresponds to the amplitude of twitch contractions (Kernell and Sjoholm, 1975; Milner-Brown et al., 1973). As the force of a reflex or voluntary contraction is increased, firing rates increase for already recruited motor units (Denny-Brown, 1929; Kernell and Sjoholm, 1975; Milner-Brown et al., 1973; Grimby and Hanerz, 1977; reviewed by Kernell, 1992) and in turn, the unfused twitch contractions progressively fuse to develop maximal tetanic forces. The minimal rate of firing of a motoneurone generally corresponds to the lower end of the steep region of the force–frequency relationship of its motor unit (Kernell, 1965c, 1992). As frequency of firing increases, there is a sharp increase in force to a maximum when tetanic contractions are completely fused. As a result, the rate gradation of force is particularly effective within a rather narrow range of discharge frequencies which correspond to the steep portion of the force–frequency curve (Figure 5.2).

The MU force–frequency relationship is therefore remarkably well matched to the firing patterns of the motoneurones which supply the muscle fibres and functionally important in the normal control of movement. Moreover, the minimal rate of motor unit firing is determined by the duration of the post-spike AHP.

It is therefore not surprising that within motoneurone pools there is a correlation between the duration of the after-hyperpolarization in the motoneurones and the contractile speed of the motor units (Gardiner and Kernell, 1990). Interestingly this correlation remains under conditions in which the contractile speed is experimentally altered. Huizar and his colleagues (1977) observed this close correlation of hyperpolarization and contractile speed in experiments in which the duration of the hyperpolarization of intact motoneurones to soleus was shortened in a partially denervated muscle. The contraction of the corresponding motor units was also faster than normal for soleus.

The steady rate of firing is generally lower than the initial rate (Kernell, 1965a), possibly due to the non-linear summation of successive conductances associated with the AHP (see Baldissera and Gustafsson, 1971; Barrett *et al.*, 1980). At very high frequencies, the firing rate of the slow motor units becomes irregular, with short intervals between spikes being followed by long intervals due to the summation of the hyperpolarizations produced by the closely spaced spikes (Baldissera and Gustafsson, 1971).

5.3.4 ACCOMMODATION AND FATIGUE

In a mixed muscle, units that have slow contractions and low susceptibility to fatigue are readily recruited and maintain force for long periods. Generally, the large, fast motor units, particularly the fatiguable units, are recruited for brief, intermittent activities such as jumping, running or lifting (Burke, 1981). The largest motor units, which are the most fatiguable, are recruited at high levels of force when the blood supply may be occluded. The metabolic characteristics of muscle fibres of these motor units appear to be well adapted to their order of recruitment (Pette and Staron, 1990).

During repetitive activity, the discharge rate in motoneurones falls significantly (Bigland-Ritchie *et al.*, 1983, 1986; Marsden *et al.*, 1983) concurrent with a slowing of the motor unit contractions and a decline of the fusion frequency (Burke *et al.*, 1973). As a result the matching of firing rate and tetanic force is well maintained in the fatiguing motor unit. This adaptation of motor unit discharge rates during muscle fatigue was therefore referred to as **muscular wisdom** by Marsden and his colleagues (Marsden *et al.*, 1983). As a result, tetanic contractions are developed at lower rates of firing which compensate, at least to some degree, for muscular fatigue.

A further illustration of the remarkable matching between motoneurones and the muscle fibres they supply is the susceptibility of different sites of the system to fatigue. There are multiple sites that may cause fatigue (see Stuart and Enoka, 1992). These include branch point failure in the intramuscular axonal branches, fatigue of the neuromuscular junction, failure of excitation–contraction coupling and, finally, depletion of energy substrates (reviewed by Stuart and Enoka, 1992). Each motor unit type seems to show a matching susceptibility to fatigue at all these sites: FF units readily show a decline in EMG as well as force after brief stimulation; FR units show both electrical and mechanical endurance (McPhedran *et al.*, 1965; Wuerker *et al.*, 1965; Burke *et al.*, 1973; Clamann and Robinson, 1985).

5.4 MOTOR UNITS, MUSCLES AND MOVEMENT

When we wish to carry out any movement, whether it is to run or to pick up our socks, it is necessary that the appropriate muscles are activated for the specific movement and that the force generated by those muscles is controlled. Henneman and Olson (1965), in their description of the differences between the three heads of the triceps surae muscles, emphasized the functional matching of the slow and fast twitch muscle properties to their mode of recruitment during movement and made the astute observation that 'three heads are better than one'. Even though the three muscles insert into a single Achilles tendon which acts at the ankle joint, the point of insertion at the ankle may be sufficiently different for the muscles possibly to control the rotation of the ankle joint differentially (Nichols *et al.*, 1993). The more obvious difference is the origin of the fibres, the gastrocnemii muscles inserting on the femur and acting on the knee joint to flex the limb during movement. The soleus muscle is a single joint muscle which stabilizes only the ankle joint. Finally the architecture of these synergistic ankle extensor muscles is quite different (for details see Lieber, 1992).

It is clear, therefore, that skeletal muscles are well designed and matched for highly specific functions and, although they may be grouped together loosely as physiological flexors and extensors around any one joint, their architectural specialization is taken into account in the orchestration of any movement by the central nervous system.

CONCLUSIONS

In anatomically defined skeletal muscles, the inclusion of many motor units varying widely in the contractile force, speed and endurance and in the excitability, size and firing properties of their motoneurones is the basis for the fine degree of control of force during movement. During postural activities and movements which require low levels of force, the low force units are readily recruited in response to low levels of input currents and they can maintain their contractions without fatigue. The progressively larger and fast motor units are recruited in response to increasingly high input currents. Rank-ordered recruitment of units allows force to increase smoothly. The remarkable matching of the firing characteristics of the 'fast' motoneurones and their progressively more fatigue-sensitive muscle units means that both motoneurones and their muscles do not sustain prolonged activity and are most suited to 'phasic' recruitment which allows large force development for relatively short periods.

The distribution of synapses and the physiological strength of the connections on motoneurones are differentially controlled such that the low force motor units are recruited before the more forceful ones.

REFERENCES

Adams, B. A. and Beam, K. G. (1990) Muscular dysgenesis in mice: a model system for studying excitation–contraction coupling. *FASEB J.* **4**, 2809–2816.

Adrian, E. D. and Bronk, D. W. (1929) The discharge of impulses in motor nerve fibres. Part II. The frequency of discharge in reflex and voluntary contractions. *J. Physiol. (Lond.)* **67**, 119–151.

Appelberg, B. and Émonet-Dénand, F. (1967) Motor units of the first superficial lumbrical muscle of the cat. *J. Neurophysiol.* **30**, 154–160.

Ariano, M. A., Armstrong, R. B. and Edgerton, V. R. (1973) Hindlimb muscle fiber populations of five mammals. *J. Histochem. Cytochem.* **21**, 51–55.

Baldissera, F. and Gustafsson, R. (1971) Regulation of repetitive firing in motoneurones by the after hyperpolarisation conductance. *Brain Res.* **30**; 431–434.

Bárány, M. (1967) ATPase activity of myosin correlated with speed of muscle shortening. *J. Gen. Physiol.* **50**, 197–216.

Barrett, E. F., Barrett, J. N. and Crill, W. I. (1980) Voltage-sensitive outward currents in cat motoneurones. *J. Physiol. (Lond.)* **304**, 251–276.

Bellamare F., Woods, J. J., Johansson, R. and Bigland-Ritchie, B. (1983) Motor-unit discharge rates in maximal voluntary contractions of three human muscles. *J. Neurophysiol.* **50**, 1380–1392.

Bessou, P., Émonet-Dénand, F. and Laporte, Y. (1963) Relation entre la vitesse de conduction des fibres nerveuse motrices et le temps de contraction de leurs unités motrices *C.r. Acad. Sci.* **256**, 5625–5627.

Bigland-Ritchie, B., Dawson, J. J., Johansson, R. S. and Lippold, O. C. J. (1986) Reflex origin for the slowing of motoneurone firing rates in fatigue of human voluntary contractions. *J. Physiol.* **379**, 451–459.

Bigland-Ritchie, B., Johansson, R., Lippold, O. C. J. and Woods, J. J. (1983) Contractile speed and EMG changes during fatigue of sustained maximal voluntary contractions. *J. Neurophysiol.* **50**, 313–324.

Binder, M. D. and Mendell L. M. (eds) (1990) *The Segmental Motor System*, Oxford University Press, New York.

Bodine, S. C., Roy, R. R., Eldred, E. and Edgerton V. R. (1987) Maximal force as a function of anatomical features of motor units in the cat tibialis anterior. *J. Neurophysiol.* **57**, 1730–1745.

Brandl, C. J., Green, M. N., Korczak, B. and MacLennan, D. H. (1986), Two Ca^{2+} ATPase genes: homologies and mechanistic implications of deduced amino acid sequences. *Cell* **44**, 597–607.

Brooke, M. H. and Kaiser, K. K. (1970a) Three 'myosin adenosine triphosphatase' systems, the nature of their pH lability and sulphydryl dependence. *J. Histochem. Cytochem.* **18**, 670–672.

Brooke, M. H. and Kaiser, K. K. (1970b) Muscle fibre types, how many and what kind? *Arch. Neurol.* **23**, 369–379.

Brown, M. C. and Ironton, R. (1978) Sprouting and regression of neuromuscular synapses in partially denervated mammalian muscles. *J. Physiol. (Lond.)* **278**, 235–48.

Burke, R. E. (1981) Motor units: anatomy, physiology and functional organization, in *Handbook of Physiology: The Nervous System. Motor Control*, vol. II, part 1, American Physiology Society, Bethesda, MD, pp. 345–422.

Burke, R. E. and Edgerton, V. R. (1975) Motor unit properties and selective involvement in movement. *Exercise and Sport Sci. Rev.* **3**, 31–81.

Burke, R. E., Levine, D. N., Tsairis P. and Zajac F. E. (1973) Physiological types and histochemical profiles in motor units of the cat gastrocnemius. *J. Physiol.* **234**, 723–748.

Burke, R. E. and Nelson P. G. (1971) Accommodation to current ramps in motoneurones of fast and slow twitch motor units. *Int. J. Neurosci.* **1**, 347–356.

Campbell, K. P. (1986) Protein components and their roles in sarcoplasmic reticulum function, in *Sarcoplasmic Reticulum in Muscle Physiology* (eds M. I. Entmann and W. B. van Winkle), CRC Press, Boca Raton, Florida, 1: 65–69.

Chamberlain, S. and Lewis, D. M. (1989) Contractile characteristics and innervation ratio of rat soleus motor units. *J. Physiol.* **412**, 1–21.

Clamann H. P. and Robinson, A. J. (1985) A comparison of electromyographic and mechanical fatigue properties in motor units of the cat hindlimb. *Brain Res.* **327**, 203–219.

Clark, B. D., Dacko, S. M. and Cope, T. C. (1993) Cutaneous stimulation fails to alter motor unit recruitment in the decerebrate cat. *J. Neurophysiol.* **70**, 1433–1439.

Close, R. (1972) Dynamic properties of mammalian skeletal muscles. *Physiol. Rev.* **52**, 129–197.

Cooper, S. and Eccles, J. C. (1930) The isometric responses of mammalian muscles. *J. Physiol.* **69**, 377–385.

Cope, C. and Clark, B. D. (1991) Motor unit recruitment in the decerebrate cat: several unit properties are equally good predictors of order. *J. Neurophysiol.* **66**, 1127–1138.

Crone, C., Hultborn, H., Kiehn, O. *et al.* (1988) Maintained changes in motoneuronal excitability by short-lasting synaptic inputs in the decerebrate cat. *J. Physiol.* **405**, 321–343.

Cullheim, S. (1978) Relations between cell body size, axon diameter and axon conduction velocity of cat sciatic alpha-motoneurones stained with horseradish peroxidase. *Neurosci. Lett.* **8**, 17–20.

Damiani, E., Betto, R., Salvatori, S. *et al.* (1981) Polymorphism of sarcoplasmic reticulum adenosine triphosphatase of rabbit skeletal muscle. *Biochem. J.* **197**, 245–248.

Dengler, R., Stein, R. B. and Thomas, C. K. (1988) Axonal conduction velocity and force of single human motor units. *Muscle & Nerve* **11**, 126–145.

Denny-Brown, D. (1929) On the nature of postural reflexes. *Proc. Roy. Soc. (Biol.)* **104**, 252–301.

Dubowitz, V. and Brooke, M. H. (1973) *Muscle Biopsy: A modern approach*, W. B. Saunders Co., Philadelphia.

Dubowitz, V. and Pearse, A. G. E. (1960) Reciprocal relationship of phosphorylase and oxidative enzymes in skeletal muscle. *Nature* **185**, 701–702.

Dum, R. P. and Kennedy, T. T. (1980) Physiological and histochemical characteristics of motor units in cat tibialis anterior and extensor digitorum longus muscles. *J. Neurophysiol.* **43**, 1615–1630.

Eccles, J. C., Eccles, R. M. and Lundberg, A. (1957) The convergence of monosynaptic excitatory afferents onto many different species of alpha motoneurones. *J. Physiol.* **137**, 22–50.

Eccles, J. C., Eccles, R. M. and Lundberg, A. (1958) The action potentials of alpha motoneurones supplying fast and slow muscle fibres. *J. Physiol. (Lond.)* **142**, 275–291.

Eccles, J. C. and Sherrington, C. S. (1930) Numbers and contraction-values of individual motor-units examined in some muscles of the limb. *Proc. Roy. Soc. Ser. B* **106**, 326–357.

Edström, L. and Kugelberg, E. (1968) Histochemical composition, distribution of fibres and fatigability of single motor units. *J. Neurol. Neurosurg. Psychiat.* **31**, 424–433.

Eisenberg, R. B. (1983) Quantitative ultrastructure of mammalian muscle, in *Handbook of Physiology* (eds L. D. Peachey, R. H. Adrian and S. R. Geiger), American Physiology Society, Bethesda, MD, pp. 73–112.

Émonet-Dénand, F., Hunt, C. C., Petit, J. and Pollin, B. (1988) Proportion of fatigue-resistant motor units in hindlimb muscles of cat and their relation to axonal conduction velocity. *J. Physiol.* **400**, 135–158.

Engel, W. K. (1962) The essentiality of histo- and cyto-chemical studies of skeletal muscle in the investigation of neuromuscular disease. *Neurology* **12**, 778–784.

Fleshman, J. W., Munson, J. B., Sypert, G. W. and Friedman, W. A. (1981) Rheobase, input resistance, and motor-unit type in medial gastrocnemius motoneurons in the cat. *J. Neurophysiol.* **46**, 1326–1338.

Franzini-Armstrong, C., Ferguson, D. G. and Champ, C. (1988) Discrimination between fast- and slow-twitch fibres of guinea pig skeletal muscle using the relative surface density of junctional transverse tubule membrane. *J. Muscle Res. Cell Motil.* **8**, 403–414.

Gardiner, P. F. and Kernell, D. (1990). The 'fastness' of rat motoneurones: time course of afterhyperpolarization in relation to axonal conduction velocity and muscle unit contractile speed. *Pflügers Arch.* **415**, 762–766

Gauthier, G. F., Burke, R. E., Lowey, S. and Hobbs, A. W. (1983) Myosin isozymes in normal and cross-reinnervated cat skeletal muscle fibers. *J. Cell Biol.* **97**, 756–771.

Gauthier, G. F. and Lowey, S. (1977) Polymorphism of myosin among skeletal muscle fiber types. *J. Cell Biol.* **74**, 760–779.

Gillespie, M. J., Gordon, T. and Murphy, P. R. (1987) Motor units and histochemistry in the rat lateral gastrocnemius and soleus muscles: evidence for dissociation of physiological and histochemical properties after reinnervation. *J. Neurophysiol.* **57**, 1175–1190.

Gordon, T. and Pattullo, M. C. (1993) Plasticity of muscle fiber and motor unit types. *Exercise. Sport Sci. Rev.* **21**, 331–362.

Gordon, T., Totosy de Zepetnek, J., Rafuse, V. and Erdbil, S. (1991) Motoneuronal branching and motor unit size after complete and partial nerve injuries, in *Motoneuronal Plasticity* (ed. A. Wernig), Springer-Verlag, Berlin, pp. 207–216.

Gorza, L. (1990) Identification of a novel type 2 fiber population in mammalian skeletal muscle by combined use of histochemical myosin ATPase and anti-myosin monoclonal antibodies. *J. Histochem. Cytochem.* **38**, 257–265.

Granit, R., Kernell, D. and Smith, R. S. (1963) Delayed depolarisation and the repetitive response to intracellular stimulation of mammalian motoneurones. *J. Physiol.* **168**, 890–910.

Granit, R., Henatsch, M. D. and Steg, G. (1956) Tonic and phasic ventral horn cells differentiated by post-tetanic potentiation in cat extensors. *Acta Physiol. Scand.* **37**, 114–126.

Granit, R., Phillips, C. G., Skoglund, S. and Steg, G. (1957) Differentiation of tonic from phasic alpha ventral horn cells by stretch, pinna, and crossed extensor reflexes. *J. Neurophysiol.* **20**, 470–481.

Greaser, M. L., Moss, R. L. and Reiser, P. J. (1988) Variations in contractile properties of rabbit single muscle fibres in relation to Troponin T isoforms and myosin light chains. *J. Physiol.* **406**, 85–98.

Green, H. J. (1991) Myofibrillar composition and mechanical function in mammalian skeletal muscle. *Sport Sci. Rev.* **1**, 43–64.

Grimby, L. and Hannerz, J. (1977) Firing rate and recruitment order of toe extensor motor units in different modes of voluntary contraction. *J. Physiol. (Lond.)* **264**, 865–879.

Grützner, P. (1884) Zür Anatomie und Physiologie der querquestreiften Muskeln. *Rec. Zool. Suisse* **1**, 665–684.

Gustafsson, B. and Jankowska, E. (1976) Direct and indirect activation of nerve cells by electrical pulses applied extracellularly. *J. Physiol. (Lond.)* **258**, 33–61.

Gustafsson, B. and Pinter, M. J. (1984a). Relations among passive electrical properties of lumbar α-motoneurons of the cat. *J. Physiol. (Lond.)* **357**, 401–431.

Gustafsson, B. and Pinter, M. J. (1984b) An investigation of threshold properties among cat spinal motoneurons. *J. Physiol. (Lond.)* **357**, 453–483.

Guth, L. and Yellin, H. (1971) The dynamic nature of so-called 'fibre types' of mammalian skeletal muscle. *Exp. Neurol.* **31**, 277–300.

Heckman, C. J. and Binder, M. D. (1990) Neural mechanisms underlying the orderly recruitment of motoneurons, in *The Segmental Motor System* (eds M. Binder and L. Mendell), Oxford University Press, New York, pp. 182–204.

Henneman, E. (1985) The size principle: A deterministic output emerges from a set of probabilistic connections. *J. Exp. Biol.* **115**, 105–112.

Henneman, E. and Mendell, L. M. (1981) Functional organization of the motoneurone pool and its inputs, in *Handbook of Physiology*, Sect. 1, vol. II (ed. V. B. Brooks), Williams and Wilkins Co. Baltimore, pp. 423–508.

Henneman, E. and Olson, C. B. (1965) Relations between the structure and function in the design of skeletal muscles. *J. Neurophysiol.* **28**, 581–598.

Hilton, S. M., Jeffries, M. G. and Vrbová, G. (1970) Functional specialization of the vascular bed of the soleus muscle. *J. Physiol. (Lond.)* **206**, 543–562.

Hudlická, O. (1984) Development of microcirculation: capillary growth and adaptation, in *Handbook of Physiology*, Section 2. The cardiovascular system (eds E. M. Renkin, C. C. Michel, and S. R. Geiger), American Physiology Society, Bethesda, MD, pp. 165–216.

Hudlická, O., Hoppeler, H. and Uhlmann, E. (1987) Relationship between the size of the capillary bed and oxidative capacity in various cat skeletal muscles. *Pflügers Arch.* **410**, 369–375.

Huizar, P., Kudo, N., Kuno, M. amd Miyata, Y. (1977) Reaction of intact spinal motoneurones to partial denervation of the muscle. *J. Physiol. (Lond.)* **265**, 175–191.

Huxley, A. F. (1974) Review lecture: Muscular contraction. *J. Physiol.* **243**, 1–43.

Jami, L. and Petit, J. (1975) Correlation between axonal conduction velocity and tetanic tension of motor units in four muscles of the cat hind limb. *Brain Res.* **96**, 114–118.

Jørgensen, A. O. and Jones, L. R. (1986) Localization of phospholamban in slow but not fast canine skeletal muscle fibres. *J. Biol. Chem.* **261**, 3775–3781.

Jørgensen, A. O., Shen, A. C. Y., MacLennon, D. H. and Tokuyasu, K. T. (1982) Ultrastructural localization of the Ca^{2+} Mg^{2+}-dependent ATPase of sarcoplasmic reticulum in rat skeletal muscle by immunoferritin labeling of ultrathin frozen sections. *J. Cell Biol.* **92**, 409–416.

Jørgensen, A. O., Shen, A. C. Y., Campbell, K. P. and MacLennon, D. H. (1983) Ultrastructural localization of calsequestrin in rat skeletal muscle by immunoferritin labeling of ultrathin frozen sections. *J. Cell Biol.* **97**, 1573–1581.

Julian, F. J., Moss, R. L. and Waller, G. S. (1979) Mechanical properties and myosin light chain composition of skinned fibres from adult and newborn rabbits. *J. Physiol.* **311**, 201–220.

Kanda, K. and Hashizume, K. (1992) Factors causing difference in force output among motor units in the rat medial gastrocnemius muscle. *J. Physiol.* **448**, 677–695.

Kandarian, S., O'Brien, S., Thomas, K. et al. (1992) Regulation of skeletal muscle dihydropyridine receptor gene expression by biomechanical unloading. *J. Appl. Physiol.* **72**, 2510–2514.

Katz, B. (1966) *Nerve, Muscle and Synapse*, McGraw-Hill, New York.

Kernell, D. (1965a) The adaptation and relation between discharge frequency and current strength of cat lumbosacral motoneurones stimulated by long-lasting injected currents. *Acta Physiol. Scand.* **65**, 65–73.

Kernell, D. (1965b) High-frequency repetitive firing of cat lumbosacral motoneurones stimulated by long-lasting injected currents. *Acta Physiol. Scand.* **65**, 74–86.

Kernell, D. (1965c) The limits of firing frequency in cat lumbosacral motoneurones possessing different time course of afterhyperpolarization. *Acta Physiol. Scand.* **65**, 87–100.

Kernell, D. (1966) Input resistance, electrical excitability, and size of ventral horn cells in cat spinal cord. *Science*, **152**, 1637–1340.

Kernell, D. (1992) Organised variability in the neuromuscular system. *Arch. Ital. de Biol.* **130**, 19–66.

Kernell, D. and Monster, A. W. (1982) Motoneurone properties and motor fatigue: An intracellular study of gastrocnemius motoneurones of the cat. *Exp. Brain Res.* **46**, 197–204.

Kernell, D. and Sjoholm, H. (1975) Recruitment and firing rate modulation of motor unit tension in a small muscle of the cat's foot. *Brain Res.* **98**, 57–72.

Kernell, D. and Zwaagstra, B. (1981) Input conductance, axonal conduction velocity and cell size among hindlimb motoneurones of the cat. *Brain Res.* **204**, 311–326.

Kugelberg, E., Edstrom, L. and Abruzzese, M. (1970) Mapping of motor units in experimentally reinnervated rat muscle. *J. Neurol. Neurosurg. Psychiat.* **33**, 310–329.

Kugelberg, E. and Lindegren, B. (1979) Transmission and contraction fatigue of rat motor units in relation to succinate dehydrogenase activity of motor unit fibres. *J. Physiol.* **288**, 285–300.

Leberer, E. and Pette, D. (1986a) Immunochemical quantitation of sarcoplasmic reticulum Ca-ATPase, of calsequestrin and of parvalbumin in rabbit skeletal muscles of defined fiber composition. *Eur. J. Biochem.* **156**, 489–496.

Leberer, E. and Pette, D. (1986b) Neural regulation of parvalbumin expression in mammalian skeletal muscle. *Biochem. J.* **235**, 67–73.

Lieber, R. L. (1992) *Skeletal Muscle Structure and Function: Implications for rehabilitation and sports medicine*, Williams & Wilkins, Baltimore.

Lowey, S., Slayter, H. S., Weeds, A. G. and Baker, H. (1969) The substructure of the myosin molecule. I. Subfragments of myosin by enzymatic degradation. *J. Mol. Biol.* **43**, 1–29.

Luff, A. R. and Atwood, H. L. (1971) Changes in the sarcoplasmic reticulum and transverse tubular system of fast and slow skeletal muscles of the mouse during postnatal development. *J. Cell Biol.* **51**, 369–383.

Marsden, C. D., Meadows, J. C. and Merton, P. A. (1983) 'Muscular Wisdom' that minimises fatigue during prolonged effort in man: Peak rates of motoneurone discharge and slowing of discharge during fatigue, in *Advances in Neurology*, Vol. 39 (ed. J. E. DeSmedt), Raven Press, New York, pp. 169–212.

Martin, T. P., Bodine-Fowler, S. C., Roy, R. R. *et al.* (1988) Metabolic and fiber size properties of cat tibialis anterior motor units. *Am. J. Physiol.* **255**, C43-C50.

McPhedran, A. M., Wuerker, R. B. and Henneman, E. (1965) Properties of motor units in a homogeneous red muscle (soleus) of the cat. *J. Neurophysiol.* **28**, 71–84.

Mendell, L. M., Collins, W. F. III and Koerber, H. R. (1990) How are 1a synapses distributed on spinal motoneurons to permit orderly recruitment?, in *The Segmental Motor System* (eds M. D. Binder and L. M. Mendell), Oxford University Press, New York, pp. 308–327.

Milner-Brown, H. S., Stein, R. B. and Yemm, R. (1973) Changes in firing rate of human motor units during voluntary isometric contractions. *J. Physiol. (Lond.)* **230**, 371–390.

Németh, P., Pette, D. and Vrbová, G. (1981) Comparison of enzyme activities among single muscle fibres within defined motor units. *J. Physiol.* **311**, 489–495.

Nichols, T. R., Lawrence, J. H. and Bonasera, S. J. (1993) Control of torque direction by spinal pathways at the cat ankle joint. *Exp. Brain Res.* **97**, 366–371.

Pette, D. and Staron, R. S. (1990) Cellular and molecular diversities of mammalian skeletal muscle fibres. *Rev. Physiol. Biochem. Pharmacol,* **116**, 1–76.

Pette, D. and Vrbová, G. (1985) Invited review: Neural control of phenotypic expression in mammalian muscle fiber. *Muscle & Nerve* **8**, 676–689.

Pette, D. and Vrbová, G. (1992) Adaptation of mammalian skeletal muscle to chronic electrical stimulation. *Rev. Physiol. Biochem. Pharmacol.* **120**, 115–202.

Pinter, M. J. (1990) The role of motoneuron membrane properties in the determination of recruitment order, in *The Segmental Motor System* (eds M. D. Binder and L. M. Mendell), Oxford University Press, New York, pp. 165–181.

Ramón y Cajal, S. (1909) *Histologie du système nerveux de l'homme et des vertébrés*, Librarie Maloin, Paris, Vol. 1.

Ranvier, L. (1874) De quelques faits relatifs à l'histologie et à la physiologie des muscles striés. *Arch. Physiol. Norm. Path.* **6**, 1–15.

Rayment, I., Holden, H. M., Whittaker, M. *et al.* (1993) Structure of the actin-myosin complex and its implications for muscle contraction. *Science* **261**, 58–65.

Reiser, P. J., Moss, R. L., Giulian, G. G. and Greaser, M. L. (1985) Shortening velocity of single fibers from adult rabbit soleus muscles is correlated with myosin heavy chain composition. *J. Biol. Chem.* **260**, 9077–9080.

Rios, E. and Brum, G. (1987) Involvement of dihydropyridine receptors in excitation–contraction coupling in skeletal muscle. *Nature* **325**, 717–720.

Romanul, F. C. A. (1964) Enzymes in muscle. I. Histochemical studies of enzymes in individual muscle fibres. *Arch. Neurol.* **11**, 355–368.

Schiaffino, S., Saggin, L., Viel, A. and Gorza, L. (1985) Differentiation of fibre types in rat skeletal muscle visualized with monoclonal antimyosin antibodies. *J. Muscle Res. Cell. Motil.* **6**, 60–61.

Sechenov, I. M. (1863) in *Reflexes of the Brain* (trans. S. Belsky, ed. G. Gibbons), MIT Press (1965).

Sherrington, C. (1939). The correlation of reflexes and the principle of the common final path. *Brit. Ass.* **74**, 728–741.

Squire, J. M. (1986) *Muscle: Design, Diversity and Disease*, Benjamin & Cummings, Menlo Park, Calif.

Sreter, F. A. and Gergely, J. (1964) Comparative studies of the Mg^{2+} activated ATPase activity and Ca^{2+} uptake of fractions of white and red muscle homogenates. *Biochem. Biophys. Res. Commun.* **16**, 438–443.

Stein, R. B., Gordon, T. and Totosy de Zepetnek, J. E. (1990) Mechanisms for respecifying muscle properties following reinnervation, in *The Segmental Motor System* (eds L. Mendell and M. D. Binder), Oxford University Press, London, pp. 278–288.

Stein, R. B., Gordon, T. and Schriver, J. (1982) Temperature dependence of mammalian muscle contractions and ATPase activities. *Biophys. J.* **10**, 97–107.

Stein, J. A. and Padykula, H. A. (1962) Histochemical classification of individual skeletal muscle fibres of the rat. *Am. J. Anat.* **110**, 103–124.

Stuart, D. G. and Enoka, R. M. (1992) Neurobiology of muscle fatigue. *J. Appl. Physiol.* **72**, 1631–1648.

Sweeney, H. L., Kushmerick, M. J., Mabuchi, K. *et al.* (1988) Myosin alkali light chain and heavy chain variations correlate with altered shortening velocity of isolated skeletal muscle fibers. *J. Biol. Chem.* **263**, 9034–9039.

Sypert, G. W. and Munson, J. B. (1981) Basis of segmental motor control: Motoneuron size or motor unit type? *Neurosurgery* **9**, 608–621.

Thomas, C. K., Johanson, R. S., Westling, G. and Bigland-Ritchie, B. (1990) Twitch properties of human thenar motor units measured in response to intraneural motor axon stimulation. *J. Neurophysiol.* **64**, 1339–1346.

Thomas, C. K., Stein, R. B., Gordon, T. *et al.* (1987) Patterns of reinnervation and motor unit recruitment in human hand muscles after complete ulnar and median nerve section and resuture. *J. Neurol. Neurosurg. Psychiat.* **50**, 259–268.

Totosy de Zepetnek, J. E., Gordon, T., Stein, R. B. and Zung, H. V. (1991) Comparison of force and EMG measures in normal and reinnervated tibialis anterior muscles of the rat. *Can. J. Physiol. Pharmacol.* **69**, 1774–1783.

Totosy de Zepetnek, J. E., Zung, H. V., Erdebil, S. and Gordon, T. (1992a) Innervation ratio is an important determinant of force in normal and reinnervated rat anterior tibialis muscle. *J. Neurophysiol.* **67**, 1385–1403.

Totosy de Zepetnek, J. E., Zung, H. V., Erdebil, S. and Gordon, T. (1992b) Motor unit categorization on the basis of contractile and histochemical properties: a glycogen depletion analysis of normal and reinnervated rat tibialis anterior muscle. *J. Neurophysiol.* **67**, 1404–1415.

van Winkle, W. B. and Schwartz, A. (1978) Morphological and biochemical correlates of skeletal muscle contractility in the cat. *J. Cell. Physiol.* **97**, 99–120.

Wuerker, R. B., McPhedran, A. M. and Henneman, E. (1965) Properties of motor units in a heterogeneous pale muscle (m. gastrocnemius) of the cat. *J. Neurophysiol.* **28**, 85–99.

Yellin, H. and Guth, L. (1970) The histochemical classification of muscle fibres. *Exp. Neurol.* **26**, 424–432.

Zengel, J. E., Reid, S. A., Sypert, G. W. and Munson, J. B. (1985) Membrane electrical properties and prediction of motor unit type of medial gastrocnemius neurones in the cat. *J. Neurophysiol.* **53**, 1323–1344.

Emergence of the mammalian motor unit 6

6.1 EARLY LAYOUT OF MOTOR UNIT TERRITORY

By definition, the motor unit is the motoneurone together with the muscle fibres it supplies. In the adult, the unique feature of such a unit is that a muscle fibre can belong only to a single motoneurone and is not shared by others. This arrangement is necessary to enable the motoneurone and the muscle fibres it supplies to function as a unit.

During most of embryonic development such a concept of the motor unit does not exist. This is due to the fact that initially there is convergence of the full complement of motor axons on to relatively few myogenic cells. Moreover, several axons share the same muscle cell. Early in development all axons of a particular motor pool are already present among the few myogenic cells that will form a particular muscle. At this stage the number of muscle fibres available for the axons to innervate is relatively small. The majority of these early muscle fibres are primary myotubes, which are relatively well developed, span the whole length of the muscle and have a large diameter (for review, see Kelly, 1983). With further development, small myotubes appear enclosed within the same basal lamina as the primary myotube. These are of small diameter, contain few myofilaments and are initially short. With time these secondary myotubes fuse with other myoblasts, increase in length and gradually reach the tendons, where they form independent attachments (Ross *et al.*, 1987).

Quantitative measurements of the numbers of primary and secondary muscle fibres have been carried out for some muscles. It was reported that in the IVth lumbrical muscle of the rat there are only about 100 primary myotubes (Ross *et al.*, 1987), which is one eighth of the total number of fibres the muscle contains in its adult state (Betz *et al.*, 1980). Similarly, in the rat sternocostalis muscle the initial number of primary myotubes present at the time when innervation of the muscle fibres starts is less than one fifth of the final number of muscle fibres (Sheard *et al.*, 1991).

The oldest, more advanced primary myotubes are contacted by several nerve terminals derived from different axons (Figure 6.1a). The new secondary myotubes are in close proximity to the primary myotubes and the axon terminals that contact them. The intimate contact between the primary and secondary myotubes is insured by gap junctions between the two populations of myocytes (Kelly, 1983). In addition, the primary and secondary myotubes often share the same axon terminal profiles (Duxson *et al.*, 1986). With further development some of these axon terminals lose contact with the primary myotube and innervate only the secondary myotube. Thus there is a transfer of terminals from the primary to the secondary myotube (Duxson *et al.*, 1986). At this stage in development, a selection is carried out as to which of the many terminals is to be shared and later transferred from the primary to the secondary myotube (Figure 6.1b, c).

What can influence this selection? An important factor in this selection process could be the compatibility of the secondary myotube with a particular terminal. Compatibility may depend on the characteristic properties of:

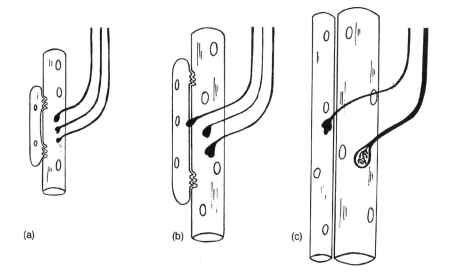

(a) (b) (c)

Figure 6.1 Nerve terminal transfer from primary to secondary myotubes during myogenesis: (a) primary myotube is contacted by three terminals of unequal size; (b) the smallest of these shares contact with both primary and small secondary myotube (note close contact between the two myotubes); (c) secondary myotube is separated from primary myotube and has the small terminal. (After Duxson *et al.*, 1986, *Neurosci. Lett.*, **71**, 147–152 taken, from Navarette and Urbová (1993), *Progress in Neurobiology*, **41**, p. 109, with permission from Pergamon Press.)

- the axon terminal that is to be transferred;
- the secondary myotube that is to accept the terminal; and
- the interaction between the terminal that is to be transferred with other terminals sharing contact with the primary myotube.

We will deal first with the characteristics of the small secondary myotubes, since more is known about these than about the axon terminals. The characteristics of the secondary myotube will be very different from those of the primary one, in that it will have a much higher input resistance, immature AChRs and less AChE (Chapter 3). Thus the secondary myotube will respond with a relatively large depolarization to even a low amount of transmitter released from the immature terminal. Its response to higher amounts of transmitter released will be disproportionately large. In addition, the immature myotube is likely to have a prolonged **endplate potential (EPP)**, since it has AChRs with long channel opening times, and little AChE (Chapter 3). Indeed, recent results confirm this and show that EPPs of young myotubes are much longer than in more mature cells (Sheard *et al.*, 1991). With increasing age the rise time and decline of depolarization becomes faster.

A recent study on the rat sternocostalis muscle shows that, at a time when the duration of the endplate potential is already shorter in the primary myotube, the less mature secondary myotube still has prolonged EPPs (Sheard *et al.*, 1991). A more mature or more active terminal would therefore produce a large prolonged response on a secondary myotube. Such a large response can be expected to lead to the retraction of the nerve ending, by mechanisms explained in section 6.1.2. For this reason the most likely candidate for transfer would be a relatively immature, inactive terminal, which will initially produce small depolarizations and only seldom fully activate the myotube. Such a terminal will have a better chance of maintaining contact with a small secondary myotube and for gradually strengthening this synaptic contact. According to this proposal, an immature terminal will be best matched to establish contact with a secondary myotube. Indeed, the earliest independent contacts between secondary myotubes and axon terminals are those which initiate an EPP of long duration, and small amplitude, indicating a low quantal content of the terminal (Sheard *et al.*, 1991).

Indirect supportive evidence for such a mechanism are findings obtained on the developing anterior and posterior latissimus dorsi muscles of the chick (**ALD** and **PLD**). In these muscles, quite early in development, slow tonic muscle fibres are segregated in the ALD and fast twitch fibres in the PLD. Most muscle fibres in the ALD are generated between embryonic days 8 and 11; by day 12, ALD has the full complement of its muscle fibres. PLD develops later than ALD and the most pronounced increase in the number of muscle fibres takes place between days 10 and 12; thereafter it continues at a slower rate until day 15 (Oppenheim and Chu-Wang, 1983). At 16 days of embryonic develop-ment, the ALD muscle fibres are larger and more mature (as judged by their content of myofibrils and by the position of the nuclei) than muscle fibres of the fast PLD (Gordon *et al.*, 1974, 1977a). At 16 days of embryonic development, the axon profiles on ALD muscle fibres are more numerous and mature than those on PLD muscle fibres, as judged by the content of vesicles and mitochondria (Srihari and Vrbová, 1978, 1980). While many axon profiles on E16 PLD muscle fibres are still surrounded by Schwann cell processes, most axon profiles on E16 ALD make good contact with the muscle fibre (Srihari and Vrbová, 1978, 1980). Figure 6.2 illustrates this point. Thus the developmentally younger PLD muscle fibres, which could be considered to be derived mainly from secondary myotubes, seem to be contacted by less mature axons.

In mixed mammalian muscles, a selective transfer of less mature nerve terminals to small myotubes would explain the checkerboard pattern of muscle fibre type distribution seen in most mammalian muscles, and could be the earliest event that helps to establish the anatomical distri-bution of motor unit territories. Recent findings showing some degree of segregation of fast and slow muscle fibres to individual motor units, even at a time when these are still supplied by several axons (Jansen and Fladby, 1990), support the possibility of an initial layout which could subsequently be changed. Later events refine this rough outline by mechanisms related to activity and environment-induced changes of muscle phenotype.

6.1.1 REORGANIZATION OF SYNAPTIC CONTACTS DURING POSTNATAL DEVELOPMENT

Even after contacts between motoneurones and muscle fibres have been made, the emergence of the motor unit in its adult form is not yet complete. In adults, each muscle fibre is supplied by a single axon and is thus activated by only one motoneurone. Each motoneurone in turn supplies several muscle fibres and the number of muscle fibres supplied by a single motoneurone is referred to as the **innervation ratio**. During late embryonic and early postnatal life the situation is very different, in that individual muscle fibres are contacted by several axons (Redfern, 1970). Figure 6.3 illustrates examples of electrophysiologically and histo-logically identified endplates from neonatal animals with polyneuronal innervation (Figure 6.3a) and an endplate with one axon contacting a muscle fibre and one axon withdrawing (Figure 6.3b). Thus in young animals each motoneurone is supplying many more muscle fibres than in the adult animal (Bagust *et al.*, 1973), i.e. it has a much higher inner-

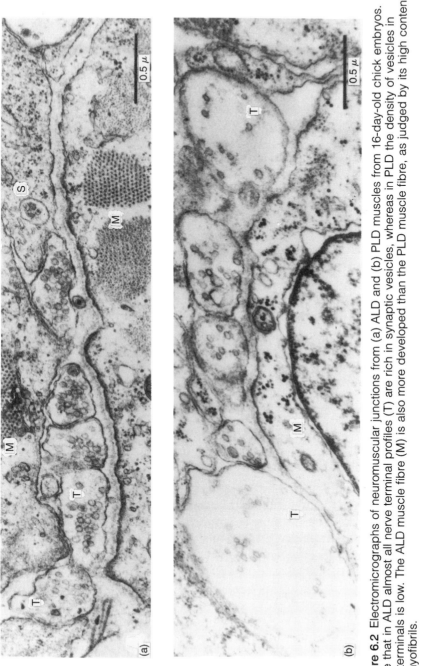

Figure 6.2 Electromicrographs of neuromuscular junctions from (a) ALD and (b) PLD muscles from 16-day-old chick embryos. Note that in ALD almost all nerve terminal profiles (T) are rich in synaptic vesicles, whereas in PLD the density of vesicles in the terminals is low. The ALD muscle fibre (M) is also more developed than the PLD muscle fibre, as judged by its high content of myofibrils.

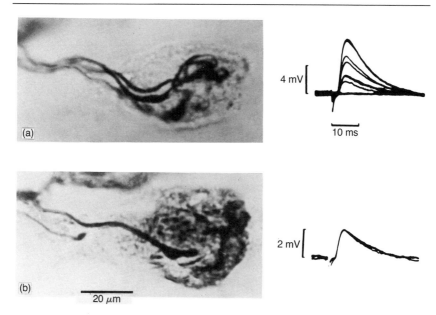

Figure 6.3 Histological and electrophysical illustration of the pattern of innervation of endplates of rat soleus muscle fibres during postnatal development. Left: cholinesterase silver-stained longitudinal sections from a muscle of (a) 9-day-old rat (note three axons contacting endplate) and (b) 14-day-old rat (note three axons in contact with endplate). Right: records of endplate potentials from soleus muscle fibres innervated by three axons (three stepwise increments) taken from 9-day-old rat (top), and mononeuronal innervation (no stepwise increment) taken from 14-day-old rat (bottom). (Reproduced from Vrbová, 1987, with permission from Manchester University Press.)

vation ratio (Figure 6.4). The role of motor unit activity in the regulation of the innervation ratio will be discussed first, since it has a profound influence on the motor unit topography.

In the adult animal the distribution of motor units within a given muscle is such that the most active motor units have the smallest innervation ratio, while the largest motor units are usually least active (Chapter 5). Consistent with this situation are results obtained in developing animals by Callaway *et al.* (1987). These authors found that motor units whose activity was reduced during the neonatal period by blocking the action potential with TTX had larger territories than those whose axons remained active. This finding is at odds with results on adult lumbrical muscles of the rat during reinnervation, where the territories of the inactive motor units were smaller than those of the active ones (Ribchester and Taxt, 1983, 1984). However, comparisons between these two sets of experiments are difficult, for they were done in the one case

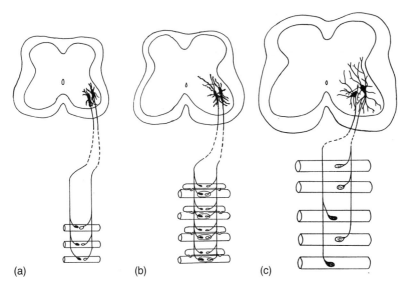

(a)   (b)   (c)

Figure 6.4 The mammalian motor unit: two motoneurones contacting muscle fibres during different stages of postnatal development: (a) the terminals of the two embryonic motoneurones converge on few primary myotubes; (b) the number of primary myotubes increases, secondary myotubes are connected via gap junctions to primary myotubes (smaller size fibres), terminal from the two motoneurones converge on all the muscle fibres; (c) in the adult, each muscle fibre is contacted by a single nerve terminal. (Reproduced from Navarette and Urbová (1993) *Progress in Neurobiology*, **41**, p. 107, with permission from Pergamon Press).

(Ribchester and Taxt, 1984) on an adult fast muscle and in the other case on a neonatal rabbit slow muscle (Callaway *et al.*, 1987).

It is likely that the cellular mechanisms that are involved in the activity-regulated maintenance of contacts between axon terminals and muscle fibres will be of great importance for the development of the motor unit.

6.1.2 MECHANISMS INVOLVED IN THE REGULATION OF SYNAPTIC CONTACTS

In rat hindlimb muscles the emergence of the adult motor unit size, achieved by elimination of polyneuronal innervation, takes place during the second and third week of life when the amount of neuromuscular activity of the animal suddenly increases (Navarrete and Vrbová, 1983, 1993). Indeed the elimination of polyneuronal innervation is critically dependent on neuromuscular activity. If neuromuscular activity is increased, the rate of loss of synaptic contacts is dramatically accelerated (O'Brien and Vrbová, 1978). Conversely, if neuromuscular activity is reduced, the rate of synapse elimination is reduced (Benoit and Changeux, 1975; Caldwell and Ridge, 1983; Callaway *et al.*, 1987). Such results could be explained in terms of a competitive interaction between the axons contacting the endplate region without an active participation of the target muscle, or it could be that the activity of the muscle influences its own innervation.

This question has been addressed in experiments where motoneurone activity was not interfered with but the activity of the target muscle was manipulated pharmacologically. This was achieved by blocking the enzyme that hydrolyses the transmitter, AChE. Under these conditions the loss of contacts between nerve terminals and muscle fibres was greater than normally seen (O'Brien *et al.*, 1982; Duxson and Vrbová, 1985; Greensmith and Vrbová, 1991). Moreover, if during the period of rapid synapse elimination the response of the muscle is prevented by blocking the AChR by curare or BTX, the rate of loss of neuromuscular contacts is temporarily arrested (Srihari and Vrbová, 1978; Duxson, 1982). The mechanism by which the target muscle regulates its synaptic inputs is not entirely clear, but there is evidence that:

- the depolarization of the endplate region is an important step in this process;
- Ca^{2+} ions are involved in regulating neuromuscular contacts; and
- a Ca^{2+} activated neutral protease (**CANP**), calpain, which is known to be involved in the breakdown of cytoskeletal proteins (Croall and DeMartino, 1991), eliminates nerve terminals (O'Brien *et al.*, 1982, 1984; for review, see Navarrete and Vrbová, 1993).

We are therefore proposing the model summarized in Figure 6.5 to account for synapse elimination. During activity, the skeletal muscle fibre releases K^+, which is known to accumulate in the synaptic cleft. (For review, see Lowrie *et al.*, 1989; Navarrete and Vrbová, 1993). Such accumulation of K^+ almost certainly causes depolarization of the nerve terminal, which in turn will lead to opening of Ca^{2+} channels and allow entry of Ca^{2+} from the extracellular compartment. If the Ca^{2+} transients are prolonged and large enough, then the concentration of Ca^{2+} in the nerve terminal may become high enough to activate the CANP present and lead to the breakdown of cytoskeletal proteins and the subsequent loss of contacts between the nerve and muscle.

According to this proposal, small axon profiles would be more vulnerable for two reasons:

- Because of their surface-to-volume ratio the Ca^{2+} transients can be expected to be greater in small than in large terminals.
- Because of their smaller size, they probably contain less cytoskeletal protein.

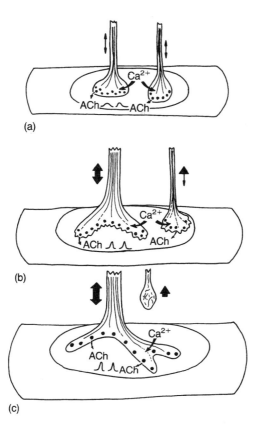

(a)

(b)

(c)

Figure 6.5 Mechanisms leading to elimination of polyneuronal innervation: (a) an endplate contacted by two terminals of unequal size (arrows indicate entry of Ca²⁺ into them); (b) in the smaller terminal, Ca²⁺ reaches high enough concentrations to disrupt neurofilaments; (c) consequently, the small terminal is withdrawn.

It has indeed been shown that, in situations where a loss of terminals is known to take place, more contacts between small terminals and muscle fibres are lost than between large terminals and muscle fibres (O'Brien *et al.*, 1984; Duxson and Vrbová, 1985). Thus any event that prolongs or increases depolarization of the post-synaptic membrane may have a deleterious effect on small nerve terminals that contact it.

The possibility that the degree of maturity of the endplate may be important for the maintenance of synaptic contact between a nerve terminal and muscle fibre is an important concept, for it indicates another dimension by which matching between motoneurones and muscle fibres can be achieved. Primary myotubes that are larger and possibly more mature than secondary myotubes will provide more favourable conditions for nerve terminal contacts, and it is indeed these fibres that initially are able to accept a much higher degree of polyneuronal innervation than the secondary, smaller fibres (Sheard *et al.*, 1991). The transfer of terminals from primary to secondary myotubes which follows (Figure 6.1) may be regulated by the mechanisms described in this section.

There are other possibilities by which muscle activity may regulate its synaptic input, such as the release of specific motoneurone (or nerve terminal) survival factors, but one would have to postulate that the less active the endplate, the more survival factors it releases. Such a mechanism, though attractive by its simplicity, has not so far gained much experimental support (for review see Jansen and Fladby, 1990).

6.2 MODIFICATIONS OF MUSCLE FIBRE PROPERTIES DURING EARLY POSTNATAL DEVELOPMENT

Until now this chapter has been concerned mainly with the outlay and site of motor unit territory. Factors that will finally determine the characteristics of muscle fibres within a motor unit have not been discussed. It is well known that one of the major influences on muscle fibre phenotype is activity (Chapter 7).

Electromyographic (**EMG**) studies in immature rats indicate that motor activity patterns to slow and fast muscles become differentiated during the first 3 weeks of postnatal life (Navarrete and Vrbová, 1983). In freely moving rats up to 8–10 days of age, the EMG activity of future slow motor units from the soleus muscle is strikingly different from that in the adult, in that they do not sustain activity for more than a few seconds. Interestingly, at this stage the muscle does not yet have a significant role in hindlimb postural support. During spontaneous or reflex movements, several motor units are activated almost synchronously at low frequencies (3–10 Hz) for brief periods, and resemble in this aspect phasic motor units of adult fast muscles. In the EDL of neonatal animals, EMG activity is also phasic but, in contrast to the situation in the adult, the firing frequency of the majority of motor units is low. Thus, during early stages of development, both types of motoneurones fire at relatively low rates and are unable to sustain their firing for long periods. With development, slow motoneurones become capable of tonic firing at low frequencies, whereas fast motoneurones continue to discharge in a phasic manner but increase their rate of firing. Moreover, an increase in the aggregate activity of both types of motor units is found during postnatal development, the increase being much greater in slow motor units (Navarrete and Vrbová, 1983). Figure 6.6a shows a dramatic increase in the amount of activity in soleus (a postural muscle). This increase occurs rapidly between the 10th–25th day of postnatal development, when the tonic EMG activity of the muscle is first apparent. Concurrent with this increase in EMG activity there is also a marked increase in the number of slow fibres (Figure 6.6b). Figure 6.6 illustrates the increase of slow myosin-containing fibres that occurs during the same period.

Before muscle fibres become committed to a single motoneurone, they combine a set of immature characteristics that are different from muscle fibre types later in life. Because of the different rates of development of the various cellular determinants of the speed of muscle contraction, the developmental changes of muscle fibre properties are often difficult to understand or describe. For instance, although embryonic and neonatal muscles contract and relax slowly, they do not contain myosin and troponin isoforms characteristic of adult slow muscles but variable mixtures

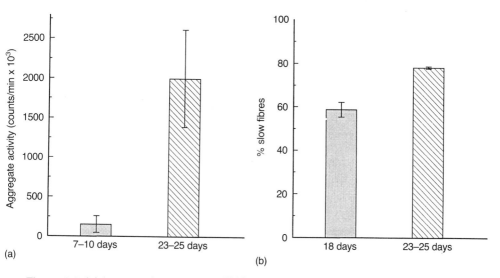

Figure 6.6 (a) Increase in aggregate EMG activity in rat soleus muscle during postnatal development; (b) percentage change of slow myosin-containing muscle fibres in the soleus muscle at two different developmental stages.

of isozymic forms, some of which are present only in immature muscles (Whalen *et al.*, 1981). This peculiar combination of a slow contractile speed with contractile protein isoforms that are not typical of slow muscles indicates that, in immature muscles, the slow rate of rise in tension is not necessarily associated with the particular form of contractile protein isoforms that the immature muscle contains. Instead, other factors that influence the speed of contraction, such as the time course of the release and uptake of Ca^{2+} ions that in turn regulate the interaction between the contractile proteins, appear to be more important (Close, 1964; Pette *et al.*, 1979). It is known, for instance, that Ca^{2+} release and re-uptake from the sarcoplasmic reticulum (SR) is slower in immature muscle fibres (Martonosi, 1982). In kittens after birth, both slow and fast muscles first become faster-contracting and then the slow muscles gradually achieve their slow rate of tension development while fast muscles continue to increase their contractile speed (Buller *et al.*, 1960). A similar situation is found in rats where, after birth, both slow and fast muscles increase their speed of contraction with age before their contractile speeds diverge (Close, 1964).

Remodelling of muscle fibre characteristics continues throughout the life of the animals and is also seen in the rat EDL and lumbrical muscles, where the number of slow fibres decreases with age. These age-related changes could be due to the shift of motor unit population, i.e. that whole motor units are transformed from a fast towards a slow type. This proposal was as originally made by Kugelberg (1976). Alternatively, transitions of muscle fibre phenotype may occur in those muscle fibres which contain a different MHC isoform during early development from that of the motor unit to which they finally belong. Although Kugelberg (1976) supported the first possibility, recent results indicate that the latter phenomenon might also contribute to the final 'sorting out' process. If the suggestion that only whole motor units are altered during development is accepted, it would have to be assumed that the original matching between a particular type of axon and the muscle fibre it supplies is indeed perfect, with no errors. This is unlikely to be the case. Recent results on the IVth lumbrical muscle of the rat show that there is a much greater heterogeneity of muscle fibres belonging to the same motor unit in younger animals than during later postnatal life, indicating that some muscle fibres became gradually transformed under the influence of innervation (Gates and Ridge, 1992). This result is not surprising in view of the ability of muscle fibres to adapt to different patterns of activity or mechanical stimuli. In the rat the soleus muscle fast fibres, which usually convert into slow ones, fail to do so if they are not exposed to their normal load and such a muscle continues to contain a relatively low proportion of slow fibres. Moreover, many muscle fibres in such an 'unloaded' soleus muscle contain both fast and slow isoforms of the myosin heavy chain (Lowrie *et al.*, 1989).

Indeed it is well known that muscle fibres change their phenotype during postnatal development (Goldspink and Ward, 1979) and the relative slowness of this process may account for the conflicting interpretations concerning the homogeneity of muscle fibres within a single motor unit. It was claimed by Edström and Kugelberg (1968), and later by Németh *et al.* (1981), that motor units are homogeneous (Chapter 5). This, in broad terms, is true, but if there are transitions with age that take a long time to be completed there may be considerable variability in the results obtained. Moreover, until recently it was possible to study only the presence but not the gene expression of a protein, and this, particularly in the case of myosin, may be very misleading because of the time lag involved between the change in gene expression and the presence of detectable amounts of the protein in the muscle fibre. It is likely that the use of more modern experimental tools, such as *in situ* hybridization and analysis of gene expression in single muscle fibres, will reveal that the second stage of refinement of motor unit territory, during postnatal development, is regulated by activity and to some extent other environmental factors, such as hormone levels and mechanical conditions during muscle activity.

CONCLUSIONS

The relationship between the axon terminal and developing myotubes during embryonic development is described. Nerves enter developing muscles at a time when only primary myotubes are present, and make contact with these. With time, secondary myotubes develop and some nerve endings are transferred from the primary to the secondary myotubes. It is argued that the least mature nerve ending is the one that is transferred, for it is better matched to the immature secondary myotube. This matching based on 'maturity' is then responsible for the initial anatomy of motor unit territory. Following this initial layout of the anatomy of the motor unit, further refinement and some redistribution of motor unit size is achieved.

Initially nerve terminals establish an excessive number of contacts with individual muscle fibres. With time many of these are eliminated and leave each muscle fibre supplied by a single axon. This elimination of excess terminals reduces the size of individual motor units and establishes the 'adult' pattern of motor unit size. The mechanisms that lead to this final refinement are discussed. The role of muscle fibre activity in the regulation of its innervation is highlighted.

Finally, mechanisms that determine the emergence of different muscle fibre types and their matching properties to motoneurones that supply them are proposed. These include neural activity, mechanical constraints and hormonal influences.

REFERENCES

Bagust, J., Lewis, D. M. and Westerman, R. A. (1973) Polyneuronal innervation of kitten skeletal muscle. *J. Physiol. (Lond.)* **229**, 241–255.

Benoit, P. and Changeaux, J.-P. (1975) Consequences of tenotomy on the evolution of multi-innervation in developing rat soleus muscle. *Brain Res.* **99**, 345–358.

Betz, H., Caldwell, J. H. and Ribchester, R. R. (1980) The effects of partial denervation at birth on the development of muscle fibres and motor units in rat lumbrical muscle. *J. Physiol.* **303**, 265–279.

Buller, A. J., Eccles, J. C. and Eccles, R. M. (1960) Differentiation of fast and slow muscles in the cat hind limb. *J. Physiol.* **150**, 399–416.

Caldwell, J. H. and Ridge, R. A. M. P. (1983) The effects of deafferentation and spinal cord transection on synapse elimination in developing rat muscles. *J. Physiol.* **339**, 145–159.

Callaway, E. M., Soha, J. M. and Van Essen, D. C. (1987) Competition favouring inactive over active motoneurones during synapse elimination. *Nature* **328**, 422–426.

Close, R. (1964) Dynamic properties of fast and slow skeletal muscles of the rat during development. *J. Physiol. (Lond.)* **173**, 75–95.

Croall, D. E. and DeMartino, G. N. (1991) Calcium activated neutral protease (calpain) system: structure, functions and regulation. *Physiol. Rev.* **71**, 813–847.

Duxson, M. J. (1982) The effect of postsynaptic block on development of the neuromuscular junction of postnatal rats. *J. Neurocytol.* **11**, 395–408.

Duxson, M. J., Ross, J. J. and Harris, A. J. (1986) Transfer of differentiated synaptic terminals from primary myotubes to new-formed muscle cells during embryonic development in rats. *Neurosci. Lett.* **71**, 147–152.

Duxson, M. J. and Vrbová, G. (1985) Inhibition of acetylcholinesterase axon terminal withdrawal *J. Neurocytol.* **14**, 337–363.

Edström, L. and Kugelberg, E. (1968) Histochemical composition, distribution of fibres and fatiguability of single motor units. Anterior tibial muscles of the rat. *J. Neurol. Neurosurg. Psychiat.* **31**, 424–433.

Gates, H.-J. and Ridge, R. M. A. P. (1992) The importance of competition between motoneurones in developing rat muscle: effects of partial denervation at birth. *J. Physiol. (Lond.)* **445**, 457–472.

Goldspink, G. and Ward, P. S. (1979) Changes in rodent muscle fibre types during postnatal growth, undernutrition and exercise. *J. Physiol.* **296**, 453–469.

Gordon, T., Perry, R., Tuffery, A. R. and Vrbová, G. (1974) Possible mechanisms determining synapse formation in developing skeletal muscle of the chick. *Cell Tiss. Res.* **155**, 13–25.

Gordon, T., Perry, R., Srihari, T. and Vrbová, G. (1977) Differentiation of slow and fast muscles in chickens. *Cell Tiss. Res.* **180**, 211–222.

Gordon, T., Purves, R. D. and Vrbová, G. (1977b) Differentiation of electrical and contractile properties of slow and fast muscle fibres. *J. Physiol.* **269**, 535–547.

Greensmith, L. and Vrbová, G. (1991) Neuromuscular contacts in developing rat soleus depend on activity. *Dev. Brain Res.* **62**, 121–129.

Jansen, J. K. and Fladby, T. (1990) The perinatal reorganization of the innervation of skeletal muscle in mammals. *Prog. Neurobiol.* **34**, 39–90.

Kelly, A. M. (1983) Emergence of specialisation of skeletal muscle, in *Handbook of Physiology*, Section 10 (ed. L. D. Peachy), William & Wilkins, Baltimore, pp. 507–537.

Kugelberg, E. (1976) Adaptive transformations of rat soleus motor units during growth. Histochemistry and contraction speed. *J. Neurol. Sci.* **27**, 269–289.

Lowrie, M. B., Moore, A. F. K. and Vrbová, G. (1989) The effect of load on the phenotype of the developing rat soleus muscle. *Pflügers Archiv.* **415**, 204–208.

Martonosi, A. (1982) Transport of calcium by sarcoplasmic reticulum, in *Calcium and Cell Function* (ed. W. Y. Cheung), Academic Press, San Diego, Vol III pp. 37–102.

Navarrete, R. and Vrbová, G. (1983) Changes of activity patterns in fast muscle during postnatal development. *Dev. Brain Res.* **8**, 11–19.

Navarrete, R. and Vrbová, G. (1993) Activity dependent interactions between motoneurones and muscles: Their role in the development of the motor unit. *Prog. Neurobiol.* **41**, 93–124.

Németh, P. M., Pette, D. and Vrbová, G. (1981) Comparison of enzyme activities among single muscle fibres within defined motor units. *J. Physiol. (Lond.)* **311**, 485–495.

O'Brien, R. A. D. and Vrbová, G. (1978) Acetylcholine synthesis in nerve endings to slow and fast muscles in developing chicks: effects of muscle activity. *Neurosci.* **3**, 1227–1230.

O'Brien, R. A. D., Östberg, A. J. C. and Vrbová, G. (1982) The reorganisation of neuromuscular junctions during development in the rat, in *Membranes in Growth and Development* (eds J. H. Hoffman, G. H. Giebisch and L. Bollis), Alan Liss, New York, pp. 247–257.

O'Brien, R. A. D., Östberg, A. J. C. and Vrbová, G. (1984) Protease inhibitors reduce the loss of nerve terminals induced by activity and calcium in developing rat soleus muscles *in vitro*. *Neurosci.* **3**, 1227–1230.

Openheim, R. W. and Chu-Wang, J. W. (1983) Aspects of naturally occurring motoneurone cell death in the chick spinal cord during embryonic development, in *Somatic and Automatic Nerve-Muscle Interactions* (eds G. Burnstock, R. A. D. O'Brien and G. Vrbová), Elsevier, Amsterdam, pp. 57–107.

Pette, D., Vrbová, G. and Whalen, R. C. (1979) Independent development of contractile properties and myosin light chains in embryonic chick fast and slow muscle. *Pflügers Arch.* **378**, 251–257.

Redfern, P. A. (1970) Neuromuscular transmission in newborn rats. *J. Physiol.* **209**, 701–709.

Ribchester, R. R. and Taxt, T. (1983) Motor unit size and synaptic competition in rat lumbrical muscle reinnervated by active and inactive motor axons. *J. Physiol.* **344**, 89–111.

Ribchester, R. R. and Taxt, T. (1984) Repression of inactive motor nerve terminals in partially denervated rat muscle after regeneration of active motor axons. *J. Physiol. (Lond.)* **347**, 497–511.

Ross, J. J., Duxson, M. J. and Harris, A. J. (1987) Formation of primary and secondary myotubes in rat lumbrical muscles. *Development* **100**, 383–394.

Sheard, P. W., Duxson, M. J. and Harris, A. J. (1991) Neuromuscular transmission to identified primary and secondary myotubes: a re-evaluation of polyneuronal innervation patterns in rat embryos. *Dev. Biol.* **148**, 459–472.

Srihari, T. and Vrbová, G. (1978) The role of muscle activity in the differentiation of neuromuscular junction in slow and fast chick muscle. *J. Neurocytol.* **7**, 529–540.

Srihari, T. and Vrbová, G. (1980) Effects of neuromuscular blocking agents on the differentiation of nerve–muscle connections in slow and fast chick muscles. *Dev. Growth Diffn.* **4**, 645–657.

Vrbová, G. (1987) Reorganization of nerve–muscle synapses during development, in *Growth and Plasticity of Neuroconnections* (eds W. Wimlow and C. R. MacCrohan), Manchester University Press, pp. 36–57.

Whalen, R. G., Schwartz, K., Bouveret, P. *et al.* (1981) Contractile protein isozymes in muscle development: identification of an embryonic form of myosin heavy chain. *Proc. Nat. Acad. Sci. USA* **76**, 5197–5201.

Plasticity of muscles and their motor units   7

The characteristic differences in the structure and function of slow tonic and fast twitch muscles in lower vertebrates have been described in Chapter 4 and the differences between fast twitch and slow twitch mammalian muscles were described in Chapter 5. These differences were first recognized more than a century ago (Ranvier, 1874). Within muscles, differences between muscle fibres are also well recognized:

- the phenotypic characteristics of all muscle fibres within a single motor unit are normally homogeneous; and
- the differing physiological and metabolic properties of different motor units are matched to the functional requirements imposed upon them by the motoneurone.

How this apparent matching of muscle fibre properties to their motoneurones is brought about has been investigated by perturbing the interaction between motoneurones and muscle fibres in specific ways and examining the consequences.

7.1 NEURAL DETERMINATION OF MUSCLE PROPERTIES

7.1.1 MUSCLE PROPERTIES AFTER CROSS-REINNERVATION

The dependence of muscle characteristics on a particular type of nerve was clearly demonstrated in the experiments of Buller *et al.* (1960a), who connected the nerve which supplies the slow soleus muscle of the cat to the fast flexor digitorum longus (**FDL**) muscle and vice versa. Some time after this 'cross-innervation' the characteristic contractile speeds of the two muscles were compared with those of the contralateral, unoperated side. Figure 7.1 shows the change in both fast FDL and slow soleus muscles produced by the alien innervation. Fast muscles supplied by a slow nerve become slow contracting and the slow muscles supplied by

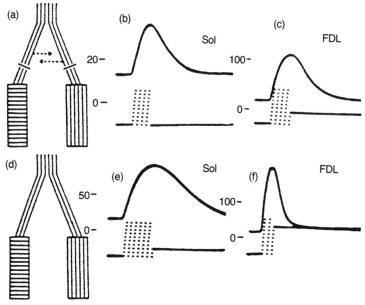

Figure 7.1 Cross-innervation in 6-day-old kittens. (a) Records showing isometric twitches of (b) the soleus and (c) flexor digitorum longus muscles recorded 45 days later. (d) Records showing isometric twitches ((e), (f)) from the same muscles on the contralateral side. (Reproduced from Buller, Eccles and Eccles, 1960, *J. Physiol.* **150**, 417–439, with permission of the authors and *Journal of Physiology.*)

a fast nerve become fast contracting. By this single series of experiments it was established that the contractile properties characteristic of a muscle are not inherent but are determined by the motor nerve. It was later found that many of the properties of fast or slow muscles could be manipulated by altering their innervation. Hence, in several species, the slow soleus muscle supplied by a fast motor nerve shows not only a decrease in time to peak twitch and tetanic tension, but also an increased tetanic-to-twitch ratio and an increase in shortening velocity of contraction (Amphlett *et al.*, 1975; Buller *et al.*, 1960a; Buller and Pope, 1977; Close, 1969; Gillespie *et al.*, 1986; Luff, 1975; Sreter *et al.*, 1974).

Cross-innervation also leads to corresponding changes in muscle metabolism reflected by a marked rearrangement of the enzyme activity patterns (Drahota and Gutmann, 1963; Prewitt and Salafsky, 1967, 1970; Guth, 1968; Golisch *et al.*, 1970) and histochemical profiles of muscle fibres (Dubowitz, 1967; Kárpáti and Engel, 1967; Romanul and Van der Meulen, 1967; Yellin, 1967; Gordon *et al.*, 1988).

In fast muscles that are reinnervated by a nerve from a slow muscle, there is an increase in oxidative enzyme activity with a simultaneous decrease in the enzyme activities of glycogenolysis, glycolysis and lactate metabolism. Myosin ATPase activity, which is related to the contraction speed (Bárány, 1967), is also altered together with the changes in contractile characteristics of cross-innervated muscles (Bárány and Close, 1971; Sreter *et al.*, 1976). In addition, there are changes in the light chain components of the myosin molecule toward slow isoforms and in the electrophoretic mobility of myosin (Hoh, 1975).

Other biochemical changes contribute to the early changes in the contractile characteristics. In the rat, ATPase activity of the sarcoplasmic reticulum in slow muscles increases after cross-innervation by a fast nerve, leading to faster accumulation of calcium, shortening of the active state and a decrease in contraction time (Margreth *et al.*, 1973).

The changes in twitch contraction speed induced by the alien innervation are also accounted for by alteration in the structure of both thin and thick filaments. Figure 7.2 shows that when the soleus muscle of a rabbit is reinnervated by the lateral popliteal nerve, which usually innervates fast muscle fibres, its contractile properties become fast. This change

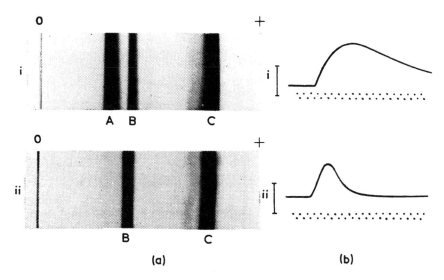

Figure 7.2 Effect of cross-innervation on the speed of contraction and the troponin I composition of rabbit soleus muscles. (a) Electrophoresis of troponin I (20 μg), isolated by affinity chromatography from rabbit soleus muscle in the presence of troponin C (50 μg) from rabbit white skeletal muscle; 6 M urea, 25 mM Tris, 80 mM glycine, pH 8.6, 8% acrylamide; (i) control soleus; (ii) soleus from the other leg of the same animal 26 weeks after cross-innervation; O = origin; A = slow troponin I-troponin C complex; B = fast troponin I-troponin C complex; C = troponin C. (b) Records of single isometric contractions of the soleus muscles used for troponin I preparations analysed in (a). The distance between successive dots represents 10 ms. Vertical scale represents 100 g tension. (Reproduced from Amphlett *et al.*, 1975, *Nature*, **257**, 602–604, with permission of the authors and *Nature*.)

of contractile properties is accompanied by a transformation of the structure of troponin I, one of the three regulatory proteins in skeletal muscles (Amphlett *et al.*, 1975, 1976).

Interestingly, cross-reinnervated muscles are not always as fast or slow as may be predicted from their altered novel innervation (Gordon and Pattullo, 1993). The cross-reinnervated soleus muscle is not usually as fast contracting as a control fast muscle. This is partially explained by the fact that most nerves contain axons from both fast and slow motoneurones. The extent of change in contraction speed probably reflects the relative proportion of fast and slow motor units that become established following cross-reinnervation (Dum *et al.*, 1985a, b; Gordon *et al.*, 1988). Buller *et al.* (1987) found that slow to fast conversion of the cat soleus was better after cross-innervation with the nerve to larger flexor hallucis longus (**FHL**) than with the smaller flexor digitorum longus (FDL), while conversion of the FHL by soleus nerve was less effective, further suggesting that the number of available fast or slow motoneurones crucially influences the results of such studies. However, this cannot entirely explain these results, for at single motor unit and muscle fibre levels too, the neural influence does not always induce a complete change in phenotype. New characteristics imposed on the existing phenotypic characteristics often result in muscle fibres co-expressing both novel and original proteins. Cross-reinnervated cat soleus muscle fibres are fast contracting but nevertheless contain both fast and slow myosin heavy chain isoforms and maintain their high oxidative capacity (Gauthier *et al.*, 1983). Figure 7.3 shows that in cross-innervated rabbit muscles, during transition from one type of muscle to another, a proportion of the muscle fibres express both fast and slow isoforms of troponin I. However, with time, complete conversion was achieved (Dhoot and Vrbová, 1991) and this is illustrated in Figure 7.4.

Despite some discrepancies, cross-innervation experiments show that the nerve has a powerful influence on the properties of adult muscle fibres. This is also seen during reinnervation (described in more detail in Chapter 9), when ingrowing nerves supply muscle fibres which formerly belonged to different motor units. After reinnervation the muscle fibres innervated by their new axon acquire the characteristic properties appropriate to their novel innervation (Kugelberg *et al.*, 1970; Albani and Vrbová, 1985; Totosy de Zepetnek *et al.*, 1992).

7.1.2 HOW DOES THE NERVE SPECIFY MUSCLE PROPERTIES?

There are several possible explanations as to how the nerve exerts its influence over the muscle fibres it supplies. Buller *et al.* (1960a) suggested that the nerve exerts a special 'trophic' influence over the muscle fibres it supplies. Another explanation of these results was put forward by

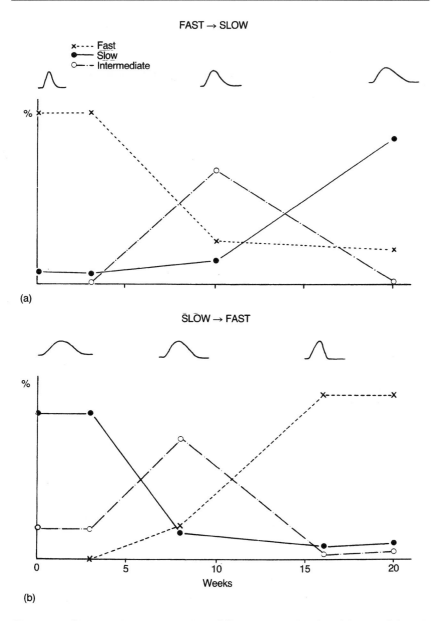

Figure 7.3 Changes in the proportion of fibres containing fast (x), slow (•) and both (○) isoforms of troponin I: (a) after cross-reinnervation of rabbit EDL muscle with a nerve from soleus muscle; (b) after cross-reinnervation of rabbit soleus muscle by the peroneal nerve. Results are plotted as percentage of initial values (ordinate) against days after operation (abscissa). Curves inserted above each panel show time course of contraction at different stages after cross-innervation.

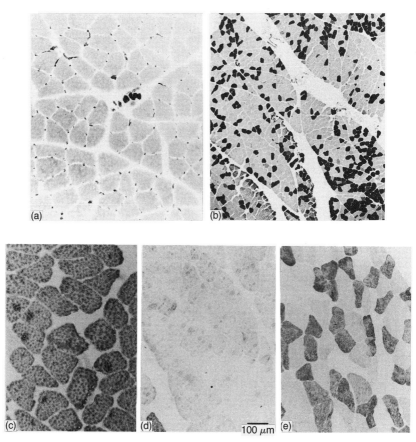

Figure 7.4 Cross-sections through cat soleus muscle stained for alkali-stable ATPase: (a) control; (b) cross-innervated with FDL nerve. Cross-sections of rabbit soleus muscles reacted with an antibody against slow MHC: (c) control; (d) cross-innervated with a peroneal nerve; (e) cross-innervated with an FDL nerve. (From Dhoot and Vrbová, 1991, *Muscle and Motility*, **2**, 125–130.)

A. F. Huxley, who suggested that the motor nerve maintains the slow time course of contraction of the soleus muscle fibres by imposing on it a slow frequency activity, which may act as a 'vibratory stress' (see Buller *et al.*, 1960a). Buller and his colleagues did not favour this interpretation and it was not until later that evidence was provided to show the crucial importance of the activity of the muscle in determining its properties.

It is known from the work of Sperry (1945) that the activity pattern of motor nerves remains unaltered when they are transposed into different muscles. After cross-innervation the nerve from a fast muscle, now supplying a slow muscle, would activate the slow muscle by the activity pattern that was originally transmitted to the fast muscle and the nerve from the slow muscle would continue activating the fast muscle by the activity received by the original slow muscle. Thus the effects produced by crossing the motor nerves can be explained by the altered activity imposed on the muscle without the need to evoke a special trophic influence.

7.1.3 THE INFLUENCE OF ACTIVITY PATTERNS ON MUSCLE PROPERTIES

The role of activity patterns in determining muscle fibre properties has been investigated using an experimental situation in which the effects of different imposed activity patterns could be tested without interfering with the muscle's own innervation. Soleus motoneurones are readily activated by 1a afferent connections from muscle spindles, while other muscles, including tibialis anterior are activated during voluntary or reflex movement. It might therefore be expected that the stretch reflex would be of greater importance for the activity of soleus than for that of muscles involved in other types of movement. After cutting the tendons (tenotomy) of the soleus muscle, preventing it from being stretched, the continuous activity of the soleus muscle is reduced to only occasional activity involving very small motor units. The contractile speed of the tenotomized soleus muscle also becomes fast (Figure 7.5). In this study (Vrbová, 1963a, b), the innervation to soleus was unaltered and only the

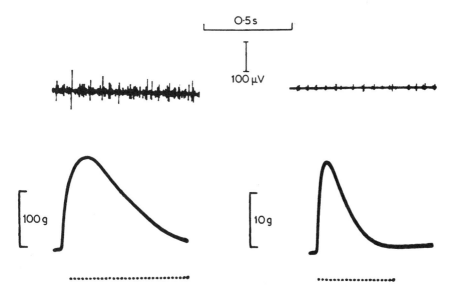

Figure 7.5 EMG recordings from (a) normal soleus muscle and (b) soleus muscle following tenotomy. Isometric single twitches of (c) control soleus and (b) soleus muscle 14 days after tenotomy; the distances between successive dots indicate 10 ms.

activity pattern of the soleus muscle was changed. Tenotomy of the other calf muscles does not appreciably alter their phasic type of EMG activity and correspondingly the contractile speed remains unaffected by tenotomy.

In order to study the effects of activity at different frequency patterns on muscle contractile properties, electrical stimulation was imposed on a muscle that had little or no activity of its own. As already mentioned, the tenotomized soleus, although quiescent, has some activity, probably due to supra spinal excitation of soleus motoneurones. After spinal cord section even this activity is abolished and the tenotomized muscle is completely 'silent' (Vrbová, 1963a). The tenotomized, quiescent soleus of a spinal rabbit is fast contracting, suggesting that its normal activity is necessary for maintaining the slow contractile speed (Vrbová, 1963b).

When silenced soleus muscles are stimulated electrically via implanted electrodes continuously at 5 or 10 Hz for 8 hours a day over a period of 2–3 weeks, they fail to become fast and remain slow. However, when bursts at higher frequencies of 20 or 40 Hz are used, the soleus muscle becomes fast contracting. These results are summarized in Figure 7.6, and clearly show that the contractile speed of the rabbit soleus is determined by the particular activity pattern imposed on the muscle (Vrbová, 1963b; Salmons and Vrbová, 1969).

Another approach to explore whether the neural influence is mediated by the pattern of neural activity is to eliminate neural activation physically by cutting the nerve supply and to replace the neural activity with direct electrical stimulation. When denervated rat soleus muscles were stimulated with a continuous low frequency pattern, the muscles remained as slow as normal (Lømo et al., 1974). In contrast, when the denervated muscle was stimulated with the same total number of stimuli but delivered in high-frequency intermittent bursts, the muscles developed force as rapidly as a fast twitch muscle (Gorza et al., 1988). The increased contractile speed and associated induction of fast myosin heavy chains was attributed to the pattern of stimulation rather than any effects of denervation alone (Al-Amood and Lewis, 1987; Gorza et al., 1988).

Whether contractile properties of the fast muscles can be influenced by activity as easily as those of slow muscles has been studied by stimulating the motor nerves to tibialis anterior (**TA**) and extensor digitorum longus (**EDL**) muscles of rabbits at 10 Hz. After 2–3 weeks of low-frequency stimulation, these fast muscles had become slow contracting (Figure 7.7). When stimulated for longer periods (2–3 months) their contractile speeds were similar to those of the normal, slow soleus muscle. Thus, even though the slow frequency activity was superimposed on the normal phasic activity of the fast muscles, it had a dramatic slowing effect (Salmons and Vrbová, 1969; Pette et al., 1973, 1976). These

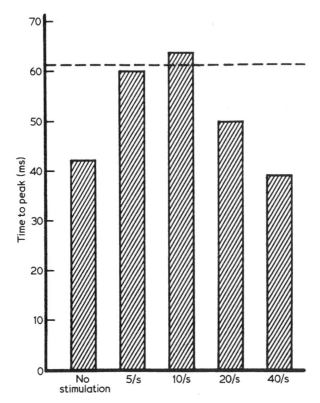

Figure 7.6 Mean values of time to peak of tenotomized soleus muscles of spinal animals under different experimental conditions. Interrupted line represents mean time to peak of control soleus muscles from a separate group of animals. Values for unstimulated tenotomized soleus muscles in first column are from previous experiments (Vrbová, 1963, *J. Physiol.* **169**, 513–526.)

results taken together indicate that the nerve exerts its influence by imposing a particular pattern of activity on the muscle.

Not only the contractile speed but also endurance and the enzyme composition of fast muscles is altered by long-term electrical stimulation (for review see Pette and Vrbová, 1992; Gordon and Mao, 1994).

Fast-to-slow transitions of skeletal muscle fibres also include changes in troponin T and myosin heavy chain isoforms, and these occur synchronously (reviewed by Pette and Vrbová, 1994). The increase of the slow isoforms of contractile proteins is accompanied by the down-regulation of their fast counterparts (Figure 7.8).

Thus, slow tonic activity induces high levels of oxidative enzymes, low levels of anaerobic enzymes, reduced rate of uptake of calcium and altered Ca^{2+} binding capacity of the sarcoplasmic reticulum and, finally, 'slow' isoforms of the myofibrillar contractile proteins are expressed

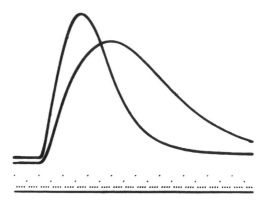

Figure 7.7 Isometric twitch of control (faster) and 14-day stimulated (slower) tibialis anterior muscle of a rabbit. Intervals between dots represent 10 ms. Tension developed by the control muscle was 650 g and by the stimulated muscle 750 g.

(reviewed by Pette and Vrbová, 1985, 1992, 1994). Fast phasic activity, on the other hand, produces muscle fibres with low oxidative and high anaerobic enzyme activities with high myosin ATPase activity (Riley and Allin, 1973; Donselaar *et al.*, 1987).

7.1.4 ADAPTIVE CHANGES OF MUSCLE TO INCREASED ACTIVITY

The finding that muscle fibre activity can profoundly and precisely affect muscle fibre properties provides fertile ground for the exploration of the mechanisms involved in the regulation of gene expression and novel proteosynthesis in skeletal muscle. The time sequence of the development of particular changes is a specific aspect of this and investigation of the timing of the different changes induced by chronic electrical stimulation, and the time course of their reversal when stimulation ceases has furthered our understanding of the mechanisms involved.

The first apparent change in response to slow frequency activity is functional hyperaemia followed by growth of new capillaries. Such increase in capillary density can be brought about earlier by electrical stimulation at low then at high frequencies of stimulation (Brown *et al.*, 1976; reviewed by Hudlická, 1990). The increase of capillary density is followed by an increase in the muscle's oxidative enzyme capacity, and a decrease of enzymes concerned with anaerobic metabolism. Increased vascularization and a denser capillary network and increased availability of oxygen may shorten the diffusion distance between the blood and working muscle cell, so that the muscle fibre functions in an environment with a higher partial pressure of oxygen. It is possible that this induces

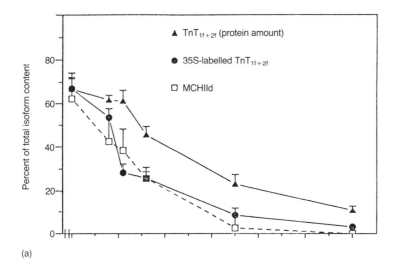

(a)

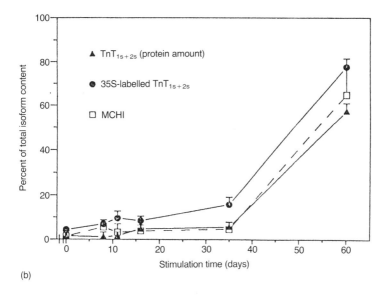

(b)

Figure 7.8 Synchronous fast-to-slow transitions of myosin heavy chain and troponin T isoforms in rabbit tibialis anterior (TA) muscle subjected to chronic low-frequency stimulation (10Hz, 12 h daily). Relative concentrations of the MHC and troponin isoforms were determined in muscles stimulated for different time periods. Values are means ± SD. MHCIId, MHCIIa = fast myosin heavy chain isoforms; MHCI = slow myosin heavy chain; TnT$_{1f+2f}$ = fast troponin T$_{1f}$ + fast troponin T$_{2f}$; TnT$_{3f}$ = fast troponin T$_{3f}$; TnT$_{1s+2s}$ = slow troponin T$_{1s}$ + slow troponin T$_{2s}$. Note the decrease of the fast isoforms of troponin T and MHC shown in (a), and the concomitant increase of the slow isoforms shown in (b).

the transformation from anaerobic to aerobic metabolism (Brown *et al.*, 1976; Hudlická *et al.*, 1977) but this is probably not related to the change in contractile characteristics or in myosin light or heavy chains, since no clear-cut relationship between high levels of oxidative enzymes and contractile speed has been established. Activities of membrane-bound enzymes increase within minutes after the onset of stimulation. The increase in the oxidative enzymes is apparent later. Another early change is that of the ATPase activity of the sarcoplasmic reticulum. Finally, after a longer period of continuous electrical stimulation, the structure of the myosin heavy and light chains also changes (reviewed by Pette and Vrbová, 1992).

The relationship of the physiological and concomitant biochemical changes has recently been evaluated. Increased resistance to fatigue is initially associated with the early changes in enzyme activities, particularly those of glucose transport, and phosphorylation, while later increases in oxidative enzymes and alterations of sarcomeric proteins take over (Pette and Vrbová, 1994). Increases in the time to peak twitch tension and relaxation are initially due to changes in Ca^{2+} sequestration and later to the conversion of contractile and regulatory proteins from fast to slow isoforms (Pette *et al.*, 1976; Klug *et al.*, 1988; reviewed by Pette and Vrbová, 1992, 1994).

The rapid changes in Ca^{2+} sequestration which account for the early changes in contractile speed are due to inactivation of the sarcoplasmic Ca^{2+}-ATPase, which is later followed by a replacement of the fast with the slow isoform of Ca^{2+}-ATPase, down-regulation of calsequestrin and parvalbumin, and expression of phospholamban (Dux *et al.*, 1990; Leberer *et al.*, 1986). There is good temporal correlation between parvalbumin down-regulation and slowing of the twitch relaxation in chronically stimulated muscles (Klug *et al.*, 1988). Transition from an anaerobic-glycolytic energy supply to a more aerobic-oxidative capacity, which correlates with increased muscle endurance, follows a similar time course (Pette *et al.*, 1976). There is a co-ordinated decline in the activity of glycolytic pathway enzymes and increased activity of enzymes involved in the citric acid cycle, the respiratory chain and fatty acid oxidation (Pette and Vrbová, 1992, 1994).

Changes in contractile proteins occur later and this is explained by findings that the fast-to-slow fibre type transition is a 'graded event' during which the number of type IIB fibres declines, type IID increases transiently, type IIA then increase and finally, in most species but not in rats and mice, type I fibres emerge.

Transition between the different fibre types involves down-regulation of mRNA for fast myosin isoforms and progressive up-regulation of intermediate and eventually slow isoforms of both myosin heavy and light chains (Brown *et al.*, 1983). Although changes in gene expression are induced within days of commencing stimulation, different rates of synthesis and degradation of the various myosin isoforms results in coexpression of different isoforms in the same fibre (Pette and Vrbová, 1992).

The nature of the coupling of electrical activity to altered gene expression is not yet understood although a rapid increase in internal Ca^{2+} concentration and/or reduced ATP relative to ADP and free inorganic phosphate have been proposed as possible triggers (Pette and Vrbová, 1994).

It has been suggested that chronic stimulation may change muscle phenotype simply by selective degeneration of fast fibres, followed by satellite cell differentiation to produce the new slow phenotype (Jolesz and Sreter, 1981). There is evidence for degeneration of the type II fibres which do not have the oxidative capacity to adapt to increased activity. Their glycogen stores are rapidly depleted and the fibres may undergo degeneration (Maier and Pette, 1987). However, selective degeneration and replacement cannot account for the dramatic muscle fibre conversions which have been demonstrated in innervated and denervated muscles (Pette and Vrbová, 1992; Gorza et al., 1988; Gundersen et al., 1988).

The potential of adult skeletal muscle fibres to alter their contractile and biochemical properties in response to particular activity patterns imposed upon them is quite remarkable. The finding that exercise or increased use achieved by overload cannot radically change the fibre composition of these muscles or their contractile characteristics (Frischknecht and Vrbová, 1991) has often been used to argue that it is not possible to transform one type of fibre into another (Burke and Edgerton, 1975). Such an argument does not take into account the fact that during exercise the motor units are still recruited in an orderly fashion and, although activated more often, the sequence of recruitment of motor units remains the same and there is no evidence that the motoneurone properties which influence the patterns of firing are changed by increased use (reviewed by Edström and Grimby, 1986). Electrical stimulation activates all motor units simultaneously with the same amount of activation. It is for this reason that the results obtained with exercise or overload must necessarily differ from those obtained by electrical stimulation.

7.2 RESPONSE TO CHRONIC ELECTRICAL STIMULATION IN DIFFERENT SPECIES

The fast-to-slow transformation of muscles has been shown to occur in various mammalian species, such as rabbit, cat, dog, goat, sheep and humans. However, it is more difficult to induce this pattern of responses in small laboratory animals such as rat and mouse. In larger animals, fast type II fibres are ultimately transformed into slow type I fibres. In small mammals, e.g. rat, the transformation of fibres proceeds mainly within the fast fibre sub-populations, following the sequence from type IIB -> type IID -> type IIA. The type IIA to type I transition, which represents the last step in the acquisition of slow characteristics, is diffi- cult to achieve in the rat (Kwong and Vrbová, 1981; Mayne et al., 1993). This difference in the responses to chronic low-frequency stimulation may be related to the initial fibre type composition of homologous muscles in small and large mammals. For example, assessment of MHC isoforms has shown that tibialis anterior (TA) muscles of rat and mouse contain a high proportion of the very fast type IIB fibres, while rabbit TA contains only traces of type IIB fibres and the majority of its fast fibres is type IID (Hämäläinen and Pette, 1993; Aigner et al., 1993). Type IID fibres, display slightly slower contractile properties than type IIB fibres (Galler et al., 1994). In rat TA muscle, the transition of the fibre types starts from type IIB, whereas in the rabbit TA it starts from type IID. Thus the fast to slow transition may be more readily achieved and complete in the rabbit, and similarly in other larger animals, than in the rat.

Thyroid hormone has an important regulatory influence on the expression of MHC isoforms (Izumo et al., 1986). Therefore, differences between fibre type composition of homologous muscles in small and large mammals may be due to different thyroid hormone levels. The responses to identical chronic low-frequency stimulation of homologous muscles may, therefore, differ in small and large mammals (Simoneau and Pette, 1988). Indeed, at low thyroid hormone levels a more complete fast-to-slow transition is achieved by low frequency stimulation in the rat and the ultimate transition from type IIA to type I fibres does occur (Kirschbaum et al., 1990). This is consistent with the finding that thyroid hormone suppresses the expression of the slower MHC isoforms, especially of MHCI (Kirschbaum et al., 1990). Effects of other hormones on myosin were less pronounced. Steroid hormones did not influence changes induced by chronic low frequency stimulation in female rabbits (Salmons, 1991). However, other workers reported that the number of androgen receptors increases rapidly in electrically stimulated gastro- cnemius muscles of male rats (Inoue et al., 1993). These discrepancies could be due to the use of different species and sex.

7.3 MUSCLE PLASTICITY IN MODELS OF DISUSE

Buller et al. (1960b) noted that soleus muscles in kittens became fast contracting after spinal cord transection or tenotomy and concluded that during development the slow nerve and its activity was required to express the slow phenotype. Since these original observations, the general trend for muscle to express fast phenotype and undergo atrophy in response to 'disuse' has been attributed to reduced neuromuscular activity. However, the term 'disuse' is more correctly defined as 'altered neuronal discharge' (Fischbach and Robbins, 1969) as activity is not always eliminated, or even greatly reduced, by such intervention (Roy *et al.*, 1992). In addition to any change in activity, paralysis by spinal cord transection results in altered muscle length and loss of weight-bearing function, as is also the case in several other models of disuse including limb immobilization, tenotomy, hindlimb suspension and space flight. Expression of fast phenotype and disuse atrophy is most pronounced in the postural weight-bearing muscles, particularly those which act upon a single joint. These observations suggest that the decreased load during muscle contractions is an important contributing factor in the trend for conversion of type I to type II fibres in many experimental models (reviewed by Gordon and Pattullo, 1993).

7.3.1 SPINAL CORD TRANSECTION

After spinal cord injury, spinal cord transection or spinal cord transection combined with deafferentation in animals, the paralysed muscles may become fast contracting and more fatiguable and may lose muscle strength (Hník *et al.*, 1985). The changes in contractile properties can be reversed when the inactive muscles are electrically stimulated (Vrbová, 1963b; Salmons and Vrbová, 1969; Stein *et al.*, 1992). The physiological changes after disuse are associated with limited conversion of type I to type II (slow to fast fibre type) (Kárpáti and Engel, 1968; Grimby *et al.*, 1976; Graham *et al.*, 1992; Hoffmann *et al.*, 1990; Jiang *et al.*, 1991; Martin *et al.*, 1992; Gordon and Mao, 1994).

Of all the changes, glycolytic enzyme activity is most closely linked with a reduction of neuromuscular activity, increasing in both slow and fast twitch muscles after spinal cord transection (Baldwin *et al.*, 1984). Paralysed fast twitch muscles become more fatiguable (Peckham *et al.*, 1973; Lenman *et al.*, 1989; Stein *et al.*, 1992; Gordon and Mao, 1994) and this is associated with a decrease in oxidative/glycolytic balance (Jiang *et al.*, 1991).

Qualitatively, the same changes are seen in hindlimb muscles after paralysis by blocking the sciatic nerve with tetrodotoxin (TTX) (Spector, 1985). Considerable atrophy also occurs after impulse blockade with maintained pressure (Kowalchuk and McComas, 1987).

7.3.2 LIMB IMMOBILIZATION

After limb immobilization, muscles may also become faster and show partial type I to type II conversion in animals (Fitts *et al.*, 1986, 1989; Kárpáti and Engel, 1968; Maier *et al.*, 1976; Mayer *et al.*, 1981; Corley *et al.*, 1984) and humans (Haggmark *et al.*, 1986) but changes are smaller than after spinal transection. Atrophic changes are particularly dramatic in slow twitch muscles which act across a single joint (Sjöstrom *et al.*, 1979). Muscle force, weight and fibre size decline rapidly within the first 3 weeks of immobilization, during which time there are dramatic alterations in protein synthesis and associated morphological changes (Baker and Matsumoto, 1988; Pachter and Eberstein, 1984). Rate of protein synthesis falls within 6 hours of immobilization (Booth and Seider, 1979), with a later enhancement of the rate of protein degradation (Goldspink, 1977). The greater atrophy of slow twitch muscles may be partly due to their more rapid protein turnover compared with fast twitch muscles (Goldberg, 1967).

Although type I to type II conversion and atrophy have been attributed to reduced activity, EMG analyses show that considerable neuromuscular activity persists after immobilization. Integrated EMG activity may be as low as 5–15% or as high as 50–80% of normal in immobilized soleus muscles (Fischbach and Robbins, 1969; Fournier *et al.*, 1983; Hnik *et al.*, 1985) and the pattern of discharge becomes more phasic, presumably due to the unloading of the muscle spindles (Hník *et al.*, 1985; Fischbach and Robbins, 1969). In the flexor muscles EDL and TA, EMG activity was unchanged by immobilization (Hník *et al.*, 1985; Pattullo *et al.*, 1992).

An important factor influencing skeletal muscle during immobilization is load or stretch (Goldspink *et al.*, 1992; Pattullo *et al.*, 1992). The importance of load on the structural integrity of slow muscles was demonstrated long ago by experiments in which stimulation was applied to unloaded slow twitch muscles of the rabbit and caused rapid fibre degeneration (McMinn and Vrbová, 1964, 1967). Consistent with this is the finding that electrical stimulation of the human quadriceps muscle under isometric conditions leads to an increase in the proportion of type I fibres, but this increase did not occur after tenotomy (Munsat *et al.*, 1976). Conditions of reduced load during muscle contraction may arise when a muscle is transposed from its original position and subjected to electrical stimulation, as in cardiomyoplasty (Chachques *et al.*, 1988).

Immobilization-induced type I to type II conversion has been attributed to a preferential type I atrophy and replacement with type II fibres which may arise through regeneration (Booth and Kelso, 1973; Fournier *et al.*, 1983; Haggmark *et al.*, 1986; Kárpáti and Engel, 1968; Spector *et al.*, 1982). However, conversion from type II to type I was found in fast twitch rabbit TA muscles immobilized in a lengthened position without any atrophic change (Pattullo *et al.*, 1992). This demonstrated that immobilization which resists contractions of elongated flexor muscles was sufficient to induce slow myosin expression. Thus, under conditions of immobilization, the expression of slow or fast myosin genes can be differentially influenced by increasing or decreasing the load of contracting muscle.

Methods of immobilization are generally traumatic themselves or are associated with traumatic injury. Procedures such as plaster-casting (Booth and Kelso, 1973) and joint-pinning (Solandt *et al.*, 1943; Fournier *et al.*, 1983) are stressful. Plaster casts are most often used to immobilize limbs during the repair of bone fracture or following tendon or ligament injury and repair (Sjöstrom *et al.*, 1979) which themselves provoke physiological responses related to pain and/or stress, including elevated levels of glucocorticoids, adrenal hypertrophy and gastric ulcers (Thomason and Booth, 1990) and which may also promote atrophy. Furthermore, glucocorticoid treatment results in increase of contractile speed (Gardiner *et al.*, 1980) and it is conceivable that increased circulating glucocorticoids could contribute to the increased proportion of type II fibres in immobilized muscle, particularly in the light of the finding that the number of glucocorticoid receptors rises in immobilized muscles (Dubois and Almon, 1980).

7.3.3 REDUCED LOAD BEARING: HINDLIMB SUSPENSION AND SPACE FLIGHT

Hindlimb suspension has been used as a model of reduced muscle activity and to simulate the weightlessness of space flight. Contraction speed increases concurrent with type I to type II fibre conversion and considerable atrophy, particularly in unloaded soleus muscles (Thomason and Booth, 1990; Fitts *et al.*, 1986, 1989; Templeton *et al.*, 1984; Winiarski *et al.*, 1987). Glycolytic enzyme activity increases in both fast and slow extensor muscles but soleus retains a high oxidative capacity in contrast to the fast extensor muscles, which also become more fatiguable (Winiarski *et al.*, 1987).

The presence of an increasing number of type II fibres may be due, at least in part, to preferential atrophy of type I fibres: up to 80% of myofibrillar proteins are lost in the soleus muscle within 28 days of hindlimb suspension, with relatively little effect on sarcoplasmic reticular proteins (Thomason and Booth, 1990). Greater degradation of slow myosin (Templeton *et al.*, 1984) contributes to the type I to type II conversion in addition to *de novo* synthesis of fast myosin (Thomason and Booth, 1990). During suspension, muscles are not prevented from length changes as in immobilization. EMG activity returns to normal after a transient fall

(Alford *et al.*, 1989). Thus atrophy and phenotypic change are more likely to be linked to unresisted contractions of the unloaded muscles.

Similar changes are observed after space flight with type I to type II fibre conversion apparent in hindlimb muscles of rats subjected to weightlessness (Martin *et al.*, 1988; Roy *et al.*, 1992), further emphasizing the importance of weight-bearing as a determinant of the slow phenotype. Muscle atrophy during 7–12 days of space flight was comparable to atrophy in the hindlimb suspension model, particularly for the soleus muscles, although there was little evidence of preferential type I atrophy within the time course of the space flight study. In both cases, considerable degenerative changes were associated with active shortening of muscles which were not resisted by any load. Flexor muscles, which are slightly stretched during suspension and weightlessness, are relatively unaffected despite their high levels of EMG activity (Alford *et al.*, 1989; Martin *et al.*, 1988).

Thus, unresisted shortening has similar adverse effects to those of eccentric contraction which produces extensive muscle damage (Faulkner *et al.*, 1989). Apparently it is not simply the reduced neural activity itself, but the contractions of muscles under abnormal mechanical conditions, that leads to type I to type II conversion and disuse atrophy. When Huxley suggested that the pattern of activity may be important in determining muscle contractile properties, he indicated that the mechanical vibration of the postural muscles under load could maintain the slow contractile speed, in contrast to the forceful contractions induced at higher frequencies in fast muscles (Buller *et al.*, 1960a). Data from chronic stimulation experiments and disuse models which take into account the different loading conditions of muscle fibres are consistent with this idea, suggesting that the pattern of muscle contraction is a very important factor in determining and maintaining mature muscle phenotype.

7.4 LIMITED PLASTICITY OF SLOW TONIC AND FAST PHASIC MUSCLES IN LOWER VERTEBRATES AND BIRDS

Surprisingly, when an attempt was made to change the contractile properties of slow tonic and fast twitch muscles of lower vertebrates by cross-innervation, different results from those observed in mammals were obtained (Hník *et al.*, 1967; Close and Hoh, 1968; Elul *et al.*, 1970). Altering the innervation of adult slow tonic and fast twitch muscles in lower vertebrates does not lead to a change in the characteristic properties of their muscles.

In the chick, when the nerve from the slow tonic anterior latissimus dorsi (ALD) was connected to the fast posterior latissimus dorsi (PLD) muscle, and vice versa, both nerves established connections with the muscles but slow tonic muscles remain slow contracting and fast twitch muscles continued to develop tension rapidly despite innervation by slow motor nerves (Feng *et al.*, 1963; Hník *et al.*, 1967). In line with this, the histochemical appearance of slow ALD muscles was not transformed by reinnervation by the fast PLD nerve but the fibres resembled denervated fibres. Similar results were obtained for the cross-innervated PLD muscle (Koenig and Fardeau, 1973).

As described in Chapter 4 at the ultrastructural level, slow tonic and fast twitch muscle fibres of lower vertebrates and birds are distinct from one another (Hník *et al.*, 1967; Page, 1969; Stefani and Steinbach, 1969). However, the ultrastructure of the frog slow tonic muscle fibres remained unaltered, even when the muscle was transplanted and connected to a nerve which formerly supplied a twitch muscle for periods longer than a year (Miledi and Orkand, 1966). Even the pattern of innervation remains that of the original muscle – the slow muscle maintains its multiple distributed innervation even when supplied by the fast motor nerve, and the fast muscle remains focally innervated by the slow nerves which normally innervate the slow muscle fibres at several points along each muscle fibre (Hník *et al.*, 1967; Bennett *et al.*, 1973; Vyskočil and Vyklický, 1974).

Membranes of slow tonic muscle fibres in these species do not generate a propagated action potential, but become depolarized by the endplate potentials and their decremental spread, while twitch fibres have a single endplate at which the action potential is initiated. The possibility that these two types of muscle fibre are inherently different has been discussed in Chapter 4 and rejected in view of new experimental evidence which clearly shows that:

- the contractile and histochemical properties of slow and fast muscles in chick embryos are similar, and differentiate only some time after innervation has taken place (Gordon and Vrbová, 1975a, b; Gordon *et al.*, 1977 a, b);
- when the nerves are crossed from one type of muscle into another during early development, considerable alterations of contractile and structural characteristics of the respective muscles can be produced (Jirmanová *et al.*, 1971; Zelená and Jirmanová, 1973; Jirmanová and Zelená, 1973);
- even after differentiation, slow chick muscles can be changed into fast muscles and vice versa during regeneration (Gordon and Vrbová, 1975a, b; Gordon *et al.*, 1977a, b).

Thus the inability of these muscles to alter their properties when innervated by an alien nerve is due to some characteristic that the muscles acquired during postnatal development that may not be changed in later life. The possibility that the different membrane properties of the two types of muscle and the different distribution of endplates along their surface may be responsible for this inability to change in later life has been considered. Although the membrane properties of slow and fast muscles are similar during early development and become different only some time after innervation has taken place (Gordon *et al.*, 1977b), their properties do not change appreciably on reinnervation by an alien nerve, or on denervation. In fact, in frogs, the persistent difference in the membrane resistance, time constant and space constant of fibres of the iliofibularis muscle, which contains both tonic and twitch fibres, allows the identification of the different fibre types even after denervation or reinnervation by an alien nerve (Elul *et al.*, 1970; Miledi and Stefani, 1970).

A possible reason for the lack of plasticity of muscles of lower vertebrates and birds could be different electrical properties of fast and slow muscle fibres: the muscles may have become incompatible with the alien nerve. Therefore in adult birds or amphibians, when fast twitch and slow tonic muscles receive inappropriate innervation, the normal pattern of activity that is characteristic of the motor nerve cannot be imposed on a different type of muscle. The finding that the pattern of innervation in slow tonic muscles cannot be altered even on reinnervation or cross-innervation further highlights their inability to alter, and is consistent with the observation in mammals that the normal pattern of innervation is restored during reinnervation by the growing axons, which preferentially innervate the old endplate sites on the muscle membrane (Chapter 9).

Stimulation of the motor nerve to tonic fibres normally elicits only local potentials in slow or fast muscles (Hník *et al.*, 1967), the quantal content being significantly smaller than that of motor nerve terminals to twitch muscles (Vyskočil *et al.*, 1971). In fast muscles supplied by a nerve from a slow tonic muscle, endplate potentials large enough to initiate an action potential can only be generated after repetitive stimulation and during post-tetanic potentiation (Hník *et al.*, 1967). Since the pattern of innervation remains that of the muscle and not the original nerve, the local potentials elicited at the single junction can spread decrementally along the twitch muscle fibre only for a short distance either side of the junction. The problem is further exacerbated by those membrane characteristics of the muscle which are not altered by the nerve, such as low resistance and short time and space constants which reduce the size of the endplate potential and limit the decremental spread from the junction.

In the case of cross-innervation of the slow tonic muscle by a fast motor nerve, the slow tonic muscle membrane receives multiple innervation by the fast motor nerve (Bennett *et al.*, 1973; Harris *et al.*, 1977). Since the ability of the ALD muscle fibres to propagate action potentials is poorly developed, local potentials are conducted along the slow muscle membrane, which has a high membrane resistance and long space and time constant. The time constant of the slow ALD muscle is too long to permit high-frequency activation of the muscle fibres and so the activity of the muscle remains that of a slow muscle. Thus, during development, the membranes develop properties to match best the activity of their nerves (Chapter 4); once the membrane becomes specialized and the muscles can be activated only by the appropriate nerve supply, its characteristic properties cannot be modified by alien innervation. The membrane of the adult muscle might therefore be regarded functionally as a gate that allows activation by the correct nerve but prevents an alien nerve from imposing its activity pattern on the contractile machinery.

CONCLUSIONS

Evidence has been presented to show that extra-fusal fibres of mammalian skeletal muscles can be transformed by changing their activity. Changes of activity brought about either by crossing the motor nerves or by imposing particular types of activity by electrical stimulation on skeletal muscles are effective at changing their properties. Since such a transformation of muscle can be brought about in adult mammals as well as in young animals, the muscle fibre evidently does not lose its ability to adjust to different activity patterns.

The finding that skeletal muscle fibres undergo complete functional and biochemical transformation has been challenged on the grounds that, in some cases, transformation of some enzymes or structural proteins does not take place, even though the functional characteristics may be altered. While this is so, many of these arguments do not take into account the time it may take for a cell to alter its biochemical composition completely or that these changes are sequential. From all existing evidence it seems reasonable to conclude that, in mammals, the contractile machinery and those biochemical properties of the muscle fibre that are associated with the contractile activity of the muscle can be altered by activity. Failure of lower vertebrates to exhibit similar plasticity is attributed to the very different membrane properties of fast twitch and slow tonic muscle fibres and their inability to respond to different activity patterns imposed by cross-innervation in adult muscles.

REFERENCES

Aigner, S., Golisch, B., Hamalainen, N. *et al.* (1993) Fast myosin heavy chain diversity in skeletal muscles of the rabbit: heavy chain 11d, not 11b predominates. *Eur. J. Biochem.* **211**, 367–372.

Al-Amood, W. S. and Lewis, D. M. (1987) The role of frequency in the effects of stimulation on denervated slow-twitch muscles in the rat. *J. Physiol.* **392**, 377–395.

Albani, M. and Vrbová, G. (1985) Physiological properties and patterns of innervation of regenerated muscles in the rat. *Neurosci.* **15**, 489–498.

Alford, E. K., Roy, R. R., Hodgson, J. A. and Edgerton, V. R. (1989) Electromyography of rat soleus, medial gastrocnemius, and tibialis anterior during hindlimb suspension. *Exp. Neurol* **96**, 635–649.

Amphlett, G. W., Perry, S. V., Syska, H. *et al.* (1975) Cross innervation and the regulatory protein system of rabbit soleus muscle. *Nature* **257**, 602–604.

Amphlett, G. W., Syska, H. and Perry, S. V. (1976) The polymorphic forms of tropomyosin and troponin I in developing rabbit skeletal muscle. *FEBS Letters* **68**, 22–26.

Baker, J. H. and Matsumoto, D. E. (1988) Adaptation of skeletal muscle to immobilization in a shortened position. *Muscle and Nerve* **11**, 231–244.

Baldwin, K. M., Roy, R. R., Sacks, B. *et al.* (1984) Relative independence of metabolic enzymes and neuromuscular activity. *J. Appl. Physiol.* **56**, 1602–1607.

Bárány, M. (1967) ATPase activity of myosin correlated with speed of shortening. *J. Gen. Physiol.* **50**, Suppl. 2, 197–218.

Bárány, M. and Close, R. I. (1971) The transformation of myosin in cross-innervated rat muscles. *J. Physiol. (Lond.)* **218**, 455–474.

Bennett, M. R., Pettigrew, A. G. and Taylor, R. S. (1973) The formation of synapses in re-innervated and cross-innervated adult avian muscle. *J. Physiol. (Lond.)* **280**, 881–7.

Booth, F. W. and Kelso, J. R. (1973) Effect of hind-limb immobilization on contractile and histochemical properties of skeletal muscle. *Pflügers Arch.* **342**, 231–238.

Booth, F. W. and Seider, M. J. (1979) Early change in skeletal muscle protein synthesis after limb immobilization of rats. *J. Appl. Physiol.* **47**, 974–977.

Brown, M. D., Cotter, M. A., Hudlická, O. and Vrbová G. (1976) The effects of different patterns of muscle activity on capillary density, mechanical properties and structure of slow and fast rabbit muscles. *Pflügers Arch.* **361**, 241–250.

Brown, W. E., Salmons, S. and Whalen, R. G. (1983) The sequential replacement of myosin subunit isoforms during muscle type transformation induced by long term electrical stimulation. *J. Biol. Chem.* **258**, 14686–14692.

Buller, A. J., Eccles, J. C. and Eccles, R. M. (1960a) Interactions between motoneurones and muscles in respect of the characteristic speeds of their responses. *J. Physiol. (Lond.)* **150**, 417–439.

Buller, A. J., Eccles, J. C. and Eccles, R. M. (1960b) Differentiation of fast and slow muscles in the cat hind limb. *J. Physiol. (Lond.)* **150**, 399–416.

Buller, A. J. and Pope, R. (1977) Plasticity in mammalian skeletal muscle. *Phil. Trans. R. Soc. Lond. Ser. B* **278**, 295–305.

Buller, A. J., Kean, C. J. C. and Ranatunga, K. W. (1987) Transformation of contraction speed in muscles following cross-reinnervation; dependence on muscle size. *J. Musc. Res. Cell Motil.* **8**, 504–516.

Burke, R. E. and Edgerton, V. R. (1975) Motor unit properties and selective involvement in movement. *Exercise and Sport Sci. Rev.,* **3**, 31–81.

Chachques, J.-C., Grandjean, P., Schwartz, K. *et al.* (1988) Effect of latissimus dorsi cardiomyoplasty on ventricular function. *Circulation* **78**, 111–203.

Close, R. (1969) Dynamic properties of fast and slow skeletal muscles of the rat after nerve cross-union. *J. Physiol. (Lond.)* **294**, 331–346.

Close, R. and Hoh, J. F. (1968) Effects of nerve cross-union on fast-twitch and slow-graded muscle fibres in the toad. *J. Physiol. (Lond.)* **198**, 103–125.

Corley, K., Kowalchuk, N. and McComas, A. J. (1984) Contrasting effects of suspension on hind limb muscles in the hamster. *Exp. Neurol.* **85**, 30–40.

Dhoot, G. K. and Vrbová, G. (1991) Conversion of cat and rabbit soleus muscle fibres by alien nerves, in *Muscle and Mobility*, Vol. 2, Intercept Ltd, Andover, pp. 125–130.

Donselaar, Y., Eerbeek, O., Kernell, D. and Verhey, B. A. (1987) Fibre sizes and histochemical staining characteristics in normal and chronically stimulated fast muscle of cat. *J. Physiol. (Lond.)* **382**, 237–254.

Drahota, Z. and Gutmann, E. (1963) Long-term regulatory influence of the nervous system on some metabolic differences in muscles of different functions. *Physiologia Bohemoslov.* **12**, 339–348.

Dubois, D. C. and Almon, R. R. (1980) Disuse atrophy of skeletal muscle is associated with an increase in number of glucocortocoid receptors. *Endocrinology* **107**, 1649–1651.

Dubowitz, V. (1967) Cross-innervated mammalian skeletal muscle: histochemical, physiological and biochemical observations. *J. Physiol. (Lond.)* **193**, 481–496.

Dum, R. P., O'Donovan, M. L., Toop, L. and Burke, R. E. (1985a) Cross re-innervated motor units in cat muscle I. Flexor digitorum longus muscle units re-innervated by soleus motoneurones. *J. Neurophysiol.* **54**, 818–836.

Dum, R. P., O'Donovan, M. L., Toop, L. *et al.* (1985b) Cross re-innervated motor units in cat muscle II. Soleus muscle reinnervated by flexor digitorum longus motoneurones. *J. Neurophysiol.* **54**, 837–851.

Dux, L., Green, H. J. and Pette, D. (1990) Chronic low frequency stimulation of rabbit fast twitch muscle induces partial inactivation of the sarcoplasmic reticulum Ca^{2+}-ATPase and changes in its tryptic cleavage. *Eur. J. Biochem.* **192**, 95–100.

Edström, L. and Grimby, L. (1986) Effect of exercise on the motor unit. *Muscle & Nerve*, **9**, 104–126.

Elul, R., Miledi, R. and Stefani, E. (1970) Neural control of contracture in slow muscle fibres of the frog. *Acta Physiol. Latino-Am.* **20**, 194–226.

Faulkner, J. A., Jones, D. A. and Round, J. M. (1989) Injury to skeletal muscles of mice by forced lengthening contractions. *Quart. J. Exp. Physiol.* **74**, 661–670.

Feng, T. P., Yang, H. W. and Wu, W. Y. (1963) The contrasting trophic changes of the anterior and posterior latissimus dorsi of the chick following denervation, in *Effect of Use and Disuse on Neuromuscular Functions* (eds E. Gutmann and P. Hník), Czechoslovak Academy of Science Publishing House, Prague.

Fischbach, G. D. and Robbins, N. (1969) Changes in contractile properties of disused soleus muscles. *J. Physiol. (Lond.)* **201**, 305–320.

Fitts, R. H., Brimmer, C. J., Heywood-Cooksey, A. and Timmerman, R. J. (1989) Single muscle fiber enzyme shifts with hindlimb suspension and immobilization. *Am. J. Physiol.* **256**, C1082-C1091.

Fitts, R. H., Metzger, J. M., Riley, D. A. and Unsworth, B. R. (1986) Models of disuse: A comparison of hindlimb suspension and immobilization. *J. Appl. Physiol.* **60**, 1946–1953.

Fournier, M., Roy, R. R., Peckham, H. *et al.* (1983) Is limb immobilization a model of disuse? *Exp. Neurol.* **80**, 147–156.

Frischknecht, R. and Vrbová, G. (1991) Adaptation of rat extensor digitorum longus to increased activity. *Pflügers Arch.* **419**, 319–326.

Galler, S., Schmitt, T. and Pette, D. (1994) Stretch activation, unloaded shortening velocity, and myosin heavy chain isoforms of rat skeletal muscle fibres. *J. Physiol. (Lond.)* **478**, 513–522.

Gardiner, P. G., Montanaro, D. S. and Edgerton, V. R. (1980) Effects of glucocorticoid treatment and food restriction on rat hindlimb muscles. *Pflügers Arch.* **385**, 147–153.

Gauthier, G. F., Burke, R. E., Lowey, S. and Hobbs, A. W. (1983) Myosin isozymes in normal and cross-reinnervated cat skeletal muscle fibers. *J. Cell Biol.* **97**, 756–771.

Gillespie, M. J., Gordon, T. and Murphy, P. R. (1986) Reinnervation of the lateral gastrocnemius and soleus muscles in the rat by their common nerve. *J. Physiol. (Lond.)* **372**, 485–500.

Gillespie, M. J., Gordon, T. and Murphy, P. R. (1987) Motor units and histochemistry in rat lateral gastrocnemius and soleus muscles: Evidence for dissociation of physiological and histochemical properties after reinnervation. *J. Neurophysiol.* **57**, 921–937.

Goldberg, A. L. (1967) Protein synthesis in tonic and phasic skeletal muscles. *Nature* **216**, 1219–1220.

Goldspink, D. F. (1977) The influence of immobilization and stretch on protein turnover of rat skeletal muscle. *J. Physiol. (Lond.)* **264**, 267–282.

Goldspink, G., Scott, A., Loughna, P. T. *et al.* (1992) Gene expression in skeletal muscle in response to stretch and force generation. *Am. J. Physiol.* **262**, R356-R363.

Golisch, G., Pette, D. and Pichlmaier, H. (1970) Metabolic differentiation of rabbit skeletal muscles as induced by specific innervation. *Eur. J. Biochem.* **16**, 110–116.

Gordon, T. and Mao, J. (1994) Muscle atrophy and procedures for training after spinal injury. *Phys. Ther.* **74**, 50–60.

Gordon, T. and Pattullo, M. C. (1993) Plasticity of muscle fiber and motor unit types. *Exercise Sport Sci. Rev.* **21**, 331–362.

Gordon, T., Perry, R., Srihari, T. and Vrbová, G. (1977a) Differentiation of slow and fast muscles in chickens. *Cell and Tissue Res.* **180**, 211–222.

Gordon, T., Purves, R. D. and Vrbová, G. (1977b) Differentiation of electrical and contractile properties of slow and fast muscle fibres. *J. Physiol. (Lond.)* **269**, 535–547.

Gordon, T., Thomas, C. K., Stein, R. B. and Erdebil, S. (1988) Comparison of physiological and histochemical properties of motor units after cross-reinnervation of antagonistic muscles in the cat hindlimb. *J. Neurophysiol.* **60**, 365–378.

Gordon, T. and Vrbová, G. (1975a) The influence of innervation on the differentiation of contractile speeds of developing chick muscles. *Pflügers Arch.* **360**, 199–218.

Gordon, T. and Vrbová, G. (1975b) Changes in chemosensitivity of developing chick muscle fibres in relation to endplate formation. *Pflügers Arch.* **360**, 349–364.

Gorza, L., Gundersen, K., Lømo, T. *et al.* (1988) Slow-to-fast transformation of denervated soleus muscles by chronic high-frequency stimulation in the rat. *J. Physiol. (Lond.)* **402**, 627–649.

Graham, S. C., Roy, R. R., Navarro, C. *et al.* (1992) Enzyme and size profiles in chronically inactive cat soleus muscle fibers. *Muscle & Nerve* **15**, 27–36.

Grimby, G., Broberg, C., Krotkiewska, I. and Krotkiewski, M. (1976) Muscle fiber composition in patients with traumatic cord lesion. *Scand. J. Rehab. Med.* **8**, 37–42.

Gundersen, K., Leberer, E., Lømo, T. *et al.* (1988) Fibre types, calcium-sequestering proteins and metabolic enzymes in denervated and chronically stimulated muscles of the rat. *J. Physiol. (Lond.)* **398**, 177–189.

Guth, L. (1968) 'Trophic' influences of nerve on muscle. *Physiol. Revs.*, **48**, 645–687.

Haggmark, T., Eriksson, E. and Jansson, E. (1986) Muscle fiber type changes in human skeletal muscle after injuries and immobilization. *Orthopedics* **9**, 181–185.

Hämäläinen, N. and Pette, D. (1993) The histochemical profiles of fast fibre types IIB, IID, and IIA in skeletal muscles of the mouse, rat and rabbit. *J. Histochem. Cytochem.* **41**, 733–743.

Harris, A. J., Ziskind, L. and Wigston, D. (1977) Spontaneous release of transmitter from 'repressed' nerve terminals in axolotl muscle. *Nature* **268**, 265–267.

Hník, P., Jirmanová, I., Vyklický, L. and Zelená, J. (1967) Fast and slow muscles of the chick after nerve cross-union. *J. Physiol. (Lond.)* **193**, 309–325.

Hník, P., Vejsada, R., Goldspink, D. F. *et al.* (1985) Quantitative evaluation of electromyogram activity in rat extensor and flexor muscles immobilized at different lengths. *Exp. Neurol.* **88**, 515–528.

Hoffmann, S. J., Roy, R. R., Bianco, C. E. and Edgerton, V. R. (1990) Enzyme profiles of single muscle fibers in the absence of normal neuromuscular activity. *J. Appl. Physiol.* **69**, 1150–1158.

Hoh, J. F. Y. (1975) Selective and non-selective reinnervation of fast-twitch and slow-twitch rat skeletal muscle. *J. Physiol. (Lond.)* **251**, 791–801.

Hudlická, O., Brown M. D., Cotter, M. *et al.* (1977) The effect of long term stimulation of fast muscles on their blood flow, metabolism and ability to withstand fatigue. *Pflügers Arch. ges. Physiol.* **369**, 141–149.

Hudlická, O. (1990) The response of muscle to enhanced and reduced activity. *Baillière's Clin. Endocrin. & Metab.* **4**, 417–439.

Inoue, K., Yamasaki, S., Fushiki, T. *et al.* (1993) Rapid increase in the number of androgen receptors following electrical stimulation of rat muscle. *Eur. J. Appl. Physiol.* **66**, 134–140.

Izumo, S., Nadal-Ginard, B. and Mahdavi, V. (1986) All members of the MHC multigene family respond to thyroid hormone in a highly tissue-specific manner. *Science* **231**, 597–600.

Jiang, B., Roy, R. R., Navarro, C. *et al.* (1991) Enzymatic responses of cat medial gastrocnemius fibers to chronic inactivity. *J. Appl. Physiol.* **70**, 231–239.

Jirmanová, I., Hník, P. and Zelená, J. (1971) Implantation of 'fast' nerve into slow muscle in young chickens. *Physiol. Bohem.* **20**, 199–203.

Jirmanová, I. and Zelená, J. (1973) Ultrastructural transformation of fast chicken muscle fibres induced by nerve union. *Z. Zellforsch. mikrosk. Anat.* **106**, 333–347.

Jolesz, F. and Sreter, F. A. (1981) Development, innervation and activity-pattern induced changes in skeletal muscle. *Ann. Rev. Physiol.* **43**, 531–552.

Kárpáti, G. and Engel, W. K. (1967) Neuronal trophic function. *Arch. Neurol.* **17**, 542–545.

Kárpáti, G. and Engel, W. K. (1968) Correlative histochemical study of skeletal muscle after suprasegmental denervation, peripheral nerve section and skeletal fixation. *Neurology* **18**, 681–692.

Kirschbaum, B. J., Kucher, H.-B., Termin, A. *et al.* (1990) Antagonistic effects of chronic low frequency stimulation and thyroid hormone on myosin expression in rat fast-twitch muscle. *J. Biol. Chem.* **265**, 13974–13980.

Klug, G. A., Leberer, E., Leisner, E. *et al.* (1988) Relationship between parvalbumin content and the speed of relaxation in chronically stimulated rabbit fast-twitch muscle. *Pflügers Arch.* **411**, 126–131.

Koenig, J. and Fardeau, M. (1973) Étude histochemique des muscles grand dorsaux antérieur et postérieur du poulet et des modifications observées après dénervation et reinnervation homologue ou croisée. *Arch. Anat. Micr. Morph. Exp.* **62**, 249–267.

Kowalchuk, N. and McComas, A. (1987) Effects of impulse blockade on the contractile properties of rat skeletal muscle. *J. Physiol. (Lond.)* **382**, 255–266.

Kugelberg, E., Edström, L. and Abruzzese, M. (1970) Mapping of motor units in experimentally reinnervated rat muscle. *J. Neurol. Neurosurg. Psychiat.* **33**, 310–329.

Kwong, W. H. and Vrbová, G. (1981) Effects of low-frequency electrical stimulation on fast and slow muscles of the rat. *Pflügers Arch.* **391**, 200–207.

Leberer, E., Seedorf, V. and Pette, D. (1986) Neural control of gene expression in skeletal muscle Ca^{2+}-sequestering proteins in developing and chronically stimulated rabbit skeletal muscles. *Biochem. J.* **239**, 295–300.

Lenman, A. J. R., Tulley, F. M., Vrbová, G. *et al.* (1989) Muscle fatigue in some neurological disorders. *Muscle and Nerve* **12**, 938–942.

Lømo, T., Westgaard, R. H. and Dahl, D. H. (1974) Contractile properties of muscle: control by pattern of muscle activity in the rat. *Proc. Roy. Soc. B (Lond.)* **187**, 99–103.

Luff, A. R. (1975) Dynamic properties of fast and slow skeletal muscles in the cat and rat following cross reinnervation. *J. Physiol. (Lond.)* **248**, 83–96.

Maier, A., Crockett, J. L., Simpson, D. R. *et al.* (1976) Properties of immobilized guinea pig hindlimb muscles. *Am. J. Physiol.* **231**, 1520–1526.

Maier, A. and Pette, D. (1987) The time course of glycogen depletion in single fibers of chronically stimulated rabbit fast twitch muscle. *Pflügers Arch.* **408**, 338–342.

Margreth, A., Salviati, G. and Carraro, U. (1973) Neural control of the activity of the calcium transport system in sarcoplasmic reticulum of rat skeletal muscle. *Nature* **241**, 285–286.

Martin, T. P., Edgerton, V. R. and Grindeland, R. E. (1988) Influence of spaceflight on rat skeletal muscle. *J. Appl. Physiol.* **65**, 2318–2325.

Martin, T. P., Stein, R. B., Hoeppner, P. H. and Reid, D. C. (1992) Influence of electrical stimulation on the morphological and metabolic properties of paralyzed muscle. *J. Appl. Physiol.* **72**, 1401–1406.

Mayer, R. F., Burke, R. E., Toop, J. *et al.* (1981) The effect of long-term immobilization on the motor unit population of the cat medial gastrocnemius muscle. *Neuroscience* **6**, 725–739.

Mayne, C. N., Mokrusch, T., Jarvis, J. C. *et al.* (1993) Stimulation induced expression of slow muscle myosin in fast muscle of the rat. Evidence of the unrestricted adaptive capacity. *FEBS Lett.* **327**, 297–300.

McMinn, R. M. H. and Vrbová, G. (1964) The effect of tenotomy on the structure of fast and slow muscles of the rabbit. *Quart. J. Exp. Physiol.* **49**, 424–430.

McMinn, R. M. H. and Vrbová, G. (1967) Motoneurone activity as a cause of degeneration in the soleus muscle of the rabbit. *Quart. J. Exp. Physiol.* **52**, 411–415.

Miledi, R. and Orkand, P. (1966) Effect of a fast nerve on slow muscle fibres in the frog. *Nature* **209**, 717–718.

Miledi, R. and Stefani, E. (1970) Miniature potentials in denervated slow muscle fibres of the frog. *J. Physiol. (Lond.)* **209**, 176–186.

Munsat, T. L., McNeal, D. and Waters, R. (1976) Effects of nerve stimulation on human muscle. *Arch. Neurol.* **33**, 608–617.

Pachter, B. R. and Eberstein, A. (1984) Neuromuscular plasticity following limb immobilization. *J. Neurocytol.* **13**, 1013–1025.

Page, S. G. (1969) Structure and some contractile properties of fast and slow muscles of the chicken. *J. Physiol. (Lond.)* **205**, 131–145.

Pattullo, M. C., Cotter, M. A., Cameron, N. E. and Barry, J. A. (1992) Effects of lengthened immobilization on functional and histochemical properties of rabbit tibialis anterior muscle. *Exp. Physiol.* **77**, 433–442.

Peckham, P. H., Mortimer, J. T. and van ter Meulen, J. P. (1973) Physiologic and metabolic changes in white muscle of cat following induced exercise. *Brain Res.* **50**, 424–429.

Pette, D., Müller, W., Leisner, E. and Vrbová, G. (1976) Time dependent effects on contractile properties, fibre population, myosin light chains and enzymes of energy metabolism in intermittently and continuously stimulated fast twitch muscles of the rabbit. *Pflügers Arch.* **364**, 103–112.

Pette, D., Smith, M. E., Staudte, H. W. and Vrbová, G. (1973) Effects of long-term electrical stimulation on some contractile and metabolic characteristics of fast rabbit muscles. *Pflügers Arch.* **338**, 257–272.

Pette, D. and Vrbová, G. (1985) Invited review: Neural control of phenotypic expression in mammalian muscle fiber. *Muscle & Nerve* **8**, 676–689.

Pette, D. and Vrbová, G. (1992) Adaptation of mammalian skeletal muscle to chronic electrical stimulation. *Rev. Physiol. Biochem. Pharmacol.* **120**, 115–202.

Pette, D. and Vrbová, G. (1994) Transformation of skeletal muscle by electrical stimulation. *Annual of Cardiac Surgery,* Current Sciences Ltd, pp. 14–22.

Prewitt, M. A. and Salafsky, B. (1967) Effect of cross-innervation on biochemical characteristics of skeletal muscle fibres. *Am. J. Physiol.* **213**, 295–300.

Prewitt, M. A. and Salafsky, B. (1970) Enzymatic and histochemical changes in fast and slow muscles after cross-innervation. *Am. J. Physiol.* **218**, 69–74.

Ranvier, L. (1874) De quelques faits relatifs à l'histologie et à la physiologie des muscles striés. *Arch. Physiol. Norm. Path.* **6**, 1–15.

Riley, D. A. and Allin, E. F. (1973) The effects of inactivity, programmed stimulation and denervation on the histochemistry of skeletal muscle fibre types. *Exp. Neurol.* **40**, 391–413.

Romanul, F. C. A. and Van der Meulen, J. P. (1967) Slow and fast muscles after cross reinnervation. Enzymatic and physiological changes. *Arch. Neurol.* **17**, 387–402.

Roy, R. R., Baldwin, K. M. and Edgerton, V. R. (1992) The plasticity of skeletal muscle: effects of neuromuscular activity. *Exercise Sports Sci. Rev.* **19**, 269–312.

Salmons, S. (1991) Myotrophic action of an anabolic steroid (nondrolone decenoale) in normal but not chronically stimulated rabbit limb muscles. *BAM* **1**, 241–251.

Salmons, S. and Vrbová, G. (1969) The influence of activity on some contractile characteristics of mammalian fast and slow muscles. *J. Physiol. (Lond.)* **201**, 535–549.

Simoneau, J. and Pette, D. (1988) Species-specific effects of chronic nerve stimulation upon tibialis anterior in mouse, rat, guinea pig and rabbit. *Pflügers Arch.* **412**, 86–92.

Sjöstrom, M., Wahlby, L. and Fugl-Meyer, A. (1979) Achilles tendon injury. III. Structure of rabbit soleus muscles after immobilization at different positions. *Acta Chir. Scand.* **145**, 509–521.

Solandt, D. Y., Partridge, R. C. and Hunter, J. (1943) The effect of skeletal fixation on skeletal muscle. *J. Neurophysiol.* **6**, 17–22.

Spector, S. A. (1985) Effects of elimination of activity on contractile and histochemical properties of rat soleus muscle. *J. Neurosci.* **5**, 2177–2188.

Spector, S. A., Simard, C. P., Fournier, M. *et al.* (1982) Architectural alterations of rat hind-limb muscles immobilized at different lengths. *Exp. Neurol.* **76**, 94–110.

Sperry, R. W. (1945). The problem of central nervous reorganisation after nerve regeneration and muscle transposition. *Quart. Rev. Biol.* **20**, 311–369.

Sréter, F. A., Gergely, T. and Luff, A. L. (1974) The effect of cross-reinnervation on the synthesis of myosin light chains characteristic of slow muscle in response to long term stimulation. *Biochem. Biophys. Res. Com.* **56**, 84–89.

Sréter, F. A., Luff, A. R. and Gergely, J. (1976) Effect of cross-reinnervation on physiological parameters and on properties of myosin and sarcoplasmic reticulum of fast and slow muscles of the rabbit. *J. Gen. Physiol.* **66**, 811–821.

Stein, R. B., Gordon, T., Jefferson, J. *et al.* (1992). Optimal stimulation of paralyzed muscle in spinal cord patients. *J. Appl. Physiol.* **72**, 1393–1400.

Stefani, E. and Steinbach, A. B. (1969) Resting potential and electrical properties of frog slow muscle fibres. Effect of different external solutions. *J. Physiol.* **203**, 383–401.

Templeton, G., Padalino, M., Manton, J. *et al.* (1984) The influence of rat suspension-hypokinesia on the gastrocnemius muscle. *Aviat. Space Environ. Med.* **55**, 381–386.

Thomason, D. B. and Booth, F. W. (1990) Atrophy of the soleus muscle by hindlimb unweighting. *J. Appl. Physiol.* **68**, 1–12.

Totosy de Zepetnek, J. E., Zung, H. V., Erdebil, S. and Gordon, T. (1992) Motor unit categorization on the basis of contractile and histochemical properties: a glycogen depletion analysis of normal and reinnervated rat tibialis anterior muscle. *J. Neurophysiol.* **67**, 1404–1415.

Vrbová, G. (1963a) Changes in motor reflexes produced by tenotomy. *J. Physiol. (Lond.)* **166**, 241–250.

Vrbová, G. (1963b) The effect of motoneurone activity on the speed of contraction of striated muscle. *J. Physiol. (Lond.)* **169**, 513–526.

Vyskočil, F. and Vyklický, L. (1974) Acetylcholine sensitivity of the chick fast muscle after cross union with the slow muscle nerve. *Brain Res.* **25**, 158–161.

Vyskočil, F., Vyklický, L. and Huston, R. (1971) Quantum content at the neuromuscular junction of fast muscle after cross-union with the nerve of slow muscle in the chick. *Brain Res.* **26**, 443–445.

Winiarski, A. M., Roy, R. R., Alford, E. K. *et al.* (1987) Mechanical properties of rat skeletal muscle after hind limb suspension. *Exp. Neurol.* **96**, 650–660.

Yellin, H. (1967) Neural regulation of enzymes in muscle fibres of red and white muscle. *J. Exp. Neurol.* **19**, 92–103.

Zelená, J. and Jirmanová, I. (1973) Ultrastructure of chicken slow muscle after nerve cross union. *Exp. Neurol.* **38**, 272–285.

Axotomy: effects on motoneurones, nerves and muscles

8

8.1 INTRODUCTION

As described in Chapters 1 and 2, motoneurones and skeletal muscle fibres are independent of each other in the early stages of embryonic development. While the alignment and fusion of myoblasts to form multinucleated cross-striated myotubes occurs in the absence of innervation, maturation and further development of muscle cells is critically dependent on the formation of functional nerve–muscle contacts. Similarly the early stages of neuronal development, with proliferation of neuro-epithelial germinal cells and early differentiation, are unaffected by ablation of peripheral target organs. Once motoneurones have sent growth cones into the periphery, however, further maturation of the cell body as well as axons and their terminals is dependent on the formation of functional connections with muscle fibres. Interaction between nerve and muscle is not only necessary in this stage of development but also critical for the survival and maintenance of the two tissues (Chapters 1, 2 and 6).

In the adult, the majority of motoneurones survive disruption of functional connections with muscles but they undergo a number of changes, some of which appear to be adaptive for regeneration and have been compared with early development, while others have been considered as degenerative. Although the return of the motoneurone to a 'growing state' has often been considered to involve de-differentiation, many properties the neurone has acquired during maturation remain unchanged and the response of the motoneurone must be considered in the general context of injury to the cell and its potential for repair by axonal regeneration. On restoration of functional nerve–muscle connections, the motoneurone recovers many of its former properties, reversing the changes due to axotomy.

Altered proteosynthetic activity and gene expression in denervated muscles has also been considered as de-differentiation involving re-enactment of early developmental stages. The trigger for such changes can be viewed as a result of muscle dysfunction since, in addition to losing contact with the motor nerve, denervated muscles are subject to very changed conditions including complete loss of neuronally elicited contractions, and all the steps that are involved in the initiation of the contractile response. In addition, altered muscle loading and length as a result of the inactivity of the muscle itself, or of synergists or antagonists when the nerve trunk to an entire limb is severed, may also influence the response of muscles to denervation.

8.2 AXOTOMY AND THE MOTONEURONE

8.2.1 THE DISTAL PERIPHERAL NERVE STUMP

Axons have very limited capacity for synthesis of proteins or lipids, containing few if any ribosomes (Zelená, 1972). It is now widely recognized that the neuronal cell body is the main site of macromolecular synthesis and that axons depend on their transport systems for the supply of materials which are synthesized in the soma (Droz et al., 1973; Lasek and Hoffman, 1976). Axons which are physically separated from the cell body after nerve injury cannot survive; they undergo degeneration. The first comprehensive description of this process (**Wallerian degeneration**) was by Waller in 1850. He proposed that the cell body is essential for the survival of axons and that connection with end-organs cannot preserve axons isolated from their cell body. Ramón y Cajal (1928) suggested that the trophic support of the cell body was of 'a dynamic and not of a material nature' because degeneration appeared to occur simultaneously along the entire isolated nerve stump (see also Donat and Wisniewski, 1973). The debate as to whether degeneration occurs in a proximodistal direction from the site of nerve injury to the target connections or in the opposite direction is reviewed by Sunderland (1978). Ljubinska (1975) made a strong case for proximodistal degeneration since isolated axons in the distal stump may continue to function for up to 7 days, depending on their length. In rat nerves, for example, it is claimed that neuromuscular failure is delayed by 2 hours for every additional centimetre of distal nerve stump (Miledi and Slater, 1968, 1970). This dependence of survival time on length suggests, but does not prove, that degeneration spreads longitudinally along the nerve stump.

The short-term functioning of the axons in the distal stump was attributed to their nutrition by continuing axonal transport in the isolated stump, and their eventual deterioration and degeneration to depletion of nutrients in the severed axons (Ljubinska, 1964, 1975). There is a minor source of axonal protein from the glial cells surrounding axons (Lasek *et al.*, 1977) but this is insufficient to maintain the mammalian axon in the absence of products transported from the cell soma. Protein synthesis in the Schwann cells of the peripheral nerves is, however, important for nerve regeneration and provides a favourable environment not only for regenerating peripheral axons, but also for CNS neurones (reviewed by Aguayo, 1985). This is discussed further in Chapter 9.

The bulk of axoplasm in the nerves is normally supplied by slow transport or axoplasmic flow and includes the cytoskeletal proteins: actin, tubulin and subunits of neurofilaments (Hoffman and Lasek, 1975). In mammalian peripheral axons, the rapid phase of axonal transport (400 mm/day) is 2 orders of magnitude faster than slow transport (1–5 mm/day) (for reviews see Grafstein and Forman, 1980; Vallee and Bloom, 1991) and transports particulate material. This consists primarily of membranous organelles for normal membrane turnover, for transmitter synthesis and secretion, and axonal metabolism (McEwen and Grafstein, 1968; Cuenod *et al.*, 1972; Droz, 1975; Grafstein, 1977). Fast but not slow transport continues in isolated nerve segments (Dahlström, 1967; Lubinska, 1964; Lasek, 1970) and neuromuscular transmission continues until supply of transmitter is depleted (Feldberg, 1943). Chemical transmission usually fails before impulse conduction (Ljubinska, 1964; Slater, 1966; Miledi and Slater, 1970). Since materials for transmission will be depleted earlier when the distal stump is shorter, a longer distal stump will maintain transmission and the function of the neuromuscular function and muscle for longer (Miledi and Slater, 1970). Consistent with this is the finding that transmission fails more rapidly if the isolated terminals are activated and made to release transmitter (Gerard, 1932; Card, 1977b).

Macrophage invasion of degenerating nerve stumps is a prominent feature, but macrophage access will depend to some extent on the type of nerve injury. Recent studies have shown that degeneration may advance from the proximal and distal ends of the stump after nerve section or from the distal regions in a retrograde fashion after a crush injury (Lunn *et al.*, 1990). At the endplate region of motor nerves, the discontinuity of the blood–brain barrier normally provided by the endo-, peri- and epineurial sheaths of the peripheral nerve allows for the infiltration of blood-borne multinucleated cells which initiate the disintegration of the axons and their eventual phagocytosis.

These findings can be reconciled with the suggestions of Ljubinska and her colleagues that the progress of degeneration is linked to depletion of source material originally supplied by the cell body. Because the nerve terminal is more metabolically demanding than the nerve fibre, the distoproximal course of Wallerian degeneration demonstrated *in vivo* would be consistent with the interplay of invading macrophages and the remaining transport of material in the isolated nerve stump. An important contribution was the finding that the time course of degeneration depends on the nature of the injury (Lunn et al., 1990). Degeneration of the distal stump of crushed nerves proceeds less rapidly than after a cut lesion, presumably because the continuity of Schwann cells in the proximal and distal stumps of crushed nerves provides additional metabolic support for the nerves over and above that provided by axonal transport alone. In cell cultures in vitro where material is artificially supplied to axons, peripheral stumps severed from growing neuroblasts do not always die but sometimes even grow and fuse again with the growing nerve (Boeke, 1950).

Degeneration of the isolated axon is probably linked to depletion of materials required for normal metabolic turnover on the one hand and to death of the mitochondria which are not replaced but accumulated at the injured end (Webster, 1962). Failure of oxidative phosphorylation leads to loss of membrane potential and disruption of chemical and electrical gradients. Calcium ion accumulation in the axons causes rapid depolymerization of microtubules and is involved in activation of calcium activated neutral protease (**CANP**). Inhibition of this enzyme prevents axonal degeneration (Schlaepfer and Freeman, 1980).

The final stages of Wallerian degeneration involve dissolution and disorganization of neurotubules and neurofilaments with final disintegration of axons, their phagocytosis and an intense proliferation of Schwann cells (reviewed by Gutmann, 1958; Altt, 1976; Sunderland, 1978; Hall, 1989; see also Salzer and Bunge, 1980; Salonen et al., 1988). Invading macrophages (Ramón y Cajal, 1928; Holmes and Young, 1942; Weinberg and Spencer, 1978) are important in the degeneration of the isolated axons, phagocytosis of axonal and myelin debris (Beuche and Friede, 1984) and the proliferative response of the Schwann cells (Williams and Hall, 1971; Baichwal et al., 1988). Delayed Wallerian degeneration is found in a mutant strain of mouse, the C57Bl/6/Ola mouse which shows a relative absence of macrophage invasion upon nerve injury (Lunn et al., 1989) suggesting an important role of macrophages in the removal of debris during Wallerian degeneration. Nonetheless, axons in the distal stump of Ola mice do eventually degenerate (Brown et al., 1992).

In addition to their proliferative action of Schwann cells, active phagocytosis by invading macrophages, removing the axonal debris and rejected myelin sheaths (Aguayo et al., 1976; Beuche and Friede, 1984; Lunn et al., 1989) is accompanied by release of cytokine L1 which triggers nerve growth factor (**NGF**) synthesis in other non-neural cells in the nerve stump (Lindholm et al., 1987; Brown et al., 1991; Matsuoka et al., 1991).

Products of degenerating nerves can be mitogenic for Schwann cells (Clemence et al., 1989; Salzer et al., 1980). Delayed Wallerian degeneration of axons in the Ola mice may not provide the stimulus for Schwann cell division and this accounts for the lack of proliferation of the Schwann cells around the intact axolemma (Thomson *et al.*, 1991). The responses of Schwann cells to nerve injury are also considered in Chapter 9, with reference to their essential role in supporting regeneration of the proximal nerve sprouts through the distal denervated nerve stump.

8.2.2 THE AXOTOMIZED NEURONE: CHROMATOLYSIS AND OTHER RESPONSES

Unlike the isolated axons in the distal nerve stump, the cell bodies of axotomized neurones maintain their synthetic ability and can, after axotomy, survive and regenerate. Observations that the emergence of growth cones from the proximal stump of transected nerves ultimately leads to axonal growth and regeneration, while growth of sprouts in severed distal nerve stumps is short-lived, indicates that continuity with the cell body is a prerequisite for regeneration (Ranson, 1912; Ramón y Cajal, 1928). The axotomized neurones undergo a number of structural, metabolic and functional changes which can be viewed as an anabolic response appropriate for repair of cellular damage and axon regeneration (Grafstein, 1977). Since many of these changes result in the expression of gene products normally associated only with developing neurones, they may also be considered as a process of de-differentiation from a fully differentiated secretory cell to a growing cell.

The cell body of axotomized neurones typically undergoes a number of characteristic morphological alterations, which were first described by Nissl (1892) and have since been extensively studied (reviewed by Hyden, 1960; Liebermann, 1971, 1974). These changes, collectively termed 'chromatolytic changes', include swelling and migration of the nucleus to an eccentric position and an increase in nuclear and nucleolar size; they refer particularly to the apparent disappearance of the prominent basophilic-staining Nissl granules. These granules are ribosome clusters and ordered arrays of rough endoplasmic reticulum which can no longer be visualized when they become disorganized, freeing polyribosomes and even ribonucleotides into the cytoplasm. Disorganization of the ribosome clusters indicates that the morphological changes in the cell body are associated with increased protein synthesis and suggests that chromatolysis is an anabolic response of the injured neurone. A schematic representation of these changes is shown in Figure 8.1.

With few exceptions, this interpretation is supported by evidence showing reduced DNA repression, conformational changes of DNA to the uncoiled form associated with metabolically active state (Barron, 1983), expression of immediate early genes (Leah *et al.*, 1991; de Filipe *et al.*, 1993), a net increase in RNA synthesis with transfer of RNA from

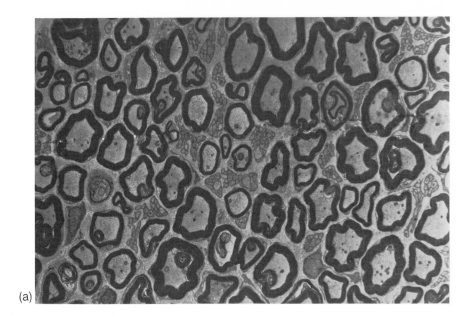

(a)

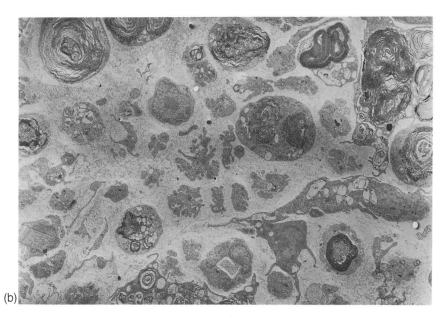

(b)

────── 5 microns

Figure 8.1 Low-power electromicrographs of transverse sections through mouse sciatic nerves 7 days after axotomy: (a) nerve taken from C57BL/601a mouse; (b) from OLA mouse. (Reproduced from Lunn *et al.*, 1989, *Eur. J. Neurosci.* 1, 27–33, with permission of the authors.)

nucleus to cytoplasm, and increased cellular protein content (Bråttgard et al., 1957, 1958; Murray, 1973; Watson, 1965, 1968, 1972, 1974). Concentrations of enzymes of the oxidative pentose phosphate shunt, which are required for RNA synthesis, are also raised (Sinicropi and Kauffman, 1979) and putrescine, a polyamine associated with actively growing or proliferating cells, is expressed (Fischer and Schmatolla, 1972; Ingoglia et al., 1977). Changes in RNA and protein synthesis often coincide with the maximal chromatolytic response observed histologically (Cragg, 1970). Specific changes in gene expression of cytoskeletal proteins including tubulin, actin and neurofilament protein, upregulation of calcitonin-gene related peptide (CGRP) (Streit et al., 1989) and induction of novel proteins including specific tubulin isoforms, T-alpha 1- and Class IIb tubulin (Miller et al., 1989), growth factor receptors (Hayes et al., 1992) and GAP-43 (Benowitz and Routtenberg, 1987; Tetzlaff et al., 1989; Bisby and Tetzlaff, 1992) have been viewed as associated with, and necessary for, regeneration. These changes have been extensively reviewed (Skene, 1989; Hoffman, 1989; Cleveland and Hoffman, 1991).

Chromatolysis is reported to be more severe and to occur earlier in neurones where axonal lesions are close to the cell body (Marinesco, 1896) and this may indicate that in motoneurones chromatolysis could be a metabolic response with a quantitative component linked to the amount of axonal growth required to re-establish connections with the target musculature.

Chromatolysis and RNA and protein synthesis are, however, not always mutually exclusive. Biochemical studies indicate that the changes in RNA and protein content which have been supposed to underlie the chromatolytic changes are actually later manifestations of early changes in RNA metabolism. These early changes, recognized as increased incorporation of uridine into mRNA and altered mobility patterns seen on gel electrophoresis (Gunning et al., 1977) occur within hours and precede chromatolytic changes (Kaye et al., 1977, reviewed by Austin and Langford, 1980). Increased mRNA expression of tubulin and actin is apparent within 12 hours of axotomy of facial and sciatic motoneurones and simultaneous down-regulation of neurofilament protein is seen (Tetzlaff et al., 1988). Thus, RNA turnover increases in axotomized neurones within hours of injury and this turnover is not necessarily associated with an increase of total cellular content of DNA, RNA or protein, or with other changes.

The signal for a change in RNA synthesis is still a matter of speculation, but it is clear that the injury alters gene expression, resulting in a change in the composition and amount of transported materials in regenerating axons (section 8.2.3). The signalling process to the cell body is extremely rapid and the rapidity of these changes suggests that a retrograde chemical signal moving at velocities of at least 140 mm/h would be required, which is very much faster than any reported velocity of axonal transport either anterograde or retrograde (Dziegielewska *et al.*, 1980). It therefore seems more likely that the signalling process involves a different mechanism such as a change in electrochemical gradient.

Some components of the chromatolytic response may be a response to the injury of the cell (Cragg, 1970; Engh and Schofield, 1972). In many neurones in the central nervous system, chromatolysis may herald cell death, but in the mature peripheral nervous system the majority of axotomized neurones which undergo chromatolysis survive. All lumbar motoneurones survive axotomy (Gordon *et al.*, 1991) but facial motoneurones are more susceptible, losing up to 20% of the neurones (Tetzlaff *et al.*, 1991; Kreutzberg, 1982).

In axotomized neurones, sequestration of intracellular organelles in membrane-bound vacuoles with ultimate degradation in secondary lysosomes (Lieberman, 1971; Mathews and Raisman, 1972; Sumner, 1975; Barron, 1983) is also seen. This lysosomal response in axotomized neurones resembles, in many respects, the changes that occur in secretory cells which follow physiological curtailment of hormonal secretion (Watson, 1976). Such responses may therefore be linked with the cessation of neurotransmitter synthesis and release, and maintenance of transmitter release may prevent these changes.

8.2.3 FAST AXONAL TRANSPORT IN AXOTOMIZED NEURONES

There are many similarities between glandular secretion and the release of neurotransmitters from neurones, and much of the neurone's metabolism is committed to elaboration of membranes for storage and transport of neurotransmitters and their secretion (reviewed by Dahlström, 1971; Heslop, 1975; Saunders, 1975). The rough endoplasmic reticulum in the neurone cell body and dendritic shafts is the major site of macromolecular synthesis, and the specialization of the axonal transport systems in neurones (Droz, 1975; Droz *et al.*, 1975; Grafstein and Forman, 1980) ensures distribution of material destined for the axolemma, the axon and its terminals. Most proteins and lipids, including transmitters and their precursors, are associated with membranes and are transported by fast axonal transport. Transport of membrane-associated materials has been considered to be primarily secretory in function in the anterograde direction and lysosomal in the retrograde direction (Droz, 1975; Schwartz, 1979).

When a peripheral nerve is cut, transmitters and their precursors initially accumulate at the site of injury and this can be used to determine rates of transport (Lubinska, 1964; Dahlström, 1965; Heslop, 1975; Saunders, 1975). However, the amount of the transmitters and their precursors which are transported falls within days after axotomy until they are barely measurable (Cheah and Geffen, 1973; Frizell and Sjöstrand, 1974). Transport rates are unaltered, suggesting that depression of synthesis is responsible for reduced transport of the neurosecretory materials. This is consistent with findings that in motoneurones, following axotomy, proteins associated with cholinergic transmission decrease (Watson, 1976). The extent and duration of the depressed synthesis appear to be determined by the type of injury. Transport of acetylcholinesterase declines in cholinergic neurones and does not recover if regeneration is impeded by nerve section (O'Brien, 1978; Heiwell et al., 1979). Reinnervation normally occurs earlier and is more successful after nerve crush than following nerve section (Gutmann and Sanders, 1943; Davis et al., 1978). This may account for the less severe effects of nerve crush compared with nerve section on synthesis of transmitter-related proteins. Synthesis and transport may be expected to recover more quickly in axons which make functional connections earlier and more readily after nerve crush injury compared with after nerve section. This supports the concept that, in the regenerating cell, synthesis and transport of neurotransmitters is suppressed until connections with the target are remade.

The anabolic response of the axotomized cell is highly specific and targeted towards increased synthesis of materials required for axonal growth. Direct evidence for quantitative and/or qualitative changes in the type of proteins manufactured has been relatively difficult to obtain. Fast transport conveys relatively small amounts of a large number of different proteins, few of which have been identified. Increased incorporation of radioactivity into fast transported proteins has been documented in pulse-labelled axotomized neurones in a number of different systems (e.g. Grafstein and Murray, 1969; Frizell and Sjöstrand, 1974; Bisby, 1980), but these experiments could not distinguish whether the increase is due to increased precursor uptake, changes in the rate of protein synthesis or alterations of the quantities of protein transported by fast axoplasmic flow during regeneration. Specific changes discovered in the spectrum of rapidly transported proteins in goldfish optic nerve after axotomy (Benowitz et al., 1981) and in regenerating rabbit nerves (Skene and Willard, 1981) led to the isolation of GAP-43 (Zwiers et al., 1987), a growth-associated protein which is up-regulated after axotomy or interruption of axonal transport (Benowitz et al., 1981; Bisby, 1988; Woolf et al., 1990; Telzlaff et al., 1991). Since GAP-43 is up-regulated mainly in neurones which regenerate, the protein was given the name **growth-associated protein** (GAP-43) (see reviews by Benowitz and Routtenberg, 1987; Skene, 1989). Although the function of the phosphoprotein remains to be determined, the axonal transport of GAP-43, its accumulation and phosphorylation in axonal growth cones (Gosling et al., 1988) are presumed to influence neurite elongation and synapse formation, possibly by controlling inositol triphosphate hydrolysis and calmodulin binding (Benowitz and Routtenberg, 1987; Willard *et al.*, 1987).

Amounts of phospholipids transported by fast transport also change after axotomy, almost doubling in quantity within 10 hours of nerve crush injury (Dziegielewska *et al.*, 1980). A burst in synthesis raises the levels of phospholipids in the axons above normal for up to 20 hours. Almost 90% of the transported phospholipid is phosphatidylcholine, a major constituent of cell membranes (Brechter and Raff, 1975) and this is presumably required for immediate sealing of damaged axons and for subsequent axon regeneration.

Although fast axonal transport has been considered primarily as secretory in function, transport systems are not necessarily linked with neurosecretion and different materials may be directed along different pathways in the same neurone. In adrenergic neurones, for example, noradrenaline-containing vesicles and associated dopamine-beta-hydroxylase, chromagranin A and ATP are transported down the axon to the nerve terminals but they are not transported in the dendrites, which have their own intrinsic transport system (Schubert *et al.*, 1972). Transport in the central process of sensory neurones in the dorsal root ganglia but not in the peripheral axon is related to neurosecretion (Anderson and McClure, 1973). GAP-43 is transported only in the axons of growing and axotomized neurones and not in the dendrites or central processes of bipolar cells (Woolf *et al.*, 1990).

Many of the changes of the axotomized cell, including up-regulation of GAP-43, are also mimicked by blocking axonal transport with colchicine in intact cells (Skene, 1989). Synaptic transmission is critically dependent on fast axoplasmic transport. When axoplasmic transport is blocked by colchicine or by a reduction in the temperature, chemical transmission fails before impulse conduction (Schwartz, 1979). The changes are unlikely to be due to loss of trophic signals or material from the target organs (cf. Cragg, 1970) because, as has been mentioned previously, changes in RNA, proteins and phospholipid synthesis are triggered by a second injury of axotomized neurones.

8.2.4 CYTOSKELETON: SLOW AXONAL TRANSPORT IN AXOTOMIZED NEURONES

A shift in metabolic activity of the axotomized cell involves a shift in proportion of the fibrillar proteins which make up the cytoskeleton. The fibrillar proteins are recognized by electron microscopy as microfilaments or actin-containing filaments of approximately 7 nm in diameter; microtubules, which are tubulin polymers and appear as hollow cylinders 20–30 nm in diameter; and neurofilaments, which are seen as rods 10 nm in diameter and are composed of neurofilament triplet protein (Hoffman and Lasek, 1975; Schwartz, 1979). The microfilaments and neurotubules are orientated longitudinally along the proximodistal axis of nerve processes with neurofilaments being the most prominent – even visible with the light microscope (Droz *et al.*, 1973). In the mature axon, the number of neurofilaments far exceeds the number of neurotubules (Friede and Samorajski, 1970). Their number is thought to determine axon size since the cross-sectional area of large axons is directly proportional to numbers of axonal neurofilaments (Forman and Borenberg, 1978; Willard *et al.*, 1979; Hoffman *et al.*, 1987).

When a mature axon is cut, the amount of neurofilament triplet protein transported down the axon is decreased so that the microtubular and microfilamental proteins increase relative to the neurofilaments (Hoffman and Lasek, 1980; Telzlaff *et al.*, 1988). This is due to specific changes in the gene expression of the cell (Hoffman and Cleveland, 1988; Miller *et al.*, 1989; Oblinger *et al.*, 1989) with corresponding changes in protein synthesis (Hall *et al.*, 1978; Greenberg and Lasek, 1988; Tetzlaff *et al.*, 1988). Within 12 hours, mRNA is increased significantly above normal for tubulin and actin simultaneously with a decline in mRNA for neurofilament protein (Tetzlaff *et al.*, 1988). The axotomized motoneurones also express alpha-tubulin, an isoform which is normally only expressed in neurones during development (Miller *et al.*, 1989).

These qualitative and quantitative changes in gene expression may be regarded as a process of de-differentiation to an immature state akin to the embryonic neurone where the cytoskeleton of the growth cone and newly formed axon is composed almost entirely of microtubules and microfilaments (Yamada *et al.*, 1971; Hoffman and Cleveland, 1988). It is only when the embryonic axon reaches a peripheral end organ that neurofilaments appear in the axons for the first time. In the adult, the proximal end of the cut axon is converted into a growth cone, which is like the embryonic growth cone. The new growth cone is characterized by an extensive network of smooth endoplasmic reticulum in its central core and the presence of actin containing filaments and microtrabeculae. Microtubules and neurofilaments are noticeably absent or few in number (reviewed by Bunge, 1986; Kuczmarski and Rosenbaum, 1974; Jockusch and Jockusch, 1981). Lasek (1981) suggested that the growth cone in regenerating axons arises specifically as a result of reorganization of the fibrillar proteins which accumulate in the cut end of the axons during the latent period before regenerative axonal elongation, and that regeneration is supported by increased transport of tubulin and actin relative to neurofilamental protein (McQuarrie and Lasek, 1989). Thus gene expression in the axotomized mature neurone appears to 'recapitulate the developmental pattern' (Hoffman and Cleveland, 1988).

Fibrillar proteins are transported distally in normal and regenerating nerves in two distinguishable waves designated slow component a and b (**SCa** and **SCb**) (Black and Lasek, 1980; Lasek and Hoffman, 1976). Tubulin and the neurofilament triplet proteins represent the bulk of the protein moving in SCa at a rate of about 0.2–2 mm/day in mammalian neurones (Hoffman and Lasek, 1975). SCb contains at least 20 major proteins, including cytomatrix proteins: actin, clathrin, myosin-like protein, fodrin and calmodulin (Black and Lasek, 1979; Hoffman and Lasek, 1975; Willard, 1977) and tubulin (Lasek and Hoffman, 1976). These also move as a coherent peak at an average velocity of 2–5 mm/day (Lasek and Hoffman, 1976; Black and Lasek, 1980), the same rate as that calculated for nerve regeneration (Gutmann *et al.*, 1942; McQuarrie and Lasek, 1989; Wujek and Lasek, 1983). The correspondence between rates of transport of SCb and regeneration at different ages (Black and Lasek, 1979) and in different neuronal processes (Wujek and Lasek, 1983) is consistent with the idea that SCb provides the cytoskeletal elements required for axonal regeneration (McQuarrie and Lasek, 1989). Moreover, the rate of SCb transport accelerates in concert with the acceleration of regeneration rate of mammalian axons after a conditioning lesion. The conditioning lesion 'primes' the neurone for accelerated SCb and correspondingly accelerated regeneration after a later test lesion at a more proximal axonal site (McQuarrie and Jacob, 1991). Even under normal conditions of regeneration, the rate of SCb transport increases (Hoffman *et al.*, 1992). The parallel between increased SCb synthesis and outgrowth also argues that the SCb proteins which are already in transit have been accelerated to contribute to the emerging growth cones.

The reduced neurofilament content of axotomized neurones accounts for the reduced fibre diameter which was first noted by Greenman in 1913 and has been confirmed many times since (e.g. Acheson *et al.*, 1942; Gutmann and Sanders, 1943; Cragg and Thomas, 1961; Gillespie and Stein, 1983; Gordon *et al.*, 1991). Down-regulation of neurofilament proteins and decline in fibre diameter rather than motoneurone death seems to account for the reduced amplitude and conduction velocity of the compound action potential of axotomized nerves since cell and axon numbers are not significantly reduced by axotomy (Gordon *et al.*, 1991). The rate of diameter decrease is most rapid immediately following axotomy and follows an exponential time course to reach a plateau value which is maintained for as long as a year if the nerves are prevented from remaking functional connections (Figure 8.2). The cell body can maintain the atrophied axons for long periods, adult neurones appear to be able to maintain this 'growing state' even if axons are prevented from elongating. Nevertheless normal axonal diameter is restored only after making functional connections between neurones and peripheral targets.

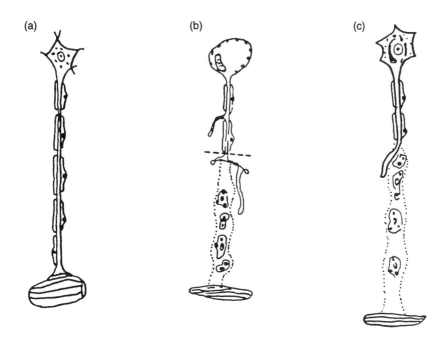

(a)   (b)   (c)

Figure 8.2 Peripheral motoneurones and muscle: (a) in the normal state; (b) during Wallerian degeneration following section of the axon at the dotted line; (c) following axonal neuropathy. Removal of the axon from the muscle causes muscle atrophy. In the case of nerve section, the nerve cell body undergoes very specific changes which include swelling, an acentrically placed nucleus and precipitation of Nissil granules. These changes in the cell body are not so prominent in the case of axonal neuropathy but, as illustrated, sprouting probably occurs in both injury and axonal degeneration. (Modified after W. G. Bradley, *Disorders of Peripheral Nerves*, Blackwell, with permission.)

8.2.5 DENDRITES AND SYNAPTIC SITES

Another feature of axotomized neurones is dendritic retraction or decline in width of the dendrites (Sumner and Sutherland, 1973). This change is reversed when the axons make peripheral connections (Sumner and Watson, 1971). However in addition to these changes, sprouts can emanate from dendrites of axotomized neurones, and their ultrastructure is similar to that of an axon (Hall, 1989). This indicates that the neurone is synthesizing proteins necessary to replace its missing axon, possibly at the expense of those that maintain its dendritic expansion.

At the ultrastructural level, the number of synaptic profiles on axotomized neurones can be seen to decline in the first few days following axon injury and this is concurrent with chromatolysis (Blitzinger and Kreutzberg, 1968; Hamberger *et al.*, 1970; Purves, 1975). Synaptic thickenings below the synaptic contacts are actually lost from the post-synaptic membrane before the boutons become detached (Sumner, 1975; Mathews and Nelson, 1975) implying that changes in the axotomized cell precede, and may be responsible for, the gradual loss of boutons. Glial cells may also play some role in the separation of boutons from the post-junctional membrane (Torvik and Skjorten, 1971; Sumner and Sutherland, 1973; Matthews and Nelson, 1975). The depression of EPSP (excitatory post-synaptic potentials) elicited by stimulation of 1a afferent fibres (Mendell *et al.*, 1974; Mendell, 1984) suggests that, in addition to loss of synaptic contacts on the axotomized motoneurones after peripheral nerve section, at least some of the reduction in efficacy is also due to atrophy of afferent terminals with decline in release of transmitter (Knyihar and Csillik, 1976).

While synaptic transmission is depressed, axotomized motoneurones are hyperexcitable, so that while monosynaptic reflexes are absent or depressed, in mammals (Eccles *et al.*, 1958; Mendell *et al.*, 1976) polysynaptic reflex discharges are reported to be larger than normal (Campbell, 1944). This can be partly accounted for by decrease in rheobase current for axotomized motoneurones and the emergence of dendritic spikes (Kuno and Llinas, 1970a).

Loss of monosynaptic reflexes and enhancement of polysynaptic reflexes may indicate selective loss of some terminals as suggested by ultrastructural studies (Sumner and Sutherland, 1973; Sumner, 1975) and by electrophysiological studies (Kuno and Llinas, 1970b; Farel, 1980).

8.2.6 PHYSIOLOGICAL CHARACTERISTICS OF AXOTOMIZED NEURONES

Reduced conduction velocity is one of several electrophysiological manifestations of the altered status of axotomized neurones (see Titmus and Faber, 1990). The molecular basis for the changed conduction velocity is the best understood. In contrast, the basis for the other changes is poorly understood and much of what we know simply describes these changes, many of which have been interpreted in the context of de-differentiation of the axotomized neurone to a more immature state (Gustaffson and Pinter, 1984; Kuno *et al.*, 1974; Huizar *et al.*, 1977; Foehring *et al.*, 1986).

Axotomy reduces but does not abolish characteristic differences in the action potentials in motoneurones supplying fast and slow muscles. The duration of the after-hyperpolarization (AHP) is normally 2 to 3 times longer in slow motoneurones and is shortened by around 35% in slow but not fast motoneurones after axotomy (Kuno *et al.*, 1974) so that axotomized slow motoneurones show similar properties to fast motoneurones (Foehring *et al.*, 1986). This change can be regarded as a process of de-differentiation of the motoneurone electrophysiological properties, since during normal development differentiation involves a change in

AHP duration in the slow but not fast motoneurones (Huizar *et al.*, 1975; Gallego *et al.*, 1978).

As a rule, AHP duration and input resistance increase and the rheobase and conduction velocity decrease (Foerhing *et al.*, 1986), further supporting the concept that motoneurones de-differentiate following axotomy. The decreased rheobase current and increased input resistance, together with an increase in the action potential overshoot, contribute to an increase in motoneurone excitability (Kuno and Llinas, 1970a; Kuno *et al.*, 1974; Gustaffson, 1979; Farel, 1980; Gustaffson and Pinter, 1984; Pinter and Van den Noven, 1989) and the development of dendritic spikes on the cell body under certain conditions (Eccles *et al.*, 1958; Mendell *et al.*, 1974; Titmus and Faber, 1990) may partially offset the effects of synaptic dysjunction in maintaining the activity of the motoneurone.

The observed changes in electrophysiological parameters after axotomy provide an interesting model system in which to study the changes in passive and active electrical membrane properties and their associated ionic channel mechanisms. These studies have stimulated extensive investigation of channel kinetics which underlie the change in the membrane properties in other neurone systems where whole-cell and single-cell patch clamp techniques can be used to characterize channel properties. For example, increased inactivation of Ca^{2+} current underlies the reduced AHP amplitude and duration in axotomized frog sympathetic neurones (Gordon *et al.*, 1987).

The nature of the changes in electrophysiological properties after axotomy also provides insights into the differential control of membrane properties. The AHP duration (but not other electrophysiological properties) is readily altered by a number of conditions which are known to induce muscle atrophy. Cordotomy in adult cats and kittens (Czeh *et al.*, 1978; Gallego *et al.*, 1978), immobilization of the muscle in a shortened position (Gallego *et al.*, 1979), partial denervation (Huizar *et al.*, 1977) and nerve conduction block (Czeh *et al.*, 1978) induce muscle atrophy and reduce the AHP duration. Other electrophysiological parameters, including rheobase current, input resistance and conduction velocity, are not affected and remain stable unless functional connectivity with muscle is disrupted either by axotomy or by functional disconnection with botulinum toxin blockade (Pinter and van den Noven, 1989). These changes may be a consequence of changed patterns of synaptic transmission which are induced by electrical stimulation and/or electrical activation of the soma. The AHP duration is the property most influenced by target connection, with respect to both sensitivity and the extent of change. Perhaps the most surprising finding is that axotomized motoneurones which grow into skin recover their longer AHP durations despite the fact that connection with skin is unable to reverse other electrophysiological changes. Functional connections with muscle are therefore thought to be essential for the motoneurone to express the full range of normal membrane properties.

While the notion of de-differentiation of the axotomized cell remains a useful descriptive concept, axotomy does not return adult neurones to the embryonic state where nerve–muscle interaction is essential for survival. Altered gene expression and changes in properties in the adult motoneurones after axotomy can provide insight into the regulatory factors which govern interactions between nerve and muscle.

8.2.7 GLIAL RESPONSE

Schwann cells and internodal length between Schwann cells surrounding the intact axolemma in the proximal nerve stump change little after axotomy (Cragg and Thomas, 1961). Furthermore, the numbers of myelin turns and myelin thickness are not changed even though the axon diameter is reduced (Gillespie and Stein, 1983).

Total fibre diameter (myelin + axon) is normally measured to assess atrophy after axotomy (reviewed by Gordon, 1983). It is also the usual parameter of fibre size that is related to conduction velocity, showing good correlation in normal nerves (Hurst, 1939; Jack, 1955; Milner et al., 1981). Conduction velocity of atrophic nerves, on the other hand, varies with axon diameter and not total diameter (Gillespie and Stein, 1983). Consequently, by measuring total fibre diameter, early histological studies of injured nerves have underestimated the atrophic changes in the axons.

The time course of the response of neuroglial cells to axotomy is similar to that of motoneurones. Watson (1972) has suggested that these glial responses may be related to changes in the dendrites of the axotomized motoneurones, since the first phase is concurrent with synaptic bouton detachment following axotomy and the second phase occurs on reinnervation when there is an expansion of the dendritic field. However, it is more likely that the appearance and activation of microglia in the vicinity of axotomized motoneurones is related to the expression of MHC on their surface (Maehlen et al., 1989). The appearance of this molecule may be the signal that the cell is damaged and may initiate the sequence of responses associated with this event.

8.2.8 TWO-STATE MODEL OF THE MOTONEURONE: GROWING OR TRANSMITTING

Neurones exhibit an early burst of RNA synthesis in response to nerve section, and a later one when the regenerating axons make functional nerve–muscle connections. A second burst of RNA activity can be observed if a second crush injury of the same nerve is carried out. Watson interpreted the second burst of RNA synthesis to nerve crush of the already axotomized nerves as the transformation of the nerve's metabolism from the 'resting non-transmitting mode' to a second 'seeking mode' of the recently axotomized neurone. This further burst of RNA synthesis after a crush of the axotomized nerve may be responsible for the enhanced regeneration after two consecutive injuries (McQuarrie and Grafstein, 1973; McQuarrie and Jacob, 1991) and for earlier findings of Gutmann (1942) that the success of regeneration increases after double crush of a peripheral nerve. The burst of RNA synthesis when axons make functional connections occurs after a single or repeated crush injuries (Watson, 1970) and suggests that increased synthetic activity is required possibly to supply transmitters and their precursors during the re-establishment of nerve–muscle contact.

Thus neurones are able to exist in either of two states, that of the regenerating or growing state when they are seeking to make connections with muscle, or in the non-growing, transmitting state when muscle connections have been formed (Watson, 1976). However, neurones also show chromatolytic changes when axons with intact peripheral contacts are induced to grow and produce terminal sprouts following intramuscular injection of botulinum toxin (Watson, 1969). Sprouting induced by botulinum toxin occurs during the course of muscle paralysis caused by the action of botulinum toxin which blocks acetylcholine release from the nerve terminals (Duchen and Strich, 1968). The chromatolytic response may reflect the axonal growth and/or re-establishment of transmission in the poisoned axons whose terminal sprouts rapidly form functional nerve–muscle contacts. Also it is not necessarily associated with the same changes in gene expression as after axotomy, since different genes may be differentially regulated, depending on their function. GAP-43 for example, which has been considered to be associated with regeneration (Skene, 1989) is not up-regulated during sprouting (Tetzlaff *et al.*, 1993). Clearly the protein is down-regulated when the motoneurone maintains contact with its target muscle fibres even when growth cones are formed during sprouting. What remains unclear is whether GAP-43 is down-regulated when the neurone is transmitting, or whether there is some retrograde inhibitory influence of the target itself.

8.3 EFFECTS OF AXOTOMY ON MUSCLE

Earlier chapters have discussed the interdependence of muscle and nerve on contact with each other during development. Nerve–muscle interactions in the adult have been studied using a number of different models and a great deal of information has been gained by examining the effects of depriving muscle of its motor innervation. The examination of changes in both form and function of muscle following experimental denervation has revealed that many aspects of muscle fibre properties are sustained and regulated by the activity imposed on muscles by their innervation and that profound changes occur in all muscle fibre properties when neural activity ceases after nerve section or blockade of nerve impulses. While some of the observed changes occur almost immediately, others develop slowly over several weeks. The prevalent features of skeletal muscle at early and late stages following denervation and their origin are discussed in this section.

8.3.1 THE NEUROMUSCULAR JUNCTION

The changes in axons that have been separated from their cell bodies have been discussed in detail in the first part of this chapter, as have some of the features associated with nerve degeneration that occur at the neuromuscular junction. Recently, detailed studies at the site of contact between nerve and muscle have revealed changes that appear to be directed towards preserving guidance mechanisms to facilitate reinner-vation (see Bloch and Pumplin, 1989; Hall and Sanes, 1992). Some of these are also discussed in Chapter 9.

Fragmentation of the nerve terminal and invasion of the endplate region by the overlying Schwann cell take place within a few hours after denervation (Birks *et al.*, 1960; Miledi and Slater, 1968; Nickel and Waser, 1969). In the frog, the axon at the neuromuscular junction is then replaced by the Schwann cell, which appears to ingest and dispose of remaining nerve terminal fragments. At later stages post-junctional changes also occur. The deep post-synaptic junctional folds so characteristic of the motor endplate become flattened and the complex post-synaptic structure is lost (Miledi and Slater, 1968; Matsuda *et al.*, 1988). Using an elegant scanning electron microscopic technique, Matsuda *et al.* (1988) described the gradual flattening of post-junctional grooves, simplification of their patterns and a flattening of the muscle fibre surface at the endplate region. By 16 weeks, no special physical structure can be discerned at the old endplate region but on reinnervation all the features of an intact endplate return.

Many other membrane-specific changes also take place at the neuro-muscular junction after denervation and these are considered in detail in section 8.3.7.

8.3.2 MUSCLE CONTRACTILE PROPERTIES FOLLOWING DENERVATION

One of the earliest changes in the functional characteristics of denervated muscle is an alteration in contractile properties with respect to both the maximum contraction force that can be developed and the rate (V_{max}) of force development (Lewis, 1972; Kean *et al.*, 1974; Gutman, 1976; Finol *et al.*, 1981). During the first few days following section of the motor nerve, the contractile force that can be developed in response to a single supramaximal electrical stimulus (twitch) is little changed but the response to repetitive (tetanic) stimuli decreases (Finol *et al.*, 1981). This is due to the fall in specific tetanic tension (tension per unit muscle-fibre area) and occurs to a similar extent and with a similar time course in slow and fast muscle fibres. Early loss of force may be followed by a transient recovery of tetanic force (Finol *et al.*, 1981). Spector (1985a, b) reports a similar reduction in the ability to develop the expected level of force per unit area in muscles inactivated by blocking impulse traffic in the motor nerve by the infusion of tetrodotoxin (TTX) (Spector, 1985a) and in denervated muscles (Spector, 1985b). As in denervated muscles, loss of the ability to develop force in inactivated muscles cannot be accounted for entirely by muscle atrophy for, although considerable atrophy does occur in denervated or inactivated muscles, the development of atrophy occurs later than the change in ability to develop force. The degree of change is generally greater in denervated than in inactivated muscles and Spector attributes this to some form of 'trophic' maintenance exerted by the nerve which remains attached to the muscle in the TTX-treated group. However, with section of the motor nerve, many other cellular changes occur which have not been reported in inactivated muscle and these may also influence the muscles' response (section 8.3.8). In addition, there is some question as to whether nerve impulse blockade with TTX can ensure total inactivity, and in other studies where botulinum toxin (BoTX) has been infused around the nerve with the aim of producing and maintaining full neuromuscular blockade other features characteristic of denervated muscles are as fully expressed as they are in surgically denervated muscles (Bambrick and Gordon, 1987). Whether there is a mechanism other than activity that can account for the differences in the contractile responses of denervated and inactivated muscles observed by Spector remains in question (Kowalchuk and McComas, 1987).

At later stages of denervation, tetanic tension falls in both fast and slow muscles by 80–90% due to gross loss of muscle bulk as well as a fall in specific tension (even when atrophy is taken into account). In long-term denervated muscles, loss of tetanic tension is generally found to be greater in slow muscles, for although they do not show such rapid early changes their properties continue to change after changes in fast muscles have apparently been completed (Finol *et al.*, 1981). At these later stages of denervation, contractile responses also become highly variable and difficult to interpret. This is not surprising in view of the observations of Schmalbruch *et al.* (1991) that muscles that have been denervated for many months have very few fibres which contain organized myofilaments and the generation of force is confined to those few fibres that retain both contractile filaments and a sarcoplasmic reticular system of sufficient complexity to allow force generation. As well as undergoing extensive atrophy, denervated muscles show alterations in all intracellular structures (section 8.3.4) as well as in their electrophysiological properties (section 8.3.6), become infiltrated by fibroblasts and other non-muscle cells (Murray and Robbins, 1982a, b) and show increased fibrous connective tissue (Garcia-Bunuel and Garcia-Bunuel, 1980), all of which make it virtually impossible to determine how far the parameters of the contractile response reflect the true state of muscle atrophy.

8.3.3 CONTRACTION SPEED

The contraction time (time to peak tension and half-relaxation time) in innervated muscle is related to the number of fast or slow motor units that each muscle contains. Within any muscle fibre the rate of force development is linked with myosin ATPase activity in fast and slow muscle fibres (Bárány, 1967; Close, 1972), its complement of myosin isoforms (Chapters 5 and 7) and the features of the sarcoplasmic reticulum (Schiaffino *et al.*, 1970; Magreth *et al.*, 1972; for review, see Frischknecht *et al.*, 1990). Slowed muscle contraction time following denervation was observed by Langley as early as 1916 and has since been reported by many others (Denny-Brown, 1929; De Smedt, 1949; Syrový *et al.*, 1971, 1972; Lewis, 1972; Kean *et al.*, 1974; Finol *et al.*, 1981). When results in different small mammalian species were compared (see Gutmann, 1976) contractile responses of rat, cat and rabbit muscles following denervation showed that, while in all species fast contracting muscles became slower, the contractile speed of slow muscles such as soleus actually became faster in rat and rabbit but not in cats (Syrový *et al.*, 1971, 1972; Gutmann *et al.*, 1972). The results in cats are in line with other studies (Eccles, 1941; Lewis, 1972; Kean *et al.*, 1974). Gutmann and his colleagues (see Syrový *et al.*, 1971) attributed the differences in response to denervation of cat and rabbit muscles to differences in their initial fibre composition, pointing out that adult cat soleus muscles are almost entirely composed of slow fibres, while those of rats and rabbits contain a proportion of fast fibres. If slow fibres were to atrophy first in these animals, this could explain the slight shortening of the contraction on denervation. However, this is not in agreement with Finol *et al.* (1981), who found that in rats, at least, early changes are greater in fast than in slow muscle fibres. The differing species, experimental protocols and varying durations of experiments make it difficult to compare these studies directly.

Prolongation of the twitch contraction is due to a longer time to peak as well as prolonged relaxation time (expressed as the time taken for the muscle to relax to half of its peak force – half-relaxation time). In rats, since prolonged contraction time is evident in the early stages of denervation (Gutmann *et al.*, 1972; Gutmann, 1976; Finol *et al.*, 1981), this may account for the smaller loss in twitch tension compared with tetanic tension at this time: the prolonged contraction represents a prolonged active state and may result in more complete activation of the contractile apparatus, counteracting direct effects of denervation on the myofilaments. Such a suggestion was in fact made by Lewis (1972) and is supported by the finding that the muscle action potential duration is also prolonged after denervation (Thesleff, 1974) (section 8.3.6).

The maximum rate of rise of the tetanic contraction is not necessarily altered and, here again, reports differ. In cats, Lewis (1972) found slower rise times in both soleus and flexor digitorum longus (FDL), while Kean *et al.* (1974) reported similar changes in maximum rate of tension development in soleus but not in the flexor muscle. However, values for control muscles in these two different experimental groups were also different and, again, this emphasizes how difficult it may be to obtain specific evidence for early change in muscle contraction properties when reported normal values can be highly variable. Also, careful studies on the length of denervated muscles indicate that sarcomeres may be added in fast (Kean *et al.*, 1974) or slow (Gorza *et al.*, 1988) muscles, possibly as a result of mechanical stretch in denervated limbs. Whether or not this is always the case, a change in sarcomere numbers or length may well be reflected in a change in the shortening velocity, giving results which vary according to the experimental design. When considered as a function of sarcomere length (Kean *et al.*, 1974), differences in values between normal and denervated slow and fast muscles were actually found to be similar.

One of the most obvious possibilities to consider in terms of altered contractile properties is an alteration in myosin ATPase activity or myosin composition. There is a correlation between myosin ATPase activity and the contractile responses of skeletal muscles (Bárány, 1967; Close, 1972), the rate of activation in fast muscles being approximately twice that in slow muscles. Changes in myosin ATPase activity accompany changes in contractile activity following denervation, according to Syrový *et al.* (1971) and Gutmann *et al.* (1972).

In spite of an apparent change in myosin ATP-ase activity, two-dimensional gel electrophoresis of whole-muscle extracts shows reduced levels of slow myosin in denervated gastrocnemius and hemidiaphragm (Carraro *et al.*, 1981) and histochemical examination of changes in the myofilament proteins or enzyme activities have produced no satisfactory explanation for the observed changes in contractile speed. Using antibody staining for fast and slow myosin, Gauthier and Hobbs (1982) noted a correlation between myosin ATPase activity and the presence of slow or fast myosin isoforms. They observed a gradual increase in fast myosin, initially often co-expressed with slow myosin in individual muscle fibres following denervation. The expression of slow myosin becomes reduced with time in both denervated (Gauthier and Hobbs, 1982; Spector, 1985b) and inactivated muscles (Spector, 1985a, b). Such changes could account for the late shortening in contraction speed seen in some slow muscles by Finol *et al.* (1981). It is of interest to note that co-expression of different myosin isoforms is also observed in some neurogenic diseases in humans (Sawchak *et al.*, 1989) where both denervated and reinnervated muscle fibres would presumably be present. In general, however, the observed alterations in muscle fibre myosin isoforms in denervated muscles do little to explain the mechanism responsible for altered contractile properties.

Similarly, the possible effects of alterations in the uptake of calcium by the sarcoplasmic reticulum with consequent changes in the duration of the active state of the muscle have been considered. Proliferation of the transverse tubular system is a well-documented feature in early denervation (Magreth *et al.*, 1972; Engle and Stonnington, 1974; Salvatori *et al.*, 1989) which accompanies early atrophic changes (Pellegrino and Franzini, 1963; Schiaffino and Settembrini, 1970; Carraro *et al.*, 1985) (section 8.3.4). Within 2 weeks of denervation, increased sarcotubular material, mostly from junctional regions and terminal cisternea rather than T-tubule regions of the sarcotubular system, can be detected. However, with this increase in membrane area there is a concomitant decrease in the density of Ca^{2+} ATPase membrane sites (Salvatori *et al.*, 1989). The net result of these contrary changes is little or no apparent change in Ca^{2+} handling until later stages of denervation, although calcium does accumulate in denervated muscle fibres (Kirby and Lindley, 1981) and impaired calcium handling could contribute to a prolonged active state.

Contraction time is a function of the properties of the membrane related to intracellular compartments of the muscle cell (see Frischknecht *et al.*, 1990 and Chapter 5) acting on a number of muscle properties which may change following denervation with differing time courses, or to differing degrees in different muscle types or in different species. Bearing this in mind, the significance of any contractile change, especially in long-term denervated muscle, must be viewed in the light of the many other extensive changes which follow denervation. Neither early nor late changes in muscle contractile properties following denervation can easily be attributed to any one factor.

8.3.4 MORPHOLOGICAL CHANGES IN DENERVATED MUSCLE

The most prominent morphological change in denervated muscle is the loss of muscle bulk. In early studies Tower (1935, 1939) describes shrinkage of muscle with the appearance of swollen nuclei centrally placed in atrophic muscle fibres. Loss of muscle bulk was attributed to reduction in muscle fibre cross-sectional area rather than to fibre degeneration (disintegration with phagocytosis of individual muscle fibres), although some muscle fibre degeneration may occur, especially in infected or injured muscles and in long-term denervated muscles (Anzil and Wernig, 1989). Atrophic changes alone can account for loss of up to 85% of muscle bulk in chronically denervated muscles (Gutmann and Zelená, 1962; Carraro et al., 1981) and although atrophy can be reversed if reinnervation is rapid, chronically denervated muscles do not necessarily return to their pevious bulk on reinnervation (Carraro et al., 1981).

During the first 3 weeks after sciatic nerve section in adult rats, both the slow contracting soleus and predominantly fast gastrocnemius muscles show similar percentage loss of wet muscle weight and reduction in muscle fibre diameter. Using electron microscopy (**EM**), Engle and Stonnington (1974) found that, as fibre diameter was reduced, myofibrillar content was also reduced such that the relationship between myofibre and myofibrillar area remained roughly constant in the earlier stages of denervation. Myofibrils start to degenerate at the periphery but central myofibrillar atrophy occurs later and at this time central nuclei with prominent Golgi bodies are common (Tower, 1935). In recent studies, Schmalbruch et al. (1991) followed light and electron microscopic changes in rat muscles denervated for between 6 and 10 months and found that, after long-term denervation, cross-sectional area of fibres was reduced to about 3% of control values and large areas of fatty infiltration were present throughout the muscle. However, detailed EM studies show that after even prolonged denervation many fibres still contain myofilaments and sarcoplasmic material, even though these are in disarray with very few regular cross-striations. Some necrotic fibres are also seen. In these studies, both satellite cells and rudimentary myotubes could be distinguished and the authors suggest that there is continuous generation of new muscle precursor cells, but these fail to produce myofibres because of lack of innervation to produce contractile activity. Their further finding (Schmalbruch et al., 1991; Al Amood et al., 1991) that, after several weeks of imposed activity achieved by electrical stimulation, many long-term denervated muscle fibres show organized myofilaments and have greater diameters than unstimulated long-term denervated muscles further indicates the importance of activity in maintaining or re-establishing viable contractile apparatus. The finding that, even after long periods of denervation, and in extremely atrophic muscles, some functional muscle tissue can be rescued by activity or re-innervation, is encouraging when considering the possibilities for restoring denervated muscle tissue in humans.

Subsarcolemmal deposits of ribosomal material indicate increased RNA activity following denervation (Engle and Stonnington, 1974; Gau-

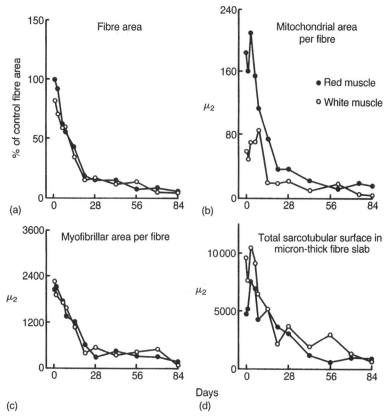

Figure 8.3 Effects of 1–84 days denervation on area and components of transversely sectioned muscle fibres: (a) mean fibre area and (b) mean fibrillar area per fibre decrease at approximately the same rate; (c) mean mitochondrial area and (d) sarcotubular surface increase absolutely during first 8 days during denervation. (Reproduced from Engle and Stonnington, 1974, *Ann. NY Acad. Sci.* **228**, 68–78.)

thier and Schaeffer, 1974). Mitochondrial area and sarcoplasmic membrane unit area also change with approximately the same time course, reaching maximum by about 20 days. Once established, all of these changes persist unless re-innervation is established. Figure 8.3 shows the relationship between changes in fibre diameter and other muscle fibre components following denervation.

Slow muscle fibres have more mitochondria than fast muscle fibres (Stonnington and Engle, 1973; Engle and Stonnington, 1974) and the total mitochondrial area in soleus muscle fibres is correspondingly greater than that in gastrocnemius. Following denervation, soleus shows a small transient increase in relative mitochondrial area followed by a later reduction with elongation of mitochondrial profiles. Distribution of mitochondria in normal muscles is relatively uniform throughout the muscle fibres, but around 7 days after denervation sub-sarcolemmal clusters begin to form. Changes in mitochondrial elements which accompany their change in shape and distribution were recently studied in more detail. In common with Hearn (1959) and Hogan *et al.* (1965), Wicks and Hood (1991) found significant reductions in mitochondrial enzyme activities which correlated with lower levels of mitochondrial-specific cardiophospholipid and reduction in muscle contractile endurance. Evaluation of mitochondrial enzyme activity in muscle sections by densitometry, using cytochrome oxidase as a marker (Nemeth *et al.*, 1980; White and Vaughan, 1991), also indicates reduced activity immediately after denervation with a rise in activity in fast fibres evident at later stages of denervation. The significance of this late change is unclear but it could be related to the late return to normal values of the nuclear controlled mRNA for cytochrome c oxidase subunit detected in the studies of Wicks and Hood (1991).

Proliferation of the sarcotubular system (SR), resulting in increased surface area, is established within a few days of denervation (Magreth *et al.*, 1972; Engle and Stonnington, 1974). An initial increase in sarcotubular area, with the most extensive proliferation taking place at junctions with the cisternea rather than in the T-tubules (Salvatori *et al.*, 1989), is followed by an overall reduction in SR area in both fibre types. Dilation of sarcotubular structures with the development of more complex and irregular outlines is commonly observed (Engle and Stonnington, 1974). In line with the observed enrichment of tubular sarcoplasmic membrane after denervation, Lehotsky *et al.* (1991) found increased incorporation of cholesterol in sarcoplasmic subcellular fractions. Together with this change, there is a decrease in a protein associated with calcium transport as well as sarcotubular ATPase activity, but the number of Ca^{2+} binding sites appears to increase, consistent with an increase in tubular membrane volume (Lehotsky *et al.*, 1991). Intracellular resting levels of calcium are increased after denervation (Kirby and Lindley, 1981) probably as a result of decreased Ca^{2+} influx and/or reduced uptake. Whether the increased calcium levels simply result from inefficient calcium uptake, and whether proliferation of sarcoplasmic membrane in tubular regions is a response to this, remains to be clarified.

8.3.5 BIOCHEMICAL CHANGES IN DENERVATED MUSCLES

A number of biochemical changes have been observed which accompany the morphological changes described above. Early observations showing increased DNA (Gutmann and Žak, 1961) and RNA (Goldspink, 1976) have been better understood in the light of subsequent findings.

There is no evidence of mitotic activity in sub-sarcolemmal nuclei following denervation. However, muscle satellite cells (Mauro, 1961) proliferate after denervation and injury (Ontell, 1974; Murray and Robbins, 1982a, b). Murray and Robbins also found substantial mitotic activity in resident connective tissue cells as well as in Schwann cells, and cellular infiltration by blood-borne mononucleated cells is pronounced between 3 and 7 days after denervation (Jones and Lane 1975). The presence of these cells could account for, or at least contribute to, the observed early increase in DNA content. It has been suggested that cellular proliferation and infiltration after denervation may be related to nerve terminal degeneration as well as to disuse (Arancio *et al.*, 1992; Jones and Vrbová, 1974; Murray and Robbins, 1982b). However, while cellular proliferation of resident non-muscle cells ceases after several days, proliferation of satellite cells continues (Murray and Robbins, 1982a) and evidence of their function in producing new myotubes has been discussed (Schmalbruch *et al.*, 1991). Presumably different mechanisms control these various cellular events and it has been suggested that long-term disuse may well be a potent signal for satellite cell proliferative activity, while nerve terminal degeneration is associated with the earlier cellular events (Jones and Vrbová, 1974; Murray and Robbins, 1982a, b).

Earlier suggestions that increased DNA content could also have been detected after long-term denervation as a result of relative changes in muscle fibre diameter and muscle bulk also require consideration since there is no satisfactory explanation as to the fate of the original sub-sarcolemmal nuclear material in long-term denervated muscles.

Increased RNA content following denervation (Goldspink, 1976; Wicks and Hood, 1991) is in line with the observed increase and sub-sarcolemmal accumulation of ribosomes (Engle and Stonnington, 1974; Gauthier and Schaeffer, 1974). More recently, Little *et al.* (1982), using acridine orange staining to visualize RNA in denervated mouse muscle, found that, during the first 4–7 days after denervation, stain was concentrated in sub-sarcolemmal areas, thereafter spreading throughout the muscle fibre, especially in very atrophic fibres. Although in the EM studies only ribosomal RNA would have been detected, it is possible that the RNA detected by acridine orange could also have included transfer and messenger RNA. Little *et al.* (1982) also noted an early increase in pentose shunt enzymes, which would indicate increased production of ribose sugars for incorporation into new RNA.

Increased RNA following denervation is related to the numerous changes in proteosynthesis which accompany the many profound intracellular and membrane changes that occur in denervated muscles and which are discussed in later sections of this chapter. However, as shown in Figure 8.4 the observation that many such denervation changes can be inhibited by actinomycin D which blocks RNA synthesis (for example,

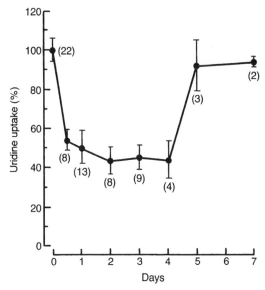

Figure 8.4 The effect of one dose of actinomycin D (0.5 mg/kg given on the day of denervation) on uridine uptake into mouse tibialis anterior muscle. Mean uptake by control muscles is expressed as 100% uptake; all other values are expressed relative to this. Figures in parentheses are numbers of experiments performed, and each point is plotted as mean ±SE. (Reproduced from Grampp, Harris and Thesleff, 1972, *J. Physiol.* **221**, 743–754, with permission of the authors.)

Grampp *et al.*, 1972) points to the importance of these early changes in RNA synthesis in denervated muscle.

Attempts to define the role of some biochemical changes that are thought to underlie muscle atrophy have been made. Hájek *et al.* (1964), finding an increase in proteolytic enzyme activity evident at 3 days after denervation and peaking at 10 days in both fast and slow rat muscles, suggested that increased enzyme activation is responsible for the processes that result in the development of atrophy. Increases in both acid and alkaline activated proteases had been detected prior to these studies (Koszalka and Miller, 1960) but the origin and role of such enzymes is still not fully understood. Acid proteases appear to have a role in muscle atrophy, and McLaughlin *et al.* (1974) detected changes in the acid protease cathepsin D activity in rat muscles reaching significance 3 days after denervation, at which time there was also loss of muscle wet weight. Results for acid proteolytic activity levels were similar to those of Hájek *et al.* (1964). Suggestions that some of these and other enzyme changes could be related to infiltration by non-muscle cells (Jones and Vrbová, 1974) or to membrane-specific denervation changes (Pollack and Bird, 1968; Tågerud and Libelius, 1984) have also been considered.

Whether muscle fibre atrophy following denervation is due to a loss of myofibrillar material resulting from an increased rate of degradation (Goldberg *et al.*, 1974) or decreased protein synthesis (Klemperer, 1972; Goldspink, 1976), or possibly both, also requires clarification. A decrease in active ribosomes in denervated muscle (Klemperer, 1972) seen both early and late in denervation indicates reduced proteosynthesis but was not specific enough to demonstrate whether this was related to fibre atrophy and myofilament loss.

Goldspink (1976) examined this question in young rats whose muscles were still growing. In slow soleus muscles, denervation resulted in loss of muscle mass in the presence of both decreased proteosynthesis and increased protein breakdown. Similar changes occurred initially in the fast extensor digitorum longus (EDL) muscle but the reduction in proteosynthesis was later reversed, resulting in restored levels of proteosynthesis which persisted. The net result was that young EDL muscles (and to a lesser extent soleus muscles) continued to grow despite being denervated, but at a much slower rate than innervated controls. Activity seems to influence this and R. Jones (unpublished) found that young denervated EDL muscles of the rat gained weight more rapidly if they were stimulated electrically (Figure 8.5). Goldspink (1976) suggests that some muscles may be subjected to greater degrees of stretch than others following section of an entire nerve trunk such as the sciatic nerve, and that factors such as degree of stretch should be taken into account when evaluating changes in proteosynthesis (also see Simard *et al.*, 1982).

Not all muscles undergo atrophy immediately following denervation. Cutting the phrenic nerve to one hemidiaphragm in the rat initially results in hypertrophy rather than atrophy (Gutmann *et al.*, 1966). The denervated hemidiaphragm may be stretched by the still active hemidiaphragm, modifying the expected changes in proteosynthesis and proteolysis following denervation in this preparation and masking atrophy. It is interesting that, even though the denervated hemidiaphragm may undergo hypertrophy, its ability to produce force was impaired in a similar manner to that expected in any denervated atrophying muscle (Gutmann *et al.*, 1966). It has also been known for some time that the multiply innervated tonic muscles of the chick respond to denervation by becoming hypertrophic (Feng *et al.*, 1963) and failure to detect any atrophy in the tonic musculature of the orbit in mammals following section of the oculomotor nerve (Assmussen and Kiessling, 1975; Christiansen *et al.*, 1992) indicates that this is a general response of tonic muscles and is not species-dependent. Inactivity induced by blocking neuromuscular transmission also produces hypertrophy in chick slow tonic muscles (Connold *et al.*, 1993) and alterations in myosin isoform expression similar to those seen following denervation (Kamel-Reid *et al.*, 1989).

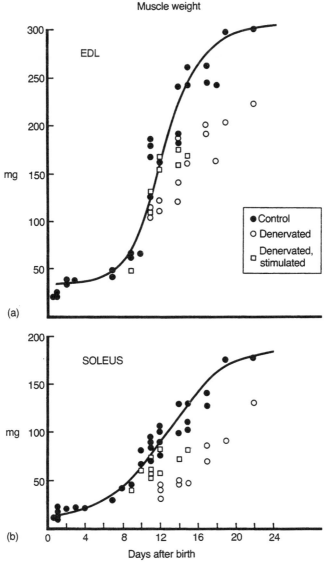

Figure 8.5 Wet weight of 1–22 days postnatal rat muscles: (a) extensor digitorum longus (EDL); (b) soleus. Normal muscles (•) are compared with denervated (○) and denervated stimulated (□) muscles. At 1–3 days, muscles were denervated by section of the sciatic nerve in the thigh. Stimulated muscles received electrical stimulation at 10 Hz for up to 6 hours daily. Denervated stimulated muscles grew as fast as normal muscles up to day 12, after which stimulation appeared to be less effective. Denervated soleus muscles did not appear to keep pace with normal muscles after day 6. Denervation did not prevent muscles from growing but denervated muscles did not grow as rapidly as innervated ones.

A rapid reduction in glycogen content in rat hindlimb muscles (Gutmann *et al.*, 1954) and other substrates for energy production (Bass, 1962) is an early consequence of inactivity following denervation in most species. It has also been shown that the compensatory 'overshoot' in glycogen content of innervated muscles following electrical stimulation fails to occur in denervated muscles (Gutmann *et al.*, 1954).

Denervation is known to induce insulin resistance (Turinsky, 1987) and, since glucose transport and glycogen synthesis are linked (Smith and Lawrence, 1984), it is possible that the changes in metabolic responses in denervated muscles are related to this change. The induction of insulin resistance is rapid, being evident by 24 hours and reaching maximum at around 3 days. Glucose transporter isoforms are also now known to be influenced by denervation and Block *et al.* (1991) have shown that expression of the muscle- and fat-specific glucose transporter GLUT–4 isoform is down-regulated on denervation, while GLUT–1, a non-specific glucose transporter, is significantly up-regulated. Insulin resistance in denervated muscle is attributed by these authors to a reduction in the expression of GLUT–4 linked to failure to suppress the expression of GLUT–1.

8.3.6 MUSCLE MEMBRANE CHANGES FOLLOWING DENERVATION

Even within the first few hours after cutting the motor nerve, sarco-lemmal membrane changes occur. The membrane properties that govern the electrical excitability of muscle fibres (DeSmedt, 1949; Beránek, 1962; Thesleff, 1974) are among the first to change, resulting in increased input resistance. There is an almost immediate fall in resting membrane potential which continues over the first few days (Albuquerque *et al.*, 1971; Card, 1977a; Kirsch and Anderson, 1986) before apparently recovering a little after approximately 3 weeks (Kirsch and Anderson, 1986). These findings are in accordance with earlier observations that metabolic alterations in denervated muscles lead to a reduced capacity to pump sodium and potassium ions across the muscle fibre membrane (Creese *et al.*, 1968). The fall in membrane potential is first seen at the endplate and spreads gradually to the rest of the muscle fibre (Albuquerque *et al.*, 1971), leading to the suggestion that acetylcholine (ACh) released by nerve terminals may be involved in regulating membrane properties. The spontaneous release of ACh, recognized as miniature endplate potentials (MEPPs), ceases after denervation with a time course shorter than that required for the development of the reduction of resting membrane potential (Card, 1977a; Rochel and Robbins, 1985). Depleting ACh in nerve terminals by stimulation (Card, 1977b) or blocking the action of released ACh on denervated muscle (Rochel and Robbins, 1985) does not alter the relative time lag between loss of MEPPs and the fall in membrane potential. This argues against the role of ACh in the regulation of membrane properties. However, it could be that non-quantal release of ACh known to occur at neuromuscular junctions may play a role. Whether or not ACh is implicated in the mechanisms of membrane change after denervation, it has regularly been observed that the length of nerve stump left attached to the muscles after denervation influences the time at which changes in membrane properties occur: the longer the nerve stump, the later such changes develop (Slater, 1966). Nerve terminal degeneration (Card, 1977b) and decline of ACh release and content are also dependent on nerve stump length.

Denervation results in a change in conductance of many ion species and, in addition to the better known alterations in Na^+ and K^+ conductance, interest has recently centred on changes in Cl^- conductance. Since Cl^- conductance accounts for about 80% of total conductance (Gm), a change in Cl^- conductance in denervated muscles might be of significance (Camerino and Bryant, 1976). The similarity between the increase in repetitive discharge of action potentials observed in denervated muscles by Lewis (1972) and the activity of myotonic muscles where low Cl^- conductance is a major feature further indicates the possible importance of changes in Cl^- conductance associated with denervation. In young rats, normal developmental changes in ion activities for both sodium and potassium (Ward and Wareham, 1985) and chloride (Conte-Camerino et al., 1989) fail to occur if the developing muscles are denervated. Observations on the complexity and diversity of chloride channels in skeletal muscles (Blatz and Magleby, 1989; Bretag, 1987) indicate the potential importance of Cl^- conductance not only for electrolyte balance and for fibre osmotic regulation but also in membrane excitability as in other excitable systems (for review, see Edwards, 1982). A further indication that an alteration in Cl^- conductance may be of significance in denervation is the finding that carbonic anhydrase activity is also greatly altered following denervation (Milot et al., 1991). Mechanisms that could influence both ion conductance and cell volume might be expected to be co-regulated.

There is a significant slowing of the rate of rise of the action potential and a reduction in overshoot (Albuquerque et al., 1971; Lewis, 1972; Redfern and Thesleff, 1971a; Thesleff, 1974; Kirsch and Anderson, 1986) with concomitant changes in electrical excitability and the development of 'anode break' responses (Marshall and Ward, 1974).

The possibility that altered Na^+ channel properties develop after denervation was explored by Kirsch and Anderson (1986) using voltage clamp techniques. They found no difference in the voltage dependence of steady state Na^+ channel activation between normal and denervated rabbit muscle in the presence of other denervation changes (e.g. reduced membrane potential, fibrillation activity) but Na^+ channels in denervated fibres were resistant to the prolonged depolarization that inactivates channels in most normal muscle fibres. There was also a difference in the rate of recovery from fast inactivation in denervated muscles. Similar studies in rat muscles (Pappone, 1980) showed a small difference in the voltage dependent activity of Na^+ channels in denervated muscles compared with control muscles. However, such studies are technically difficult, and slight methodological differences or species differences could explain some discrepancies between these studies.

Denervation results in the appearance of low-affinity binding sites for sodium channel markers similar to those seen in developing muscle (Frelin *et al.*, 1984; Hansen-Bay and Strichartz, 1980; Bambrick and Gordon, 1987). According to Bambrick and Gordon (1987), this accompanies the low density of high-affinity sites also seen following denervation.

The development of changes in Na^+ channel characteristics following denervation has been studied using specific channel blockers. Tetrodotoxin (TTX), a toxin which is known to block the action potential in normal muscles by its action on Na^+ channels, does not block action potential generation in denervated muscles (Redfern and Thesleff, 1971b; Harris and Thesleff, 1971) (Figure 8.6).

This was attributed to the development of TTX-resistant action potentials on the extrajunctional regions of the muscle fibre membrane. TTX-sensitive sodium channels contain subunits typical for the isoform now designated SKM1, and TTX-insensitive channels for the isoform SKM2. Both isoforms are expressed or re-expressed in denervated muscles (see also Colquhoun *et al.*, 1974) but recent studies have shown that, while SKM2 is barely detectable in innervated muscles, the presence of mRNA for SKM2 rises rapidly on denervation and that for SKM1 declines temporarily (Yang *et al.*, 1991), which is consistent with a decline in sodium channel density. Production and incorporation of the new sodium channel isoforms into the muscle fibre membrane takes place over the first few days after denervation, mirroring the development of TTX-resistant action potentials.

These recent studies support the implication of earlier findings that actinomycin D, which prevents the appearance of mRNA, also greatly delays the development of TTX resistance (Grampp *et al.*, 1972) and inhibits the onset of fibrillation activity (Muchnik *et al.*, 1973). However, TTX will block fibrillation activity later (Redfern and Thesleff, 1971b; Harris and Thesleff, 1971), implying that fibrillation is linked to new SKM1 rather than SKM2 channels.

It is a very old clinical observation that denervated muscles show intermittent contractile activity known as **fibrillation** (Denny-Brown and Pennybacker, 1938). Like the changes in membrane excitability, fibrillation in denervated muscles appears within the first few days (Tower,

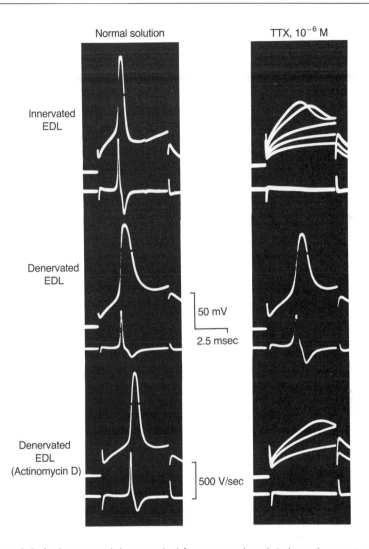

Figure 8.6 Action potentials recorded from normal and 4-days-denervated mouse EDL muscles and the effect of tetrodotoxin (TTX) 10^{-6M}. In the presence of TTX, action potential generation is blocked in normal muscles and in denervated muscles from animals treated by actinomycin D. Action potential generation in denervated muscles from untreated animals in only partially blocked. Actinomycin D was given 1 day after denervation. (Reproduced from Grampp, Harris and Thesleff, 1972, *J. Physiol.* **221**, 743–754, with permission of the authors.)

1939; Cannon and Rosenblueth, 1949; Hník and Skorpil, 1962; Thesleff, 1963; Belmar and Eyzaguirre, 1966; Kirsch and Anderson, 1986). Using intracellular recordings, Purves and Sackmann (1974) and Thesleff and

Ward (1975) observed spontaneous biphasic membrane potential oscillations, each muscle action potential being followed by a small after-hyperpolarization recovery which acts as a pre-potential to the next spontaneous fibrillation potential, thus setting up trains of spikes. Like Belmar and Eyzaguirre (1966), Purves and Sackmann (1974) found that some of the potentials seemed to be initiated at the site of old endplate regions, but also observed action potentials originating at extrajunctional sites. The failure of repolarization following denervation was ascribed to less efficient potassium transport, with prolonged duration of the outward potassium current during action potential generation but with higher conductance during after-hyperpolarization (Thesleff, 1962). The relationship between membrane changes and the development of fibrillation potentials is still in question, but Thesleff and Ward (1975) found that the membrane characteristics of denervated muscles were well correlated with the critical level for the generation of action potentials whose rate of rise reached maximum at the critical level for the generation of fibrillation potentials. They suggested that this was because sodium channel inactivation was also maximal at maximal rate of rise of the action potential. Kirsch and Anderson (1986) found that, while steady state and fast inactivation closing times of channels were not significantly altered, the rate of opening of fast channels was prolonged. They suggest that altered sodium gate activities may also be implicated in the development of fibrillation activity, especially since removing sodium or blocking sodium channels with TTX abolishes fibrillation activity (Purves and Sackmann, 1974).

8.3.7 CHANGES IN THE DISTRIBUTION AND TYPE OF ACETYLCHOLINE RECEPTORS AND ACETYLCHOLINESTERASE FOLLOWING DENERVATION

Since the initial discovery by Dale *et al.* (1936) that the influence of acetlycholine (ACh), which had previously been observed to cause contractions in skeletal muscle (Langley, 1907), was due to its action as the chemical transmitter between nerve and muscle, numerous studies have clarified the complex interactions between motor nerves and the regulation of muscle chemosensitivity.

One of the earliest observations concerning the influence of the nerve on the regulation of chemosensitivity in mammalian skeletal muscles was the finding by G. L. Brown (1937) that, while muscles with intact nerves do not respond to acetylcholine injected close arterially, following section of the motor nerve such injections cause strong contractions of nearby denervated muscles. *In vitro*, graded contractions in response to different concentrations of topically applied ACh appear (Elmquist and Thesleff, 1960) and the time course of the development of the response to topically applied ACh has been established (Jones and Vrbová, 1974). Application of drugs to cell surface membranes using small pipettes (Nastuk, 1953) enabled many to confirm the proposal of Ginetzinski and Shamarina (1942) that denervation hypersensitivity corresponds to a change in the distribution of ACh receptors (AChR). Normally ACh receptors are confined to the endplate region of adult muscle fibres. After denervation they are found over the entire muscle surface (Figure 8.7) (Axelsson and Thesleff, 1957; Miledi, 1960a; Albuquerque and McIsaac, 1969; Dreyer and Peper, 1974). Albuquerque and McIsaac (1969) suggest that the increase in extrajunctional ACh receptors may start near the endplate and, with time, spread to cover the entire muscle surface. Precise mapping of the distribution and density of ACh receptors – using an agent that binds to the AChR, alpha-bungarotoxin (α Butx), isolated from the venom of the krait snake – has been used to provide information on the distribution and turnover rate of ACh receptors in denervated muscles (see Bambrick and Gordon, 1994). Fambrough and others (Fambrough, 1970, 1979; Hartzel and Fambrough, 1973; Brockes and Hall, 1975; Merlie et al., 1984) confirmed that extrajunctional ACh receptors were newly synthesized and incorporated into the non-endplate regions of the muscle fibre membrane. The turnover rate of these new receptors was faster than that at the normal endplate region (Berg and Hall, 1974; Chang and Huang, 1974). Recent studies have identified nicotinic ACh receptors with a turnover rate of approximately 8 days (so-called 'slow' or sAChR) and others with a very rapid half-life of 1 day (rAChR). Both are present in denervated muscle fibres, but slow turnover receptors are probably associated with the endplate region of the muscle fibres (Salpeter et al., 1992).

Control of AChR is linked to the expression of regulatory proteins associated with the commitment of developing mesenchymal cells to muscle phenotype. The expression of myogenin mRNA and MyoD (see also Chapters 1 and 2) in denervated muscles indicates that denervation permits the re-expression of features characteristic of immature muscles. MyoD mRNA is increased as early as 16 hours following denervation, and is maximal at 48 hours. This time course closely follows the observed course for the increase in sensitivity to ACh and the appearance of extrajunctional AChR. On the other hand, My–5, does not change until some time after the peak increase in AChR and is thus not thought to participate in the regulation of ACh receptors. Myogenin and MyoD bind directly to AChR alpha and gamma receptor subunit enhancers, further implicating them in the regulation of the gene that controls the synthesis of new receptors. As shown in Figure 8.7 both innervation and electrical muscle stimulation dramatically down-regulate their expression (Eftimie et al., 1991; Witzemann and Sackmann, 1991; see Salpeter, 1992) and this presumably accounts for the observation that development of increased chemosensitivity can be reversed in denervated muscle by electrical stimulation (Jones and Vrbová, 1971, 1974; Lømo and Rosenthal, 1972; Drachmann and Witzke, 1972).

Figure 8.7 illustrates that the high concentration of AChR normally present at the neuromuscular junction persists after denervation in the rat (Lømo and Rosenthal, 1972; Frank *et al.*, 1976) and it has been suggested that junctional sub-sarcolemmal nuclei have a special role in maintaining endplate chemosensitivity as well as other post-synaptic features (Brenner *et al.*, 1990). However, the turnover rate of receptors at the neuromuscular junction is affected by denervation (Stanley and Drachmann, 1981; Bevan and Steinbach, 1983). Both rapid and slow turnover AChR are present at the endplate region in denervated muscle fibres, but the half-life of slow turnover receptors at the neuromuscular junction is also reduced compared with that at innervated endplates (Salpeter *et al.*, 1992). Shyng *et al.* (1991) have found that cyclic AMP stabilizes junctional ACh receptors in denervated muscles probably by reversing the increased turnover rate of sAChRs (see Levitt and Salpeter, 1981; Salpeter *et al.*, 1992). The presence of fast turnover receptors at the denervated neuromuscular junction may represent the presence at the endplate region of newly synthesized receptors with a fast turnover rate similar to those that appear throughout denervated muscles. However, there may also be a fast turnover population of receptors at the normally innervated neuromuscular junction (Stanley and Drachmann, 1983) and those seen after denervation could simply represent a continuing process of the production and incorporation of different receptor types. Sub-sarcolemmal nuclei close to the junctional region appear to have some more specific role than other muscle nuclei in regulating the production and incorporation of a range of synapse-associated molecules.

The physical anchoring of ACh receptors at the neuromuscular junction may explain the persistence of high density of receptors with fast channel opening times at the endplate for some time after denervation (Levitt and Salpeter, 1981). Although stimulation of denervated muscles reduces the level of extrajunctional ACh receptors (Jones and Vrbová, 1971, 1974; Lømo and Rosenthal, 1972; Drachman and Witzke, 1972), endplate receptors appear unaffected by stimulation (Figure 8.7), retaining their high density (Lømo and Rosenthal, 1972). Some extracellular matrix molecules have been implicated in the control of the distribution of AChR. One such molecule, agrin, appears to play a vital role in accumulation ACh receptors at very high density at the neuromuscular junction (see review by Nastuk and Fallon, 1993). This mechanism apparently persists for some time after section of the motor nerve, since muscles as well as neurones can synthesize agrin (Fallon and Gelfman, 1989).

It has long been known that acetylcholinesterase (AChE) activity is regulated by neural activity in skeletal muscles and that denervation

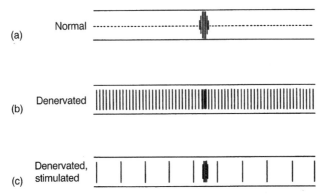

(a) Normal

(b) Denervated

(c) Denervated, stimulated

Figure 8.7 Distribution of acetylcholine (ACh) receptors in: (a) normal muscles; (b) denervated muscles; (c) denervated stimulated muscles. The density of the vertical lines represent the degree of sensitivity. The high density of ACh receptors seen at normal endplates is preserved following denervation, but in denervated muscles ACh receptors are also present along the entire muscle fibre length. Electrical stimulation of denervated muscles reduces the number of extrajunctional ACh receptors but does not affect the high density of receptors at the original endplate region.

results in a significant fall in AChE activity (Guth *et al.*, 1981). Different molecular forms of AChE can be determined in normal muscles (reviewed by Massoulie *et al.*, 1993) and asymmetric forms, especially 16S, are prominent at the endplate in both fast and slow muscles, the levels being higher in slow than in fast muscle endplates. Levels of 16S AChE fall precipitously on denervation (Sketelj *et al.*, 1994) although differences in levels of 16S AChE between fast and slow muscles are preserved. In the fast EDL muscle of rats, the 10S form of AChE becomes the predominant extrajunctional form following denervation, but soleus continues to express the 16S form at both junctional and extrajunctional sites (Sketelj *et al.*, 1994). Similar changes are seen in inactivated muscles, indicating that both cholinesterase level and type is regulated by activity.

8.3.8 NEUROMUSCULAR ACTIVITY AND CELLULAR RESPONSES IN DENERVATED MUSCLES

It has been repeatedly proposed that contractile properties of muscles are regulated by a 'trophic' influence exerted by the motor nerve, mediated independently of activity (Miledi, 1960b; Thesleff, 1960; Guth *et al.*, 1981), and it is also suggested that a similar mechanism may control surface membrane properties. The concept of a source of 'trophic' material within the nerve seems to be supported by the still unexplained finding that, following denervation, membrane changes occur sooner if the nerve is cut close to the muscle than if nerve section is more distant (Harris and

Thesleff, 1971; Guth *et al.*, 1981). In both circumstances nerve activity ceases upon nerve section, but it has been postulated that a longer nerve stump may contain more 'trophic' substance which would maintain normal muscle properties for longer. However, it is known that the time at which neuromuscular transmission fails also depends on the length of the peripheral nerve stump. Thus transient changes of membrane potential at the neuromuscular junction would be present for longer when the peripheral stump is longer, and these may be responsible for the results obtained on muscles with nerve stumps of different sizes. Moreover, no specific trophic substance has ever been isolated. Although some experiments seem to indicate that soluble extracts from nerve can help to maintain some muscle properties (e.g. Davis and Kienan, 1981), these studies did not test the influence of cytosolic extracts from any other source.

In a different group of studies, attempts were made to resolve the question by blocking the activity in axons, nerve terminals or muscle fibres without disconnecting them from muscles. Cuffs or pellets containing local anaesthetics (Roberts and Oester, 1970; Blunt and Vrbová, 1974; Lorkovic, 1975), tetrodotoxin (Pestronk *et al.*, 1976) or injections of botulinum toxin (Thesleff, 1960; Bambrick and Gordon, 1987) or bungarotoxin (Berg and Hall, 1974) were used to inactivate innervated muscles. All of these strategies produced hypersensitivity to ACh but, in general, this was usually less than following denervation. Roberts and Oester (1970) failed to induce increased sensitivity to ACh in muscles blocked by local anaesthetics, but their results were explained by the finding that local anaesthetics either diffuse away rapidly or cause nerve damage and subsequent denervation (Blunt and Vrbová, 1975; Lorkovic, 1975). Bambrick and Gordon (1987) have shown that in muscles whose nerves were blocked by botulinum toxin, where inactivity equivalent to that produced by denervation was induced, membrane changes equivalent to those seen following denervation could be obtained. These studies support the thesis that it is neurally induced muscle activity that maintains at least some normal features of adult muscle fibre membrane.

In a different approach, attempts were carried out to block the transport along the motor nerve of a putative 'trophic' factor while not arresting impulse activity. Colchicine or vincristine, which disrupt neurotubules, were applied to the nerve in cuffs and when ACh receptors appeared at extrajunctional sites following this treatment it was argued that this provided evidence for chemical trophic influence in muscle (Albuquerque *et al.*, 1972). However, such drugs are also known to have a direct effect on muscle fibres and transmitter release (Cangiano and Fried, 1977), throwing into question the earlier interpretations on nerve transport blockade. Moreover, by blocking axoplasmic transport, the maintenance of the integrity of the nerve terminal is likely to be compromised and many muscle fibres will be denervated. Thus it is impossible from these studies to resolve the question.

Partial mechanical inactivation by fixation or suspension also causes increased sensitivity to ACh (Solandt and Magladery, 1942; Johns and Thesleff, 1961; Fischback and Robins, 1971) and these findings point to the loss of muscle activity as a factor in denervation hypersensitivity. Many studies have now shown that electrical stimulation can reduce the extrajunctional sensitivity of denervated muscles (Jones and Vrbová, 1971, 1974; Drachman and Witzke, 1972; Lømo and Rosenthal, 1972) but very high rates of activity were reported to be needed by Lomo and Rosenthal (100 Hz) and prolonged activity at such rates cannot be regarded as physiological. Jones and Vrbová (1974) failed to reduce the early development of hypersensitivity by stimulation and attempted to resolve the debate by postulating that at least two (and possibly more) factors interact to provide the membrane signals of denervation changes, loss of muscle activity and factors associated with the presence of degenerating nerve. It is well documented that denervated muscles undergo profound cellular changes which include the proliferation of satellite cells (Ontell, 1974; Murray and Robbins, 1982a,b) infiltration by blood-borne cells (Jones and Lane, 1975; Murray and Robbins, 1982a,b) and accumulation of phagocytic and fibroblast-like cells at the endplate region (Connor and McMahan, 1987). Involvement of such diverse cell types probably indicates that the significance of the proliferation or activation of each cell group is related to differing factors which may be directly or indirectly linked to other denervation changes. As an example, cellular infiltration throughout the muscle and accumulation at the endplate region appear to be associated with nerve stump degeneration since the timing is dependent on nerve stump length (Arancio et al., 1992) and the time course is closely linked to the time course of nerve degeneration (Jones and Vrbová, 1974; Murray and Robbins, 1982a,b; Connor and McMahan, 1987), while satellite cells continue to proliferate actively as long as the muscle is inactive (Murray and Robbins, 1981b; see also Schmalbruch et al., 1991). Such findings lead to the suggestion that other post-denervation changes such as the development of extrajunctional chemosensitivity may also be linked to these cellular events (Jones and Vrbová, 1974). This possibility was supported by findings (illustrated in Figure 8.8) that a section of degenerating nerve placed on the surface of an innervated and active muscle induces a local area of high ACh sensitivity (Vrbová, 1970; Jones and Vrbová, 1974; Jones and Vyskočil, 1975).

In addition, Cangiano and Lutzemberger (1977) showed that the innervated muscle fibres of partially denervated muscles undergo temporary denervation-like changes at a time when the cellular responses are at their height. Brown et al. (1991) found that in the C578L/Ola

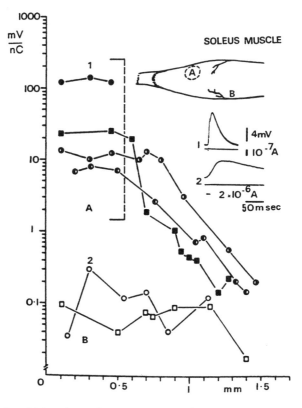

Figure 8.8 Sensitivity of extrajunctional areas of rat soleus muscle fibres to locally applied acetylcholine (ACh). Three days prior to recording, a small piece of brachial nerve was placed on the surface of the innervated soleus muscle (see A in insert) and allowed to degenerate; on the graph, the area under the degenerating nerve is indicated by the broken vertical line (A). As the recording and ACh application pipettes were moved through the area under the degenerating nerve, high levels of sensitivity to ACh were recorded. Outside this area, on the same fibres, sensitivity was the normal low extrajunctional sensitivity characteristic of the soleus muscle, as is also illustrated by the lower two plots on the graph (B). The rise time of potentials recorded under the degenerating nerve is also much faster than that recorded outside this region (see lower insert). Abscissa represents distance along the muscle fibres; ordinate is ACh sensitivity expressed as mV/nC. (After Jones and Vyskočil (1975), *Brain Res.*, **88**, 309–17.)

mouse, where nerve degeneration following nerve section is retarded, the development of denervation changes is also retarded.

The relationship between nerve stump length and the timing of the onset of denervation changes has long been regarded as the most powerful argument for the presence of a trophic factor released by the nerve, since it was postulated that the later development of denervation changes associated with a longer nerve stump indicated that more trophic factor would be left to maintain muscle properties. However, it is also clear that the timing of nerve stump degeneration changes are related to the level of nerve section (Arancio *et al.*, 1992). The interesting relationship between nerve degeneration, inflammatory cellular infiltration and surface membrane changes in both nerve and the muscle injury deserves further consideration (see Gordon, Jones and Vrbová, 1976).

As to the significance of these changes, it has been suggested that some may be related to the maintenance of features essential for the restoration of neuromuscular connections. The establishment of synapses in developing muscles requires the development of membrane sites containing high densities of ACh receptors, the expression of adhesion molecules and specialization of the basal lamina (for reviews see Bloch and Pumplin, 1988; Hall and Sanes, 1993). Following denervation the elaboration of the post-synaptic membrane is retained for a short time, as is a high density of AChR at the endplate region. However, there is thought to be a requirement for guidance of ingrowing axon terminals to the original endplate sites (Sanes *et al.*, 1980) and to provide suitable substrates for adhesion and possibly to signal to ingrowing axons that they have reached their target. Those adhesion molecules thought to be related to the early development of synapses such as N-CAM, Li, Nile and others, as well as specialized macromolecules of the basal lamina such as laminin, fibronectin and heparin sulphate proteoglycans, are also expressed in denervated muscles (for reviews see Sanes *et al.*, 1986; Hall and Sanes, 1993; Covault *et al.*, 1986). The potential importance of such molecular reorganization is suggested by the finding that ingrowing axons can form connections and initiate synaptic transmission even when underlying muscle fibres are destroyed, provided that the basal lamina remains intact and is able to express appropriate potential signalling markers such as laminin (Sanes *et al.*, 1980; Sanes and Chui, 1983).

Although these surface molecules in denervated muscle may have a role in the recovery process, their expression appears to be a more general response of the muscle fibre, similar to that produced by injury in all cells. In addition to the adhesion molecules described above, the muscle after denervation also expresses MHC1 class on ligands (Maehlen *et al.*, 1989) and it could be that this molecule is responsible for the accumulation of blood-borne mononucleated cells. These cells in turn are likely to release a number of cytokines that affect the muscle membrane and induce the observed changes described in this chapter.

What preserves or initiates the re-expression of these surface membrane and basal lamina markers is still unclear but it has been suggested that the accumulation of fibroblast-like cells at the denervated endplate could represent the sort of mechanism that would be required, since fibroblasts are known to produce such adhesion molecules (e.g. see

Connor and McMahan, 1987). Although there are still many unanswered questions concerning the precise mechanisms involved in cellular and molecular responses to denervation, there are now many clues as to how apparently unreconcilable findings may eventually turn out to be linked. Further, these findings indicate that denervated muscles, far from being passively atrophying tissues, seem to undergo a number of adaptive changes designed to provide the best possible environment to regain innervation and restore function.

CONCLUSIONS

When the axon of a motor nerve is cut extensive changes occur in the distal and proximal stumps of the axon, as well as in the motoneurone cell body. The distal stump undergoes Wallerian degeneration, and the cellular changes associated with this process are described. The role of Schwann cells, macrophages and various molecules that typically appear during Wallerian degeneration are considered, as well as the importance of interrupted supply of nutrients from the cell body for the survival and function of the separated distal nerve stump. In the adult vertebrate cell bodies of axotomized motoneurones maintain their ability to synthesize proteins and can survive the insult for long periods of time. The cell body undergoes a number of important changes, all directed towards the conversion of a cell specialized for synthesizing transmitter related proteins to a cell specializing in growth and synthesizing molecules associated with this new function. Thus the synthesis of molecules associated with growth is up-regulated and that of proteins associated with transmitter synthesis and release is down-regulated. The motoneurone cell body as well as its dendrites are also affected by axotomy so that both their morphology and physiological properties change. These changes are accompanied by alterations of synaptic inputs. The response of the motoneurone to axotomy is discussed in terms of a two-state model, i.e. the transition from a transmitting to a growing cell.

The muscle too responds dramatically to the separation from its nerve supply. The most apparent change is the rapid atrophy, which occurs after denervation and cessation of impulse traffic, i.e. muscle activity. Degenerative changes at the neuromuscular junction and the cellular response initiated by the degenerating terminals are accompanied by a number of alterations of the muscle fibre membrane, and other structural and metabolic characteristics of the muscle fibre. The mechanisms that lead to the profound changes of denervated muscle fibres are discussed and the role of loss of contractile and impulse activity for the induction of these changes evaluated.

REFERENCES

Acheson, G. H., Lee, E. S. and Morison, R. S. (1942) A deficiency in the phrenic respiratory discharges parallel to retrograde degeneration. *J. Neurophysiol.* **5**, 269–273.

Aguayo, A. J., Epps. J., Charron, L. and Bray, G. M. (1976) Multipotentiality of Schwann cells in cross-anastomosed and grafted myelinated and unmyelinated nerves: quantitative microscopy and radioautography. *Brain Res.* **104**, 1–20.

Aguayo, A. J. (1985) Axonal regeneration from injured neurons in the adult mammalian central nervous system, in *Synaptic Plasticity* (ed. C. W. Cotman), Guildford, New York, pp. 457–484.

Al-Amood, W. S., Lewis, D. M. and Schmalbruch, H. (1991) Effects of chronic electrical stimulation on contractile properties of long term denervated rat skeletal muscle. *J. Physiol. (Lond.)* **441**, 243–256.

Albuquerque, E. X. and Thesleff, S. (1968) A comparative study of membrane properties of innervated and chronically denervated fast and slow skeletal muscles of the rat. *Acta Physiol. Scand.* **73**, 471–480.

Albuquerque, E. X. and McIsaac, R. J. (1969) Early development of acetylcholine receptors on fast and slow mammalian skeletal muscle. *Life Sciences* **8**, 409–416.

Albuquerque, E. X. and McIsaac, R. J. (1970) Fast and slow muscles after denervation. *Exp. Neurol.* **26**, 183–202.

Albuquerque, E. X., Schuh, F. T. and Kauffman, F. C. (1971) Early membrane depolarisation of fast mammalian muscle after denervation. *Pflügers Archiv.* **328**, 36–50.

Albuquerque, E. X., Warnick, J. E., Tasse, J. R. and Sansone, F. M. (1972) Effects of vinblastine and colchicine on neural regulation of the fast and slow skeletal muscles of the rat. *Exp. Neurol.* **37**, 607–634.

Allt, G. (1976) Pathology of the peripheral nerve, in *The Peripheral Nerve* (ed. D. N. Landon), Chapman & Hall, London, pp. 666–739.

Anderson, L. L. and McClure, W. O. (1973) Differential transport of protein in axons: comparison between the sciatic nerve and dorsal columns of cats. *Proc. Natl. Acad. Sci. USA* **70**, 1521–1525.

Anzil, A. P. and Wernig, A. (1989) Muscle fibre loss and reinnervation after long term denervation. *J. Neurocytol.* **18**, 833–845.

Arancio, O., Buffelli, M., Cangiano, A. and Pasino, D. (1992) Nerve stump effects in muscle are independent of synaptic connections and are temporarily correlated with nerve degeneration phenomena. *Neurosci. Lett.* **146**, 1–4.

Assmussen, G. and Kiessling, A. (1975) Hypertrophy and atrophy of mammalian extraocular muscle fibres following denervation. *Experientia* **31**, 1186.

Austin, A. and Langford, C. J. (1980) Nerve regeneration: a biochemical view. *TINS* **3**, 130–132.

Axelsson, J. and Thesleff, S. (1957) A study of supersensitivity in denervated mammalian skeletal muscle. *J. Physiol.* **149**, 178–193.

Baichwal, R. R., Bigbee, J. W. and DeFries, G. H. (1988) Macrophage-mediated myelin-related mitogenic factor for cultured Schwann cells. *Proc. Natl. Acad. Sci. USA* **85**, 1701–1705.

Bambrick, L. and Gordon, T. (1987) Acetylcholine receptors and sodium channels in denervated and botulinum-toxin-treated adult rat muscle. *J. Physiol. (Lond.)* **382**, 69–86.

Bambrick, L. and Gordon, T. (1994) Neurotoxins in the study of neural regulation of membrane proteins in skeletal muscle *J. Pharmacol. Methods* (in press).

Bárány, M. (1967) ATP-ase activity of myosin correlated with speed of shortening. *J. Gen. Physiol.* **50**, Suppl. 2, 197–218.

Barron, K. D. (1983) Axon reaction and central nervous system regeneration, in *Nerve, Organ and Tissue Regeneration: research perspectives* (ed. F. J. Seil), Academic, New York, pp. 3–36.

Bass, A. (1962) Energy metabolism in denervated muscle, in *The Denervated Muscle* (ed. E. Gutmann), Czechoslovak Academy of Science Publishing House, Prague, pp. 203–264.

Belmar, J. and Eyzaguirre, C. (1966) Pacemaker site of fibrillation potentials in denervated mammalian muscle. *J. Neurophysiol.* **29**, 425–441.

Benowitz, L. I. and Routtenberg, A. (1987) A membrane phosphoprotein associated with neural development, axonal regeneration, phospholipid metabolism and synaptic plasticity. *Trends Neurosci.* **10**, 527–532.

Benowitz, L. I., Shashoua, V. E. and Yoon, M. G. (1981) Specific changes in rapidly transported proteins during regeneration of the goldfish optic nerve. *J. Neurosci.* **1**, 300–307.

Beránek, R. (1962) Electrophysiology of denervated muscle, in *The Denervated Muscle* (ed. E. Gutmann), Czechoslovak Academy of Science, Prague, pp. 123–133.

Berg, D. K. and Hall, Z. W. (1974) Fate of alpha-bungarotoxin bound to acetylcholine receptors of normal and denervated muscles. *Science* **184**, 473–475.

Beuche, W. and Friede, R. L. (1984) The role of non-resident cells in Wallerian degeneration. *J. Neurocytol.* **13**, 767–796.

Bevans, S. and Steinbach, J. H. (1983) Denervation increases the degradation rate of acetylcholine receptors and endplates *in vivo* and *in vitro*. *J. Physiol.* **336**, 159–177.

Birks, R., Katz, B. and Miledi, R. (1960) Physiological and structural changes at the amphibian myoneural junction in the course of nerve degeneration. *J. Physiol. (Lond.)* **150**, 145–168.

Bisby, M. (1980) Changes in the composition of labelled protein transported in motor axons during regeneration. *J. Neurobiol.* **11**, 435–455.

Bisby, M. A. (1988) Dependence of GAP–43 ((B50, F1) transport on axonal regeneration in rat dorsal root ganglion neurons. *Brain Res.* **458**, 157–161.

Bisby, M. A. and Tetzlaff, W. (1992) Changes in cytoskeletal protein synthesis following axon injury and during axon regeneration. *Mol. Neurobiol.* **6**. 107–123.

Black, M. M. and Lasek, R. J. (1979) Changes in cytoskeletal protein synthesis following axon injury and during axon regeneration. *Exp. Neurol.* **63**, 108–119.

Black, M. M. and Lasek, R. J. (1980) Slow components of axonal transport: two cytoskeletal networks. *J. Cell Biol.* **86**, 616–623.

Blatz, A. L. and Magleby, K. L. (1989) Adjacent interval analysis distinguishes among gating mechanisms for the fast chloride channel from rat skeletal muscle. *J. Physiol.* **410**, 561–586.

Blitzinger, K. and Kreutzburg, G. (1968) Displacement of synaptic terminals from regenerating motoneurones by microglial cells. *Z. Zellforsch. Mikrosk. Anat.* **85**, 145–147.

Bloch, R. J. and Pumplin, D. W. (1988) Molecular events in synaptogenesis: nerve–muscle adhesion and postsynaptic differentiation. *Am J. Physiol.* **23**, C354–364.

Block, N. E., Melnick, D. R. Robinson, K. A. and Buse, M. G. (1991) Effect of denervation on expression of 2-glucose transporter isoforms in rat hind limb muscles. *J. Clin. Invest.* **88**, 1546–1552.

Blunt, R. J. and Vrbová, G. (1975) The use of local anaesthetics to produce prolonged motor nerve block in the study of denervation hypersensitivity. *Pflügers Archiv.* **357**, 187–199.

Boeke, J. (1950) Nerve regeneration, in *Int. Conf. on the Development, Growth, and Regeneration of the Nervous System. Genetic Neurology* (ed. P. Weiss), University of Chicago Press, Chicago, pp. 78–91.

Bradley, W. G. (1974) *Disorders of Peripheral Nerves*, Blackwells, Oxford.

Bråttgard, S. O., Edström, J. E. and Hyden, H. (1957) The chemical changes in regenerating neurones. *J. Neurochem.* **1**, 316–325.

Bråttgard, S. O., Edström, J. E. and Hyden, H. (1958) The productive capacity of the neurone in retrograde reaction. *Expl. Cell. Res. Suppl.* **5**, 185–200.

Brechter, M. S. and Raff, M. C. (1975) Mammalian plasma membranes. *Nature* **258**, 43–49.

Brenner, H. R., Witzeman, V. and Sackmann, B. (1990) Imprinting of acetylcholine receptors messenger RNA accumulation in mammalian neuromuscular synapses. *Nature* **344**, 544–547.

Brenner, H. R. and Rudin, W. (1989) On the effects of muscle activity on the endplate membrane in denervated mouse muscle. *J. Physiol. (Lond.)* **410**, 501–512.

Bretag, A. H. (1987) Muscle chloride channels. *Physiol. Rev.* **67**, 618–724.

Brockes, J. P. and Hall, Z. W. (1975) Acetylcholine receptors in normal and denervated rat diaphragm muscle. Comparison of junctional and extrajunctional receptors. *Biochem. New York* **14**, 2100–2106.

Brown, G. L. (1937) The actions of acetylcholine on denervated mammalian and frog muscle. *J. Physiol.* **89**, 438–461.

Brown, M. C., Perry, V. H., Lunn, E. R. *et al.* (1991) Macrophage dependence of peripheral sensory nerve regeneration: Possible involvement of Nerve Growth Factor. *Neuron* **6**, 359–370.

Brown, M. C., Lunn, E. R. and Perry, V. H. (1992) Consequences of slow Wallerian degeneration for regenerating motor and sensory axons. *J. Neurobiol.* **23**, 521–536.

Bunge, M. B. (1986) The axonal cytoskeleton: Its role in generating and maintaining cell form. *Trends Neurosci.* **9**, 477–482.

Camerino D. and Bryant, S. H. (1976) Effects of denervation on the chloride conductance of rat skeletal muscle fibres. *J. Neurobiol.* **7**, 221–228.

Campbell, B. (1944) The effects of retrograde degeneration upon reflex activity of ventral horn neurones. *Anat. Rec.* **88**, 25–37.

Cangiano, A. and Fried, J. A. (1977) The production of denervation like changes in rat muscle by colchicine without interference with axonal transport or muscle activity. *J. Physiol.* **265**, 63–84.

Cangiano, A. and Lutzemberger, L. (1977) Partial denervation affects denervated and innervated fibres in mammalian muscle. *Science* **196**, 542–544.

Cannon, W. B. and Rosenblueth, A. (1949) *Supersensitivity of Denervated Structures*, Macmillan, London.

Card, D. J. (1977a) Physiological alterations of rat extensor digitorum longus motor nerve terminals as a result of surgical denervation. *Exp. Neurol.* **54**, 478–488.

Card, D. J. (1977b) Denervation: sequence of neuromuscular degenerative changes in rats and the effect of stimulation. *Exp. Neurol.* **54**, 251–265.

Carraro, U., Catani, C. and Dalla Libera, L. (1981) Myosin light and heavy chains in rat gastrocnemius and diaphragm muscle after chronic denervation or reinnervation. *Exp. Neurol.* **72**, 401–412.

Carraro, U., Morale, D., Mussini, I. *et al.* (1985) Chronic denervation of rat hemidiaphragm: maintenance of fibre heterogeneity with associated increasing uniformity of myosin isoforms. *J. Cell Biol.* **100**, 161–174.

Chang, C. C. and Huang, M. (1974) Turnover of junctional and extrajunctional acetylcholine receptors of the rat diaphragm. *Nature* **253**, 643–644.

Cheah, T. B. and Geffen, L. B. (1973) Effects of axonal injury on nor-epinephrine, tyrosine hydroxylase and monoamine oxidase levels in sympathetic ganglia. *J. Neurobiol.* **4**, 443–452.

Christiansen, S. P., Baker, R. S., Maher, M. and Terrell, B. (1992) Type-specific changes in fibre morphology following denervation of canine extraocular muscles. *Exp. Mol. Path.* **56**, 87–95.

Clemence, A., Mirsky, R. and Jessen, K. R. (1989) Non-myelin-forming Schwann cells proliferate rapidly during Wallerian degeneration in the rat sciatic nerve. *J. Neurocytol.* **18**, 185–192.

Cleveland, D. W. and Hoffman, P. N. (1991) Neuronal and glial cytoskeletons. *Cur. Op. Neurobiol.* **1**, 346–353.

Close, R. (1972) Dynamic properties of mammalian skeletal muscles. *Physiol. Revs.* **52**, 129–197.

Colquhoun, D., Rang, H. P. and Ritchie, J. M. (1974) The binding of tetrodotoxin and α-bungarotoxin to normal and denervated mammalian muscle. *J. Physiol.* **240**, 199–226.

Connold, A. L., Kamel-Reid, S., Vrbová, G. and Zak, R. (1993) Inactivity induces hypertrophy and redistribution of myosin isoenzymes in chicken anterior latissimus dorsi muscle. *Pflügers Arch.* **423**, 34–40.

Connor, E. A. and McMahan, U. J. (1987) Cell accumulation in the junctional region of denervated muscle. *J. Cell Biol.* **104**, 109–120.

Conte-Camerino D., DeLuca, A., Mambrini, M. and Vrbová, G. (1989) Membrane ion conductances in normal and denervated skeletal muscle of the rat during development. *Pflügers Arch. Euro. J. Physiol.* **413**, 568–570.

Covault, J., Merlie, J. P., Goridis, C. and Sanes, J. R. (1986) Molecular forms of N-CAM and its RNA in developing and denervated skeletal muscle. *J. Cell Biol.* **102**, 731–739.

Cragg, B. G. (1970) What is the signal for chromatolysis? *Brain Res.* **23**, 1–21.

Cragg, B. G. and Thomas, P. K. (1961) Changes in conduction velocity and fibre size proximal to peripheral nerve lesions. *J. Physiol.* **157**, 315–327.

Creese, R., El-Shafie, A. L. and Vrbová, G. (1968) Sodium movements in denervated muscle and the effects of antimycin A. *J. Physiol.* **197**, 279–294.

Cuenod, M., Boesch, J., Marko, P. *et al.* (1972) Contributions of axoplasmic transport to synaptic structures and functions. *Int. J. Neurosci.* **4**, 77–87.

Czeh, G., Gallego, R., Kudo, N. and Kuno, M. (1978) Evidence for the maintenance of motoneurone properties by muscle activity. *J. Physiol.* **281**, 239–252.

Dahlström, A. (1965) Observations on the accumulations of noradrenaline in the proximal and distal parts of peripheral adrenergic nerves after compression. *J. Anat.* **99**, 677–689.

Dahlström, A. (1967) The transport of nor-adrenaline between two simultaneously performed ligations of the sciatic nerves of rat and cat. *Acta Physiol. Scand.* **69**, 158–166.

Dahlström, A. (1971) Axoplasmic transport (with particular respect to adrenergic neurones). *Phil. Trans. Roy. Soc. London, Ser. B* **261**, 325–358.

Dale, H. H., Feldberg, W. and Vogt, M. (1936) Release of acetylcholine at voluntary motor nerve endings. *J. Physiol.* **86**, 353–380.

Davis, H. L. and Kienan, J. A. (1981) Effects of nerve extract on atrophy of denervated or immobilised muscles. *Exp. Neurol.* **72**, 582–591.

Davis, L. A., Gordon, T., Hoffer, J. A. *et al.* (1978) Compound action potentials recorded from mammalian peripheral nerves following ligation or resuturing. *J. Physiol.* **285**, 543–559.

deFilipe, C., Jenkins, R., O'Shea, R. *et al.* (1993) The role of intermediate early genes in the regeneration of the central nervous system. *Adv. Neurol.* **59**, 263–271.

De Smedt, J. E. (1949) Les propriétés electrophysiologiques du muscle squelettique au cours de la dégénerescence Wallérienne, et dans le cas d'une atrophie non-Wallérienne (resection tenoineuse). *Archs. Int. Physiol.* **57**, 98–101.

Denny-Brown, D. (1929) On the nature of postural reflexes. *Proc. Roy. Soc. (Biol.)* **104**, 252–301.

Denny-Brown, D. and Pennybacker, J. B. (1938) Fibrillation and fasciculation in voluntary muscle. *Brain* **61**, 311–334.

Donat, J. R. and Wisniewski, H. M. (1973) The spatio-temporal pattern of Wallerian degeneration in mammalian peripheral nerves. *Brain Res.* **53**, 41–53.

Drachman, D. B. and Witzke, F. (1972) Trophic regulation of acetylcholine in muscle: Effect of chronic stimulation. *Science* **176**, 514–516.

Dreyer, F. and Pepper, K. (1974) The acetylcholine sensitivity in the vicinity of the neuromuscular junction in the frog. *Pflügers Arch.* **348**, 273–286.

Droz, B. (1975) Synthetic machinery and axoplasmic transport: maintenance of neuronal connectivity, in *The Nervous System* (ed. D. B. Tower), Vol. 1. The Basic Neurosciences, Raven Press, New York, pp. 111–127.

Droz, B., Koenig, H. C. and Di Giamberardino, L. (1973) Axonal migration of protein and glycoprotein in nerve endings. I. Radioautographic analysis of renewal of protein in nerve endings of chicken ciliary ganglion after intracerebral injection of [3H]-lysine. *Brain Res.* **60**, 93–127.

Droz, B., Rambourg, A. and Koenig, H. L. (1975) The smooth endoplasmic reticulum: structure and role in the renewal of axonal membrane and synaptic vesicles by fast axonal transport. *Brain Res.* **93**, 1–13.

Duchen, L. W. and Strich, S. J. (1968) The effects of botulinum toxin on the pattern of innervation of skeletal muscle in the mouse. *Quart. J. Exp. Physiol.* **53**, 84–89.

Dziegielewska, K. M., Evans, C. A. N. and Saunders, N. R. (1980) Rapid effect of nerve injury upon axonal transport of phospholipids. *J. Physiol.* **304**, 83–98.

Eccles, J. C. (1941) Changes in muscle produced by nerve degeneration. *Med. J. Aust.* **1**, 573–575.

Eccles, J. C., Libet, B. and Young, R. R. (1958) The behavior of chromatolysed motoneurones studied by intracellular recording. *J. Physiol.* **143**, 11–40.

Edwards, C. (1982) The selectivity of ion channels in nerve and muscle. *Neurosci.* **7**, 1335–1366.

Eftimie, R., Brenner, H. R. and Buonanno, A. (1991) Myogenin and MyoD join a family of skeletal muscle genes regulated by electrical activity. *Proc. Nat. Acad. Sci.* **88**, 1349–1353.

Elmquist, D. and Thesleff, S. (1960) A study of acetylcholine induced contractures in denervated mammalian muscle. *Acta Pharmacol. Toxicol.* **17**, 84–93.

Engh, C. A. and Schofield, B. H. (1972) A review of the central response to nerve injury and its significance in nerve regeneration. *J. Neurosurg.* **37**, 195–203.

Engle, A. G. and Stonnington, H. H. (1974) Morphological effects of denervation of muscle. A quantitative ultrastructural study. *Ann. NY Acad. Sci.* **228**, 68–78.

Fallon, J. R. and Gelfman, C. E. (1989) Agrin related molecules are concentrated at acetylcholine receptor clusters in normal and aneural muscle. *J. Cell Biol.* **108**, 1527–1535.

Fambrough, D. M. (1970) Acetylcholine sensitivity of muscle fibre membranes: mechanism of regulation by motoneurones. *Science* **168**, 372–373.

Fambrough, D. M. (1979) Control of acetylcholine receptors in skeletal muscle. *Physiol. Rev.* **59**, 165–227.

Farel, P. B. (1980) Selective synaptic changes following spinal motoneuron axotomy. *Brain Res.* **158**, 331–341.

Feldberg, W. (1943) Synthesis of acetylcholine in sympathetic and cholinergic nerve. *J. Physiol.* **101**, 432–435.

Feng, T. P., Yang, H. W. and Wu, W. Y. (1963) The contrasting trophic changes of the anterior and posterior latissimus dorsi of the chick following denervation, in *Effect of Use and Discuse on Neuromuscular Functions* (eds E. Gutmann and P. Hník), Czechoslovak Academy of Science Publishing House, Prague.

Finol, H. J., Lewis, D. M. and Owens, R. (1981) The effects of denervation on contractile properties of rat skeletal muscle. *J. Physiol. (Lond.)* **319**, 81–92.

Fischbach, G. D. and Robbins, N. (1971) Effect of chronic disuse of rat soleus neuromuscular junctions on post-synaptic membrane. *J. Neurophysiol.* **34**, 562–569.

Fischer, H. A. and Schmatolla, E. (1972) Axonal transport of tritium-labelled putrescine in the embryonic visual system of zebrafish. *Science* **176**, 1327–1329.

Foehring, R. C. and Munson, J. B. (1990) Motoneuron and muscle-unit properties after long-term direct innervation of soleus muscle by medial gastrocnemius nerve in cat. *J. Neurophysiol.* **64**, 847–861.

Foehring, R. C., Sypert, G. W. and Munson, J. B. (1986) Properties of self-reinnervated motor units of medial gastrocnemius of cat. II. Axotomized motoneurons and time course of recovery. *J. Neurophysiol.* **55**, 947–965.

Forman, D. S. and Borenberg, R. A. (1978) Regeneration of motor axons in the rat sciatic nerve studied by labelling with axonally transported radioactive proteins. *Brain Res.* **156**, 213–225.

Frank, E., Gautvik, K. and Sommerschild, H. (1976) Persistence of junctional acetylcholine receptors following denervation. *Cold Spring Harbor Symp. Quant Biol.* **XL**, 275–281.

Frelin, C., Vijerberg, H. P. M. and Romey, G. (1984) Different functional states of tetrodotoxin sensitive and tetrodotoxin resistant Na^+ channels occur during *in vitro* development of rat skeletal muscle. *Pflügers Arch.* **402**, 121–128.

Friede, R. L. and Samorajski, T. (1970) Axon caliber related to neurofilaments and microtubules in sciatic nerve fibres of rats and mice. *Anat. Rec.* **167**, 379–388.

Frischknecht, R., Navarrete, R. and Vrbová, G. (1990) Introduction to the functional anatomy of the mammalian motor unit. *Ballière's Clinical Endocrinology and Metabolism*, **4**, 401–415.

Frizell, M. and Sjöstrand, J. (1974) Transport of proteins, glycoproteins, and cholinergic enzymes in regenerating hypoglossal neurons. *J. Neurochem.* **22**, 845–850.

Gallego, R., Huizar, P., Kudo, N. and Kuno, M. (1978) Disparity of motoneurone and muscle differentiation following spinal transection in the kitten. *J. Physiol.* **281**, 253–265.

Gallego, R., Kuno, M., Nunez, R. and Snider, W. D. (1979) Dependence of motoneurone properties on the length of immobilized muscle. *J. Physiol.* **291**, 179–189.

Garcia-Bunuel, L. and Garcia-Bunuel, V. M. (1980) Connective tissue metabolism in normal and atrophic skeletal muscle. *J. Neurol. Sci.* **47**, 69–77.

Gauthier, G. F. and Hobbs, A. W. (1982) Effects of denervation on the distribution of isoenzymes in skeletal muscle fibres. *Exp. Neurol.* **76**, 312–346.

Gauthier, G. F. and Schaeffer, S. F. (1974) Ultrastructural and cytochemical manifestations of protein synthesis in the peripheral sarcoplasm of denervated and newborn skeletal muscle fibres. *J. Cell Sci.* **14**, 113–137.

Gerard, R. W. (1932) Nerve metabolism. *Physiol. Rev.* **12**, 469–592.

Gillespie, M. J. and Stein, R. B. (1983) The relationship between axonal diameter, myelin thickness and conduction velocity during atrophy of mammalian peripheral nerves. *Brain Res.* **59**, 41–56.

Ginetzinski, A. and Shamarina, N. M. (1942) The tonomotor phenomenon in denervated muscles. *Usp. Sovrem. Biol.* **15**, 283–294.

Goldberg, A. L., Jablecki, C. and Li, J. B. (1974) Effects of use and disuse on amino acid transport and protein turnover in muscle. *Ann. NY Acad. Sci.* **228**, 190–201.

Goldspink, G. (1976) The effect of denervation on protein turnover of rat skeletal muscle. *Biochem J.* **156**, 71–80.

Gordon, T. (1983) Dependence of peripheral nerves on their target organs, in *Somatic and Autonomic Nerve–Muscle Interactions* (eds G. Vrbová, G. Burnstock and R. D. O'Brien), Elsevier, Amsterdam, pp. 289–325.

Gordon, T., Gillespie, J., Orozco, R. and Davis, L. (1991) Axotomy-induced changes in rabbit hindlimb nerves and the effects of chronic electrical stimulation. *J. Neurosci.* **11**, 2157–2169.

Gordon, T., Jones, R. and Vrbová, G. (1976) Changes in chemosensitivity of skeletal muscle as related to endplate formation. *Prog. Neurobiol.* **6**, 103–136.

Gordon, T., Kelly, M. E. M., Sanders, E. J. *et al.* (1987) The effects of axotomy on bullfrog sympathetic neurones. *J. Physiol. (Lond.)* **392**, 213–229.

Gorza, L., Gundersen, K., Lømo, T. *et al.* (1988) Slow-to-fast transformation of denervated soleus muscles by chronic high frequency stimulation in the rat. *J. Physiol. (Lond.)* **402**, 627–649.

Gosling, K., Schreyer, D. J., Skene, J. H. P. and Banker, G. (1988) Developmental polarity: GAP-43 distinguishes axonal from dendritic growth cones. *Nature* **336**, 672–675.

Grafstein, B. (1977) Axonal transport: the intracellular traffic of the neuron, in *The Handbook of Physiology, Section 1: The Nervous System*, Vol. 1. (eds J. M. Brookhart and V. B. Mountcastle), American Physiological Society, Washington, DC, pp. 697–717.

Grafstein, B. and Forman, D. S. (1980) Intracellular transport in neurons. *Physiol. Rev.* **60**, 1167–1283.

Grafstein, B. and Murray, M. (1969) Transport of protein in goldfish optic nerve during regeneration. *Exp. Neurol.* **25**, 494–508.

Grampp, W., Harris, J. B. and Thesleff, S. (1972) Inhibition of denervation changes in skeletal muscle by blockers of protein synthesis. *J. Physiol.* **221**, 743–754.

Greenberg, S. G. and Lasek, R. J. (1988) Neurofilament protein synthesis in DRG neurons decreases more after peripheral axotomy than after central axotomy. *J. Neurosci.* **8**, 1739–1746.

Greenman, M. J. (1913) Studies on the regeneration of the peroneal nerve of the albino rat: number and sectional area of fibers: area relation of axis to sheath. *J. Comp. Neurol.* **23**, 479–513.

Gunning, P. W., Kaye, P. L. and Austin, L. (1977) *In vivo* synthesis of rapidly-labelled RNA within the rat nodose ganglia following vagotomy. *J. Neurochem.* **28**, 1245–1248.

Gustaffson, B. (1979) Changes in motoneurone electrical properties following axotomy. *J. Physiol. (Lond.)* **293**, 197–215.

Gustaffson, B. and Pinter, M. (1984) Effects of axotomy on the distribution of passive electrical properties of motoneurones. *J. Physiol. (Lond.)* **356**, 433–442.

Guth, L., Kemperer, V. E., Samuras, T. A. *et al.* (1981) Roles of use and loss of trophic function in denervation atrophy of skeletal muscle. *Exp. Neurol.* **73**, 20–36.

Gutmann, E. (1942) Factors affecting recovery of motor function after nerve lesions. *J. Neurol. Neurosurg. and Psychiat.* **5**, 81–95.

Gutmann, E. (1958) *Die functionelle Regeneration der peripheren Nerven*, Akademie-Verlag, Berlin.

Gutmann, E. (1976) Neurotrophic relations. *Ann. Rev. Physiol.* **38**, 177–216.

Gutmann, E. and Sanders, F. K. (1943) Recovery of fibre numbers and diameters in the regeneration of peripheral nerves. *J. Physiol.* **101**, 489–518.

Gutmann, E. and Žák, R. (1961) Nervous regulation of nucleic acid levels in cross striated muscle. Changes in denervated muscle. *Physiol. Bohemoslov.* **10**, 493–500.

Gutmann, E. and Zelená, J. (1962) Morphological changes in denervated muscle, in *The Denervated Muscle* (ed. E. Gutmann), Czechoslovak Academy of Science Publishing House, Prague, pp. 57–98.

Gutmann, E., Guttmann, L., Medawar, P. B. and Young, J. Z. (1942) The rate of regeneration of nerve. *J. Exp. Biol.* **19**, 14–44.

Gutmann, E., Hanzlíková, M., Hájek, I. *et al.* (1966) Post denervation hypertrophy of the diaphragm. *Physiol. Bohemoslov.* **15**, 508–524.

Gutmann, E., Melichna, J. and Syrový, I. (1972) Contraction properties and ATP-ase activity in fast and slow muscle of the rat during denervation. *Exp. Neurol.* **36**, 488–497.

Gutmann, E., Vodličká, Z. and Vrbová, G. (1954) Nervous regulation of super compensation of glycogen in skeletal muscle. *Physiol. Bohemoslov.* **3**, 182–185.

Gutmann, E., Vodičká, Z. and Zelená, Z. (1955) Changes in striated muscle after nerve section related to the length of the peripheral stump. *Physiol. Bohemoslov.* **4**, 200–206.

Hajek, I., Gutmann, E. and Syrový, I. (1964) Proteolytic activity in denervated and reinnervated muscle. *Physiol. Bohemoslov.* **13**, 32–38.

Hall, M. E., Wilson, D. L. and Stone, G. C. (1978) Changes in synthesis of specific proteins following axotomy: detection with two-dimensional gel electrophoresis. *J. Neurobiol.* **9**, 353–366.

Hall, S. M. (1989) Regeneration in the peripheral nervous system. *Neuropathol. App. Neurobiol.* **15**, 513–529.

Hall, Z. W. and Sanes, J. (1993) Synaptic structure and development. The neuromuscular junction. *Cell*, **72**/*Neuron* **10** (Suppl.) 99–121.

Hamberger, A., Hansson, H. A. and Sjöstrand, J. (1970) Surface structure of isolated neurones. Detachment of more terminals during axon regeneration. *J. Cell. Biol.* **47**, 319–331.

Hansen-Bay, C. and Strichartz, G. K. (1980) Saxitoxin binding to sodium channels of rat skeletal muscles. *J. Physiol.* **300**, 89–103.

Harris, J. B. and Thesleff, S. (1971) Studies on tetrodotoxin-resistant action potentials in denervated skeletal muscle. *Acta Physiol. Scand.* **83**, 382–388.

Hartzell, H. C. and Fambrough, D. M. (1973) Acetylcholine receptor production and incorporation into membranes of developing muscle fibres. *Dev. Biol.* **30**, 153–165.

Hayes, R. C., Wiley, R. G. and Armstrong, D. M. (1992) Induction of nerve growth factor receptor (p75[NGFr]) mRNA within hypoglossal motoneurons following axonal injury. *Mol. Brain Res.* **15**, 291–297.

Hearn, G. R. (1959) Succinate-cytochrome c reductase and aldoase activities of denervated skeletal muscle. *Am. J. Physiol.* **196**, 465–466.

Heiwell, P. O., Dahlström, A., Larsson, P. A. and Booj, S. (1979) The intra-axonal transport of acetylcholine and cholinergic enzymes in rat sciatic nerve during regeneration after various types of axonal trauma. *J. Neurobiol.* **10**, 119–136.

Heslop, J. P. (1975) Axonal flow and fast transport in nerves. *Adv. Comp. Physiol. Biochem.* **6**, 75–163.

Hník, P. and Skorpil, V. (1962) Fibrillation activity in denervated muscle, in *The Denervated Muscle* (ed. E. Gutmann), Czechoslovak Academy of Science, Prague, pp. 123–125.

Hoffman, P. N., Lopata, M. A., Watson, D. F. and Luduena, R. F. (1992) Axonal transport of Class II and III beta-tubulin: Evidence that the slow component wave represents the movement of only a small fraction of the tubulin in mature motor axons. *J. Cell Biol.* **119**, 595–604.

Hoffman, P. N. (1989) Expression of GAP–43, a rapidly transported growth-associated protein, and class II beta tubulin, a slowly transported cytoskeletal protein, are co-ordinated in regenerating neurons. *J. Neurosci.* **9**, 893–897.

Hoffman, P. N. and Cleveland, D. W. (1988) Neurofilament and tubulin expression recapitulates the developmental pattern during axonal regeneration: Induction of a specific B-tubulin isotype. *Proc. Nat. Acad. Sci.* **9**, 893–897.

Hoffman, P. N., Cleveland, D. W., Griffith, J. W. *et al.* (1987) Neurofilament gene expression: a major determinant of axonal calibre. *Proc. Natl. Acad. Sci. USA* **84**, 3472–3476.

Hoffman, P. N. and Lasek, R. J. (1975) The slow component of axonal transport. Identification of major structural polypeptides of the axon and their generality among mammalian neurons. *J. Cell Biol.* **66**, 351–366.

Hoffman, P. L. and Lasek, R. J. (1980) Axonal transport of the cytoskeleton in regenerating motor neurons: constancy and change. *Brain Res.* **202**, 317–333.

Hogan, E. L., Dawson, D. M. and Romanul, F. C. A. (1965) Enzymic changes in denervated muscle. *Arch. Neurol.* **13**, 274–282.

Holmes, W. and Young, J. Z. (1942) Nerve regeneration after immediate and delayed suture. *J. Anat.* **77**, 63–96.

Huizar, P., Kuno, M. and Miyata, Y. (1975) Differentiation of motoneurones and skeletal muscles in kittens. *J. Physiol.* **252**, 465–479.

Huizar, P., Kudo, N., Kuno, M. and Myata, Y. (1977) Reaction of intact spinal motoneurones to partial denervation of the muscle. *J. Physiol. (Lond.)* **265**, 175–191.

Hurst, J. B. (1939) Conduction velocity and diameter of nerve fibres. *Am. J. Physiol.* **27**, 131–139.

Hyden, H. (1960) The Neuron, in *The Cell* (eds J. Brachet and A. E. Mirsky), Academic Press, New York and London, pp. 215–323.

Ingoglia, N. A., Stutman, J. A. and Eisner, R. A. (1977) Axonal transport of putrescine, spermidine and spermine in normal and regenerating goldfish optic nerves. *Brain Res.* **130**, 433–445.

Jack, J. J. B. (1955) Physiology of peripheral nerve fibres in relation to their size. *Br. J. Anaesth.* **47**, 173–182.

Jockusch, H. and Jockusch, B. M. (1981) Structural proteins in the growth core of cultured spinal cord neurons. *Exp. Cell. Res.* **131**, 345–352.

Johns, J. T. and Thesleff, S. (1961) Effects of motor inactivation on the chemical sensitivity of skeletal muscle. *Acta Physiol Scand.* **51**, 136–141.

Jones, R. and Lane, J. T. (1975) Proliferation of mononuclear cells in skeletal muscle after denervation or muscle injury. *J. Physiol.* **246**, 60–61P.

Jones, R. and Vrbová, G. (1971) Can denervation hypersensitivity be prevented? *J. Physiol.* **210**, 144–145P.

Jones, R. and Vrbová, G. (1974) Two factors responsible for the development of denervation hypersensitivity. *J. Physiol.* **236**, 517–538.

Jones, R. and Vyskočil, F. (1975) An electrophoretic examination of the changes in skeletal muscle fibres in response to degenerating nerve tissue. *Brain Res.* **88**, 309–317.

Kamel-Reid, S., Kennedy, J. M., Shimizu, N. *et al.* (1989) Regulation of expression of avian slow myosin heavy chain. *Biochem. J.* **260**, 449–454.

Kaye, P. L., Gunning, P. W. and Austin, L. (1977) In vivo synthesis of stable RNA within rat nodose ganglia following vagotomy. *J. Neurochem.* **28**, 1241–1243.

Kean, C. J. G., Lewis, D. M. and McGarrick, J. D. (1974) Dynamic properties of denervated fast and slow twitch muscle of the cat. *J. Physiol. (Lond.)* **237**, 103–113.

Kirby, A. C. and Lindley, B. D. (1981) Calcium content of rat fast and slow muscles after denervation. *Comp. Biochem. Physiol.* **70A**, 583–586.

Kirsch, G. E. and Anderson, M. F. (1986) Sodium channels in normal and denervated rabbit muscle membrane. *Muscle and Nerve* **9**, 738–747.

Klemperer, H. G. (1972) Lowered proportion of polysomes and increased amino acid incorporation by ribosomes from denervated muscle. *FEBS Let.* **28**, 169–173.

Knyihar, E. and Csillik, B. (1976) Effect of peripheral axotomy on the fine structure and histochemistry of the rolando substance: degenerative atrophy of central processes of pseudounipolar cells. *Exp. Brain Res.* **26**, 73–87.

Koszalka, T. R. and Miller, L. L. (1960) Proteolytic activity of rat skeletal muscle. I. Evidence of an enzyme optimally active at pH 8.5 to 9. *J. Biol. Chem.* **235**, 665–669.

Kreutzberg, G. W. (1982) Acute neuronal reaction to injury, in *Repair and Regeneration of the Nervous System* (ed. J. G. Nicholls), Springer, Berlin, pp. 57–69.

Kuczmarski, E. R. and Rosenbaum, J. L. (1974) Studies on the organization and localization of actin and myosin in neurones. *J. Cell. Biol.* **80**, 356–371.

Kuno, M. and Llinas, R. (1970a) Enhancement of synaptic transmission by dendritic potentials in chromatolysed motoneurones of the cat. *J. Physiol.* **210**, 807–821.

Kuno, M. and Llinas, R. (1970b) Alterations of synaptic action in chromatolysed motoneurones of the cat. *J. Physiol.* **210**, 823–838.

Kuno, M., Miyata, Y. and Munoz-Martinez, E. J. (1974) Differential reactions of fast and slow alpha-motoneurones to axotomy. *J. Physiol.* **240**, 725–739.

Langley, J. N. (1916) Observations on denervated muscle. *J. Physiol. (Lond.)* **50**, 335–344.

Lassek, R. J. (1970) Protein transport in neurones. *Int. Rev. Neurobiol.* **13**, 232–289.

Lassek, R. J. (1981) The dynamic ordering of neuronal skeletons. *Neurosci. Res. Prog. Bull.* **19**, 7–32.

Lassek, R. J., Grainer, H. and Baker, J. L. (1977) Cell to cell transfer of glial proteins to the squid axon. The glial cell neuronal protein transfer hypothesis. *J. Cell Biol,* **74**, 501–523.

Lassek, R. J. and Hoffman, P. N. (1976) The neuronal cytoskeleton, axonal transport and axonal growth, in *Cell Motility. Book C, Microtubules and Related Proteins* (eds R. Goldman, T. Pollard and J. Rosenbaum), Cold Spring Harbor Laboratory, New York, pp. 1021–1049.

Leah, J. D., Herdegen, T. and Bravo, R. (1991) Selective expression of Jun proteins following axotomy and axonal transport block in peripheral nerves of the rat evidence for a role in the regeneration process. *Brain Res.* **566**, 198–207.

Lehotsky, J., Drgova, A., Dobrota, D. and Mezesova, V. (1991) Effects of denervation on the contents of cholesterol and membrane systems involved in muscle contraction in rabbit fast sarcotubular system. *Gen. Physiol. Biophys.* **10**, 175–188.

Levitt, T. A. and Salpeter, M. M. (1981) Denervated endplates have a dual population of junctional acetylcholine receptors. *Nature* **291**, 239–241.

Lewis, D. M. (1972) The effect of denervation on the mechanical and electrical responses of fast and slow mammalian twitch muscle. *J. Physiol. (Lond.)* **222**, 51–75.

Lieberman, A. R. (1971) The axon reaction: a review of the principal features of perikaryal responses to injury. *Int. Rev. Neurobiol.* **14**, 49–124.

Lieberman, A. R. (1974) Some factors affecting retrograde neuronal responses to axonal lesions, in *Essays on the Nervous System. A Postscript for Professor J. Z. Young* (eds R. Bellairs and E. G. Gray), Oxford University Press, Oxford, pp. 71–105.

Lindholm, D., Heumann, R., Meyer, M. and Thoenen, H. (1987) Interleukin–1 regulates synthesis of nerve growth factor in non-neuronal cells of rat sciatic nerve. *Nature (Lond.)* **330**, 658–659.

Little, B. W., Barlow, R. and Perl, D. P. (1982) Denervation modulated changes in mouse skeletal muscle RNA concentration *Virchows Arch. (Cell Pathol.)* **39**, 1–7.

Ljubinska, L. (1964) Axoplasmic streaming in regenerating and in normal nerve fibres, in *Progress in Brain Research, Mechanisms of Neural Regeneration,* Vol. 13 (eds M. Singer and J. R. Schade), Elsevier, Amsterdam, pp. 1–71.

Ljubinska, L. (1975) On axoplasmic flow. *Int. Rev. Neurobiol.* **17**, 241–296.

Lømo, T. and Rosenthal, J. (1972) Control of ACh sensitivity by muscle activity in the rat. *J. Physiol.* **221**, 493–513.

Lorkovic, H. (1975) Supersensitivity to ACh in muscles after prolonged nerve block. *Arch. Int. Physiol. et Biochemie* **83**, 771–781.

X Lunn, E. R., Perry, V. H., Brown, M. C. *et al.* (1989) Absence of Wallerian degeneration does not hinder regeneration in peripheral nerve. *Eur. J. Neurosci.* **1**, 27–33.

X Lunn, E. R., Brown, M. C. and Perry, V. H. (1990) The pattern of axonal degeneration in the peripheral nervous system varies with different types of lesion. *Neurosci.* **35**, 157–165.

Maehlen, J., Nennesmo, I., Olsson, T. *et al.* (1989) Peripheral nerve injury causes transient expression of MHC Class 1 antigens in rat neurones and skeletal muscles. *Brain Res.* **481**, 368–372.

McEwen, B. and Grafstein, B. (1968). Fast and slow components in axonal transport of protein. *J. Cell Biol.* **38**, 494–508.

McLaughlin, J., Abood, L. G. and Bosman, H. B. (1974) Early elevations of glucosidase, acid phosphatase, and acid proteolytic enzyme activity in denervated muscle. *Exp. Neurol.* **42**, 541–554.

McQuarrie, I. G. and Grafstein, B. (1973) Axonal outgrowth enhanced by a previous injury. *Arch. Neurol.* **29**, 43–55.

McQuarrie, I. G. and Lasek, R. J. (1989) Transport of cytoskeletal elements from the parent axons into regenerating daughter axons. *J. Neurosci.* **9**, 436–446.

McQuarrie, I. G. and Jacob, J. M. (1991) Conditioning nerve crush accelerates cytoskeletal protein transport in sprouts that form after a subsequent crush. *J. Comp. Neurol.* **305**, 139–147.

Margreth, A., Carraro, U. and Salviati, G. (1976) Effects of denervation on protein synthesis and on properties of myosin of fast and slow muscles, in *Pathogenesis of Human Dystrophies*, Excepta Medica, pp. 161–170.

Margreth, A., Salviati, G., Di Mauro, S. and Turati, G. (1972) Early biochemical consequences of denervation in fast and slow skeletal muscles and their relationship to neural control and muscle differentiation. *Biochem J.* **126**, 1099–1110.

Marinesco, G. (1896) Sur les phénomènes de réparation dans les centres nerveux après la section des nerfs périphériques. *Compt. rendu Soc. de biol.* **111**, 930.

Marshall, M. W. and Ward, M. R. (1974) Anode break excitation in denervated rat skeletal muscle fibres. *J. Physiol.* **236**, 413–420.

Massoulié, J., Pezzementi, L., Bon, S. *et al.* (1993) Molecular and cellular biology of cholinesterases. *Prog. Neurobiol.* **41**, 31–91.

Matsuda, Y., Nojimia, M. and Desaki, J. (1988) Scanning EM study of denervated and reinnervated neuromuscular junction. *Muscle & Nerve* **11**, 1266–1271.

Matsuoka, I., Meyer, M. and Thoenen, H. (1991) Cell-type-specific regulation of nerve growth factor synthesis in non-neural cells: Comparison with other cell types. *J. Neurosci.* **11**, 3165–3177.

Matthews, M. R. and Raisman, G. (1972) A light and electron microscopic study of the cellular response to axonal injury in the superior cervical ganglion of the rat. *Proc. Roy. Soc. B.* **181**, 43–79.

Matthews, M. R. and Nelson, V. H. (1975) Detachment of structurally intact nerve endings from chromatolytic neurones of rat superior cervical ganglion during the depression of synaptic transmission induced by post-ganglionic axotomy. *J. Physiol.* **245**, 91–135.

Mauro, A. (1961) Satellite cell of skeletal muscle fibres. *J. Biophys. Biochem. Cytol.* **9**, 493.

Mendell, L. M. (1984) Modifiability of spinal synapses. *Physiol. Rev.* **64**, 260–324.

Mendell, L. M., Munson, J. G. and Scott, J. G. (1974) Connectivity changes of 1a afferents on axotomized motoneurons. *Brain Res.* **73**, 338–342.

Mendell, L. M., Munson, J. G. and Scott, J. G. (1976) Alterations of synapses on axotomised motoneurones. *J. Physiol.* **255**, 67–79.

Merlie, J. P., Isenberg, K. E., Russell, S. D. and Sanes, J. R. (1984) Denervation supersensitivity in skeletal muscle: analysis with a cloned cDNA probe. *J. Cell Biol.* **99**, 332–335.

Miledi, R. (1960a) Properties of regenerating neuromuscular synapses in the frog. *J. Physiol.* **154**, 190–205.

Miledi, R. (1960b) Junctional and extrajunctional acetylcholine receptors in skeletal muscle fibres. *J. Physiol.* **151**, 24–30.

Miledi, R. and Slater, C. R. (1968) Electrophysiology and electron-microscopy of rat neuromuscular junctions after nerve degeneration. *Proc. Roy. Soc. B.* **169**, 289–306.

Miledi, R. and Slater, C. R. (1970) On the degeneration of rat neuromuscular junctions after nerve section. *J. Physiol.* **207**, 507–528.

Miller, F. D., Tetzlaff, W., Bisby, M. A. *et al.* (1989) Rapid induction of the major embryonic a-tubulin mRNA, in adults following neuronal injury. *J. Neurosci.* **9**, 1452–1463.

Milner, T. E., Stein, R. B., Gillespie, J. and Hanley, B. (1981) Improved estimates of conduction velocity distributions using single unit action potentials. *J. Neurol. Neurosurg. Psychiat.* **44**, 476–484.

Milot, J., Fremont, P., Cote, C. and Tremblay, R. R. (1991) Differential modulation of carbonic anhydrase (CA 111) in slow- and fast-twitch skeletal muscles of rat following denervation and reinnervation. *Biochem. Cell Biol.* **69**, 702–710.

Muchnik, S., Ruarte, A. C. and Kotsias, B. A. (1973) The effect of actinomycin D on fibrillation activity in denervated skeletal muscle of the rat. *Life Sci.* **13**, 1763–1770.

Murray, M. A. and Robbins, N. (1982a) Cell proliferation in denervated muscle: Time course, distribution, and relation to disuse. *Neurosci.* **7**, 1817–1822.

Murray, M. A. and Robbins, N. (1982b) Cell proliferation in denervated muscle: Identity and origin of dividing cells. *Neurosci.* **7**, 1823–1833.

Murray, M. (1973) 3H-uridine incorporation by regenerating retinal ganglion cells of goldfish. *Exp. Neurol.* **39**, 489–497.

Nastuk, W. L. (1953) Membrane potential changes at a single muscle endplate produced by transitory applications of acetylcholine with an electrically controlled microjet. *Fed. Proc.* **12**, 102.

Nastuk, M. A. and Fallon, J. R. (1993) Agrin and the molecular choreography of synapse formation. *TINS* **16**, 72–76.

Németh, P. M., Myer, D. and Kark, R. A. P. (1980) Effects of denervation and simple disuse on rates of oxidation and on activities of four mitochondrial enzymes in type 1 muscle. *J. Neurochem.* **35**, 1351–1360.

Nickel, E. and Waser, P. G. (1969) An electron microscope study of denervated motor endplates after zinc iodide-osmium impregnation. *Brain Res.* **13**, 168–176.

Nissl, F. (1892) Über die Veränderungen der Ganglienzellen am Facialis-kern des Kaninchens nach Aureissung der Nerven. *Allg. Z. Psychiat.* **48**, 197–198.

Oblinger, M. M., Wong, J. and Parysek, L. M. (1989) Axotomy-induced changes in the expression of a type 111 neuronal intermediate filament gene. *J. Neurosci.* **9**, 3766–3775.

O'Brien, R. A. D. (1978) Axonal transport of acetylcholine, choline acetyl-transferase and cholinesterase in regenerating peripheral nerve. *J. Physiol.* **282**, 91–103.

Ontell, M. (1974) Muscle satellite cells: A validated technique for light-microscopic identification and a quantitative study of changes in their population following denervation. *Anat. Rec.* **178**, 211–228.

Pappone, P. A. (1980) Voltage clamp experiments in normal and denervated mammalian skeletal muscle fibres. *J. Physiol.* **306**, 377–410.

Pellegrino, C. and Franzini, C. (1963) An electron microscope study of denervation atrophy in red and white skeletal muscle fibres. *J. Cell Biol.* **17**, 327–349.

Pestronk, A., Drachmann, D. B. and Griffin, J. W. (1976) Effects of muscle disuse on acetylcholine receptors. *Nature* **260**, 352–353.

Pinter, M. J. and van den Noven, S. (1989) Effects of preventing reinnervation on axotomized spinal motoneurones in the cat. I. Motoneuron electrical properties. *J. Neurophysiol.* **62**, 311–324.

Pollack, M. S. and Bird, J. W. C. (1968) Distribution and particle properties of acid hydrolase in denervated muscle. *Am. J. Physiol.* **215**, 716–722.

Purves, D. (1975) Functional and structural changes of mammalian sympathetic neurones following interruption of their axons. *J. Physiol.* **252**, 429–463.

Purves, D. and Sackmann, B. (1974) Membrane properties underlying spontaneous activity of denervated muscle fibres. *J. Physiol.* **239**, 125–153.

Ramón y Cajal, S. R. (1928) *Degeneration and Regeneration of the Nervous System*, vol. II (trans. and ed. R. M. May), Oxford University Press.

Ranson, S. W. (1912) Degeneration and regeneration of nerve fibres. *J. Comp. Neurol.* **22**, 487–546.

Redfern, P. and Thesleff, S. (1971a) Action potential generation in denervated rat skeletal muscle. 1. Quantitative aspects. *Acta Physiol. Scand.* **81**, 557–564.

Redfern, P. and Thesleff, S. (1971b) Action potential generation in denervated rat skeletal muscle. 2. The action of tetrodotoxin. *Acta Physiol. Scand.* **82**, 70–78.

Richardson, P. M., Issa, V. M. K. and Aguayo, A. J. (1984) Regeneration of long spinal axons in the rat. *J. Neurocytol.* **13**, 161–182.

Roberts, E. D. and Oester, Y. T. (1970) Absence of supersensitivity to acetylcholine in innervated muscle subjected to a prolonged pharmacological nerve block. *J. Pharm. Exper. Ther.* **174**, 133–140.

Rochel, S. and Robbins, N. (1985) Is a nicotinic influence involved in denervation-induced depolarisations of muscle? *J. Neurosci.* **5**, 2331–2335.

Salonen, V., Aho, H., Roytta, M. and Peltonen, J. (1988) Quantitation of Schwann cells and endoneurial fibroblast-like cells after experimental nerve trauma. *Acta Neuropathol. (Berl.)* **75**, 331–336.

Salpeter, M., Buonanno, A., Eftimie, R. *et al.* (1992) Regulation of molecules at the neuromuscular junction, in *Neuromuscular Development and Disease* (eds A. M. Kelly and H. M. Blau), Raven Press, New York, pp. 251–283.

Salvatori, S., Damiani, E., Zorzato, F. *et al.* (1989) Denervation-induced proliferative changes of triads in rabbit skeletal muscle. *Muscle and Nerve* **11**, 1246–1259.

Salzer, J. L. and Bunge, R. P. (1980) Studies of Schwann cell proliferation. I. An analysis in tissue culture of proliferation during development, Wallerian degeneration, and direct injury. *J. Cell Biol.* **84**, 739–752.

Salzer, J. L., Williams, A. K., Glaser, L. and Bunge, R. P. (1980) Studies of Schwann cell proliferation. II. Characterization of the stimulation and specificity of the response to a neurite membrane fraction. *J. Cell Biol.* **84**, 753–766.

Sanes, J. R. and Chui, A. Y. (1983) The basal lamina of the neuromuscular junction. *Cold Spring Harbor Symp. Quant. Biol.* **48**, 667–678.

Sanes, J. R., Marshall, L. M. and McMahan, U. J. (1980) Reinnervation of muscle fibre basal lamina after removal of myofibres. Differentiation of regenerating axons at original synaptic sites. *J. Cell Biol.* **78**, 176–198.

Sanes, J. R., Schachner, M. and Covaolt, J. (1986) Expression of several adhesive macromolecules (N-CAM, L1, J1, NILE, uvomorulin, laminin, fibronectin, and a heparan sulphate proteoglycan) in embryonic, adult and denervated adult skeletal muscles. *J. Cell Biol.* **102**, 420–431.

Saunders, N. R. (1975) Axonal transport of acetylcholine, in *Cholinergic Mechanisms* (ed. P. G. Waser), Raven Press, New York, pp. 177–185.

Sawchak, J. A., Lewis, S. and Shafiq, S. A. (1989) Comparison of myosin isoforms in muscle of patients with neurogenic disease. *Muscle and Nerve* **12**, 679–689.

Schiaffino, S. and Settembrini, P. (1970) Studies on the effects of denervation in developing muscle. *Virchows Arch.* **B4**, 345–356.

Schlaepfer, W. W. and Freeman, L. A. (1980) Calcium dependent degradation of mammalian filaments by soluble tissue factors from rat spinal cord. *Neuroscience*, **5**, 2305–2314.

Schmalbruch, H., Al-Amood, W. S. and Lewis, D. M. (1991) Morphology of long-term denervated rat soleus muscle and the effect of chronic electrical stimulation. *J. Physiol. (Lond.)* **441**, 233–241.

Schubert, P., Kreutzberg, G. W. and Lux, H. D. (1972) Neuroplasmic transport in dendrites: effect of colchicine on morphology and physiology of moto-neurones in the cat. *Brain Res.* **47**, 331–343.

Schwartz, J. H. (1979) Axonal transport: components, mechanisms and specificity. *Ann. Rev. Neurosci*, **2**, 467–504.

Shyng, S.-L., Xu, R. and Salpeter, M. M. (1991) Cyclic AMP stabilises the degradation of original junctional acetylcholine receptors in denervated muscle. *Neuron* **6**, 469–475.

Simard, C. P., Spector, S. A. and Edgerton, V. R. (1982) Contractile properties of rat hind limb muscles immobilised at different lengths. *Exp. Neurol.* **77**, 467–482.

Sinicropi, D. V. and Kauffman, F. C. (1979) Retrograde alteration of 6-phosphogluconate dehydrogenase in axotomised superior cervical ganglia of the rat. *J. Biol. Chem.* **254**, 3011–3017.

Skene, J. H. P. (1989) Axonal growth-associated proteins. *Ann. Rev. Neurosci.* **12**, 127–156.

Skene, J. H. P. and Willard, M. (1981) Axonally transported proteins associated with axon growth in rabbit central and peripheral nervous systems. *J. Cell Biol.* **89**, 96–103.

Sketelj, J., Crne-Finderle, N. and Dolenc, I. (1994) *Factors influencing acetylcholinesterase regulation in slow and fast skeletal muscles* (in press).

Slater, C. R. (1966) The time course of failure of neuromuscular transmission after nerve section. *Nature* **209**, 305–306.

Smith, R. L. and Lawrence, J. C. (1984) Insulin action in denervated rat hemidiaphragms. Decreased hormonal stimulation of glycogen synthesis involves both glycogen synthase and glucose transport. *J. Biol. Chem.* **259**, 2201–2207.

Solandt, D. Y. and Magladery, J. W. (1942) A comparison of the effects of upper and lower motoneurone lesions on skeletal muscle. *J. Neurophysiol.* **5**, 373–380.

Spector, S. A. (1985a) Effects of elimination of activity on contractile and histochemical properties of rat soleus muscle. *J. Neurosci.* **5**, 2177–2188.

Spector, S. A. (1985b) Trophic effects on the contractile and histochemical properties of rat soleus muscle. *J. Neurosci.* **5**, 2189–2196.

Stanley, E. F. and Drachmann, D. B. (1981) Denervation accelerates the degradation of junctional acetylcholine receptors. *Exp. Neurol.* **73**, 390–396.

Stanley, E. F. and Drachmann, D. B. (1983) Evidence for the post-insertional stabilisation of acetylcholine receptors at the neuromuscular junction. *Soc. Neurosci.* **9**, 884.

Stonnington, H. H. and Engle, A. G. (1973) Normal and denervated muscle. A morphometric study of fine structure. *Neurology,* **23**, 714–724.

Streit, W. J., Dumoulin, F. L., Raivich, G. and Kreutzberg, G. W. (1989) Calcitonin gene-related peptide increases in rat facial motoneurons after peripheral nerve transection. *Neurosci. Lett.* **101**, 143–148.

Sumner, B. E. H. (1975) A quantitative analysis of the response of presynaptic boutons to post synaptic motor axotomy. *Expl. Neurol.* **46**, 605–615.

Sumner, B. E. H. and Sutherland, F. I. (1973) Quantitative electron microscopy on the injured hypoglossal nucleus in the rat. *J. Neurocytol.* **2**, 315–328.

Sumner, B. E. H. and Watson, W. E. (1971) Retraction and expansion of the dendritic tree of motor neurones of adult rats induced *in vivo. Nature* **233**, 273–275.

Sunderland, S. (1978) *Nerves and Nerve Injuries,* Livingstone, Edinburgh and London.

Syrový, I., Gutmann, E. and Melichna, J. (1971) Differential response of myosin ATP-ase activity and contraction properties of fast and slow rabbit muscle following denervation. *Experientia* **27**, 1426–1427.

Syrový, I., Gutmann, E. and Melichna, J. (1972) The effect of denervation on contraction and myosin properties of fast and slow rabbit and cat muscles. *Physiol. Bohemoslov.* **21**, 353–359.

Tågerud, S. and Libelius, R. (1984) Lysosomes in skeletal muscle following denervation. Time course of horseradish peroxidase uptake and increase of lysosomal enzymes. *Cell. Tiss. Res.* **236**, 73–79.

Tetzlaff, W., Alexander, S. W., Miller, F. D. and Bisby, M. A. (1991) Response of facial and rubrospinal neurons to axotomy: changes in mRNA expression for cytoskeletal proteins and GAP–43 *J. Neurosci.* **11**, 2528–2544.

Tetzlaff, W., Bisby, M. A. and Brown, M. C. (1993) Comparison of GAP–43 mRNA in axotomised and colaterally sprouting motoneurones in the mouse. *J. Physiol.* **459**, 294P.

Tetzlaff, W., Bisby, M. A. and Kreutzberg, G. W. (1988) Changes in cytoskeletal protein changes in the rat facial nucleus following axotomy. *J. Neurosci.* **8**, 3181–3189.

Tetzlaff, W., Zwiers, H., Lederis, K. *et al.* (1989) The axonal transport and localization of B50/GAP–43-like immunoreactivity in regenerating sciatic and facial nerves of the rat. *J. Neurosci.* **9**, 1303–1313.

Thesleff, S. (1960) Supersensitivity of skeletal muscle produced by botulinum toxin. *J. Physiol.* **151**, 598–607.

Thesleff, S. (1962) Spontaneous electrical activity in denervated rat skeletal muscle, in *The Effect of Use and Disuse on Neuromuscular Functions* (ed. E.

Gutmann, P. Hník and J. C. Eccles), Publishing House of the Czechoslovak Academy of Science, Prague.

Thesleff, S. (1974) Physiological effects of denervation on muscle. *Ann. NY Acad. Sci.* **228**, 89–103.

Thesleff, S. and Ward, M. R. (1975) Studies on the mechanism of fibrillation potentials in denervated muscle. *J. Physiol.* **244**, 313–324.

Thomson, C. E., Mitchell, L. S., Griffiths, I. R. and Morrison, S. (1991) Retarded Wallerian degeneration following peripheral nerve transection in C57BL/6/Ola mice is associated with delayed down-regulation of the PO gene. *Brain Res.* **538**, 157–160.

Titmus, M. J. and Faber, D. S. (1990) Axotomy-induced alterations in the electrophysiological characteristics of neurons. *Prog. Neurobiol.* **35**, 1–51.

Torvik, A. and Skjorten, F. (1971) Electron microscopic observations on nerve cell regeneration and degeneration after axon lesions. II. Changes in the glial cells. *Acta Neuropath.* **17**, 265–282.

Tower, S. S. (1935) Atrophy and degeneration in skeletal muscle. *Am. J. Anat.* **56**, 1–12.

Tower, S. S. (1939) The reaction of muscle to denervation. *Physiol. Rev.* **19**, 1–48.

Turinsky, J. (1987) Dynamics of insulin resistance in denervated slow and fast muscles *in vivo*. *Am J. Physiol.* **252**, R531–537.

Vallee, R. B. and Bloom, G. S. (1991). Mechanisms of fast and slow axonal transport. *Ann. Rev. Neurosci.* **14**, 59–92.

Vrbová, G. (1970) Control of chemosensitivity at the neuromuscular junction, in *Proc. 4th Int. Congress on Pharmacology*, Schwabe & Co., Basel, pp. 158–179.

Waller, A. V. (1850) Experiments on the section of the glossopharyngeal and hypoglossal nerves of the frog, and observations of the alterations produced thereby in the structure of their primitive fibres. *Phil. Trans. Roy. Soc. London B* **140**, 423.

Ward, K. M. and Wareham, A. C. (1986) Effects of denervation on Na^+ and K^+ activities of skeletal muscle. *J. Physiol.* **371**, 269P.

Watson, W. E. (1965) An autoradiographic study of the incorporation of nucleic acid precursors by neurones and glia during nerve regeneration. *J. Physiol.* **180**, 741–753.

Watson, W. E. (1968) Observations on the nucleolar and total cell body nucleic acid of injured nerve cells. *J. Physiol.* **196**, 655–676.

Watson, W. E. (1969) The response of motor neurones to intramuscular injection of botulinum toxin. *J. Physiol.* **202**, 611–630.

Watson, W. E. (1970) Some metabolic responses of axotomized neurones to contact between their axons and denervated muscle. *J. Physiol.* **210**, 321–343.

Watson, W. E. (1972) Some qualitative observations upon the responses of neuroglial cells which follow axotomy of adjacent neurones. *J. Physiol.* **225**, 415–435.

Watson, W. E. (1974) The binding of actinomycin D to the nuclei of axotomized neurones. *Brain Res.* **65**, 317–322.

Watson, W. E. (1976) *Cell Biology of Brain*, Chapman & Hall, New York.

Webster, H. de F. (1962) Transient focal accumulation of axonal mitochondria during the early stages of Wallerian degeneration. *J. Cell Biol.* **12**, 361–383.

Weinberg, J. J. and Spencer, P. S. (1978) The fate of Schwann cells isolated from axonal contact. *J. Neurocytol.* **7**, 555–569.

White, K. K. and Vaughn, D. W. (1991) Age effect of cytochrome oxidase activities during denervation and recovery of three muscle types. *Anat. Rec.* **230**, 460–467.

Wicks, K. L. and Hood, D. A. (1991) Mitochondrial adaptations in denervated muscle: relationship to muscle performance. *Am. J. Physiol.* **260** PC 841–850.

Willard, M. (1977) The identification of two intra-axonally transported polypeptides resembling myosin in some respects in the rabbit visual system. *J. Cell. Biol.* **75**, 1–11.

Willard, M., Meiri, K. F. and Johnson M. I. (1987) The role of GAP–43 in axon growth, in *Axonal Transport* (eds R. A. Smith and M. A. Bisby), Alan R. Liss, New York, pp. 407–420.

Willard, M., Wiseman, M., Levine, J. and Skene, P. (1979) Axonal transport of actin in rabbit retinal ganglion cells. *J. Cell Biol.* **81**, 581–591.

Witzemann, V. and Sackmann, B. (1991) Differential regulation of MyoD and myogenin mRNA levels by nerve induced muscle activity. *FEBS Lett.* **282**, 259–264.

Woolf, C. J., Reynolds, M. L., Molander, C. *et al.* (1990) The growth-associated protein GAP–43 appears in dorsal root ganglion cells and in the dorsal horn of the rat spinal cord following peripheral nerve injury. *Neurosci.* **34**, 465–478.

Wujek, J. R. and Lasek, R. J. (1983) Correlation of axonal regeneration and slow component b in two branches in a single axon. *J. Neurosci.* **3**, 243–251.

Yamada, K. M., Spooner, B. S. and Wessels, N. K. (1971) Ultrastructure and function of growth cones and axons of cultured nerve cells. *J. Cell Biol.* **48**, 614–635.

Yang, J. S.-J., Sladsky, J. T., Kallen, R. G. and Barchi, R. L. (1991) TTX-sensitive and TTX-insensitive sodium channel mRNA transcripts are independently regulated in adult skeletal muscle after denervation. *Neuron* **7**, 421–427.

Zelená, J. (1972) Ribosomes in myelinated axons of dorsal root ganglia. *Z. Zellforsch. Mikcrosk. Anat.* **124**, 217–229.

Zwiers, H. A., Oestreicher, A. B., Bisby, M. A. *et al.* (1987) Protein kinases C substrate B–50 in adult and developing rat brain is identical to axonally transported GAP–43 in regenerating rat nerve, in *Axonal Transport* (eds R. A. Smith and M. A. Bisby), Alan R. Liss, New York, pp. 421–433.

Nerve regeneration and muscle reinnervation

9

The regeneration of the motor nerve and subsequent reinnervation of skeletal muscle are of practical as well as theoretical importance. The motor axon is an extension of the cell body; when part of it becomes separated from the cell body and degenerates, the cell body attempts to restore its original shape and size by replacing its severed portion. Reinnervation of denervated muscle fibres by regenerating nerves ultimately depends on:

- how many motor nerves are successful in growing back to the muscle and forming functional connections, namely the number of functional motor units; and
- how many muscle fibres are reinnervated by each regenerated nerve, namely the motor unit size or innervation ratio (IR).

How many nerves return to the muscle depends on a number of factors, including:

- the nature of the injury and repair;
- the capacity of the axotomized nerves to grow;
- the growth environment of the distal nerve stump up to and within the denervated muscles; and
- the capacity of the muscle fibres to accept reinnervation.

Similarly there are several factors which determine how many muscle fibres each regenerated nerve supplies. These include:

- the capacity of the motor nerves to branch;
- the growth environment provided by the intramuscular nerve sheaths and the denervated muscles;
- the capacity of the denervated muscles to accept innervation; and

- the competition between regenerating nerve fibres in maintaining contacts at the reinnervated neuromuscular junctions.

In addition to the number and size of motor units, the recovery of neuromuscular function also requires reversal of the atrophic effects of axotomy and denervation on the nerves and muscles respectively.

Although considerable progress had been made in elucidating many of the events that take place during nerve regeneration and muscle reinnervation, many fundamental issues are still obscure. This chapter reviews the present knowledge of the different aspects of motor nerve regeneration and muscle reinnervation and how these influence the number and size of functional motor units in the reinnervated muscles.

The ability of the peripheral nerve to restore its functional connections after injury was already recognized in the eighteenth century (Fontana, 1781). The question as to whether this regeneration was the result of a peripheral process whereby the severed axon would be reconstituted by peripheral mechanisms, or whether regeneration was the result of an outgrowth of nerve fibres from the central stump into the periphery, was resolved by Ranvier (1874) and Vanlair (1882). They provided indisputable morphological evidence that the new nerve fibres seen during regeneration grow out from the central stump of the nerve and are therefore probably synthesized by the cell body.

Although the changes in the peripheral stump and their importance in the regenerative process were often used as an argument against a primary involvement of the neuronal cell body in nerve regeneration (Bethe, 1901), the role of the peripheral nerve stump was eventually greatly overshadowed by the importance given to the cell body in the replacement of its missing axon. Nevertheless, the changes in the peripheral stump including the multiplication of the Schwann cells and their formation of long, multinucleated bands of cells surrounded by basal lamina – the so-called 'bands of Büngner' (Büngner, 1891) – have great importance for the subsequent growth of nerve fibres from the central stump (Ramón y Cajal, 1928; Nathaniel and Pease, 1963a, b; Thomas, 1964a, b) and for their guidance to their destination (Weiss, 1943; Weiss and Taylor, 1944). Thus the argument as to the origin of the regenerating nerve fibres was resolved in favour of the nerve cell but the essential role of the distal nerve stump and its resident non-neural cells in regeneration is now appreciated and is the subject of intense investigation.

The process of regeneration of motor nerves can be divided into three major events:

1. Nerve growth.
2. Establishment of connections with the muscle fibres.
3. Maturation of the regenerated axon and reversal of the effects of axotomy and muscle denervation.

9.1 NERVE GROWTH

9.1.1 THE NEURONAL RESPONSE, REGENERATION AND MYELINATION

Following nerve section or nerve crush, several changes are seen at the place of injury, both in the central and peripheral stumps. Swelling of the central stump near the injured site was noted and considered to be an expression of 'damming' of substances usually transported from the cell body down the axon, towards the muscle (Weiss and Hiscoe, 1948). Accumulation of enzymes, cytoskeletal proteins and organelles is also thought to be caused by deposition of materials which continue to be transported by slow and fast axoplasmic transport despite the injury (Zelená et al., 1968; Heslop, 1975; Friede and Bischhausen, 1980). Local reaction to the lesion may also contribute to the changes but the swelling is due mainly to accumulation of transported material from the cell body. For example, if a nerve is crushed in two places, the swelling occurs only at the proximal crush site from whence thin nerve fibres emerge a few hours later (Gutmann, 1958; McQuarrie, 1985). Transport of material does not require continuity between axon and soma; it still carries on in the isolated nerve stump, resulting in a small accumulation of material at its distal end (e.g. Ljubinska and Niemierko, 1971).

The cell body reacts within a few hours to the injury to its axon (for details see Chapter 8).

The signal for the switch of the cellular machinery from transmitting to growth mode is poorly understood and puzzling. Cragg (1970) suggested that a possible signal for chromalysis may be the arrest of axoplasmic transport by the lesion so that the flow of material down the axon is dammed. Interestingly, the signal is sufficient to induce growth of other neuronal processes, whether or not they are injured. Injury to motoneurones can even induce growth of additional axon-like processes from motoneurones in the spinal cord, which are not normally there (Linda et al., 1985; Havton and Kellerth, 1987).

For many years, it was believed that a few days elapsed (latent period) before the injured nerve shows signs of regeneration (Gutmann et al., 1942). Using the 'pinch test' introduced by Young and Medawar (1940), the distance travelled by the regenerating sensory nerves was determined by observing the point on the distal nerve stump at which a response is elicited by a pinch to the regenerated nerves. Observations made of the distance of regeneration in a number of different animals at progressively longer intervals allowed the regeneration rate to be calculated. The latent period was obtained by extrapolation of plots of regeneration distance vs time to zero outgrowth distance and was based on the assumption that the rate of regeneration is linear from the start of axonal growth.

However, Wyrwicka (1950) and later Sjoberg and Kanje (1990), using the same pinch test, showed that there is little delay in the emergence of regenerating sprouts from the central nerve stump. Rate of elongation is initially slow and accelerates during the first days to reach a constant velocity by 3 days. It is this early period of regeneration during the latent period which varies according to the type of injury, age and species of animal (Gutmann *et al.*, 1942; Black and Lasek, 1979; Pestronk *et al.*, 1980). After nerve crush, where the continuity of the nerve sheath and basement membranes are preserved (Thomas, 1964a), the latent period is shorter than after total separation of the central and peripheral stumps by cutting. It is also shorter in young animals and varies in different species (Gutmann *et al.*, 1942).

The emergence of regenerating nerves within hours of the injury indicates that growth cone formation occurs without direct support from the cell body and depends on materials which are locally available in the axon (McQuarrie, 1985). Pre-existing cytoskeletal elements in transit are transported into the daughter axons emerging from the proximal stump (McQuarrie and Lasek, 1989). The microtubules provide a bridge from the parent to daughter axons for transfer of other selected materials which, in turn, support axonal growth (Miller *et al.*, 1987). Continued transport even in isolated axons is sufficient to allow growth cone formation *in vitro* (Bray *et al.*, 1978; Shaw and Bray, 1977; Wessells *et al.*, 1978).

Axons in the central stump 'die back' to the first node of Ranvier ('traumatic' degeneration: Ramón y Cajal, 1928), from whence sprouting daughter nerve branches arise (Ramón y Cajal, 1928; Friede and Bischhausen, 1980; McQuarrie, 1985). Ranson (1912) claimed that often a single axon splits and branches into as many as 50–100 branches. Peroncito (1907) saw that these young branches often formed spirals, or ringlets, which disappeared during the later stages of regeneration. If, however, the growing nerve fibres do not find any 'guiding' structures they produce a neuroma which is a mixture of immature nerve fibres and connective tissue. Young (1948) considered such a neuroma to be an expression of the residual growth capacity of the neurone which is disorganized in the absence of guidance by Schwann cells.

When the numerous fine nerve fibres emanating from the parent axon grow within the supportive growth environment of the distal nerve stump, an average of five daughter axons per axon regenerate (Greenman, 1913; Gutmann and Sanders, 1943; Aitken *et al.*, 1947; Bray and Aguayo, 1974; Jenq and Coggeshall, 1985; Toft *et al.*, 1988). The axonal sprouts that advance distally comprise a 'regenerating unit' (Morris *et al.*, 1972), which remains in the stump if the nerves do not make target connections (Figure 9.1a). However, only one remains once connections with the target are made (Figure 9.1c; Aitken *et al.*, 1947; Mackinnon *et*

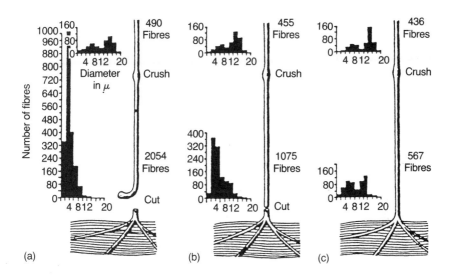

Figure 9.1 Experiments to show the effect of presence and absence of peripheral connections on the size and number of nerve fibres reached during regeneration in rabbits. The nerve to the medial head of gastrocnemius is crushed on both sides, and on one side is also cut lower down. In (a) it is prevented from reaching the muscle; in (b) it is allowed to re-establish contact; in (c) the nerve is left undamaged below the crush. The histograms show the conditions of the nerves above and below the crushed points 100 days after the operation. On the side where connection with the periphery was prevented, the number and size of the nerve fibres remained abnormal. (Reproduced from Aitken, Sharman and Young, 1937 *J. Anat.* 81, 1–22.)

al., 1991). The number of daughter axons is generally higher if the nerves have been severed and the proximal and distal nerve stumps sutured together, compared with the number after a crush injury that maintains the integrity of the basement membranes (Jenq and Coggeshall, 1985; Giannini *et al.*, 1989). Although the regenerating axons in the distal nerve stump grow in diameter, and become myelinated (Figure 9.1b), they do not approach normal size unless they make functional connections (Aitken *et al.*, 1947; Gordon and Stein, 1982a). The conduction velocity and the diameter of regenerating axons is always proportional to the conduction velocity of the parent axons from the earliest time when sprouts first grow out from the proximal stump (Devor and Govrin-Lipmann, 1979a). This suggests that the parent axons may determine the size of the regenerating axons. Less specific experiments of Simpson and Young (1945) and Evans (1947) also pointed to this conclusion. They found that somatic nerves which regenerated in the narrow tubes of degenerating sympathetic fibres attained larger diameters than the nonmyelinated fibres that previously occupied the tubes. The nerve and not the tube appeared to determine the diameter of regenerating fibres.

Nevertheless, remyelination of the regenerating axons is initiated by contact of the axolemma with the Schwann cells. The interaction induces galactocerebroside synthesis (Mirsky and Jessen, 1988) and expression of myelin-associated proteins in the Schwann cells (Lemke, 1988; LeBlanc and Poduslo, 1990) and initiates the spiralling of the myelin sheath (Bunge *et al.*, 1986, 1990). The strong relationship between the size of axons and their myelination was also demonstrated in an animal model where the capacity of Schwann cells to myelinate axons is defective, i.e. the Trembler mouse (Aguayo *et al.*, 1977). Using grafted nerve segments the authors found that, when a graft taken from a Trembler mouse was placed into a normal host, the regenerating axons of the host were abnormally small and poorly myelinated whilst in the grafted nerve segment but returned to normal once they reached the distal nerve sheath of the host mouse.

Remyelination also depends on the formation of basal lamina and sufficient numbers of Schwann cells. *In vitro* experiments show that axonal contact is essential for Schwann cells to assemble basal lamina (Bunge *et al.*, 1982; Clark and Bunge, 1989). The axonal contact stimulates the synthesis and secretion of the glycoproteins of the basal lamina, which include laminin, entactin, type IV and V collagen and heparan sulphate proteoglycan (reviewed by Bunge and Bunge, 1983; Bunge *et al.*, 1990). In turn, the basal lamina is essential for the Schwann cell to ensheath the axon and form myelin (Bunge *et al.*, 1986, 1989). Axonal contact is also mitogenic for Schwann cells (Salzer and Bunge, 1980; Salzer *et al.*, 1980a, b) with the result that Schwann cells contacted by the ingrowing axons undergo further mitosis (Pellegrino and Spencer, 1985). Thus *in vitro* experiments indicate that contact with the axon leads to Schwann cell division, up-regulation of myelin-associated proteins and synthesis of molecules for the basal lamina. These changes are concurrent with the down-regulation of the expression of growth-associated proteins, including cell adhesion molecules and growth factors and their receptors. These conditions being met, the Schwann cells ensheath the axons and form myelin, while fibroblasts elaborate the perineurium (Scaravilli, 1984).

9.1.2 SCHWANN CELLS AND THE GROWTH ENVIRONMENT

In order to reach its destination, the growing nerve fibre must find its peripheral stump, or have some form of guidance towards the structures it is to reinnervate. The changes in the peripheral stump are initially degenerative and associated with the breakdown and removal of axonal material and myelin debris (Chapter 8; Gutmann, 1958; Perry and Brown,

1992). Schwann cells proliferate during this degenerative phase and co-operate with the infiltrating macrophages in the active phagocytosis of degenerating nerv. and myelin debris (Chapter 8; Abercrombie and Johnson, 1946; Aguayo *et al.*, 1976; Allt, 1976; Baichwal *et al.*, 1988; Beuche and Friede, 1984; Salonen *et al.*, 1988; Clemence *et al.*, 1989; Perry and Brown, 1992). Schwann cells also play a vital role in the mechanical guidance of nerve fibres in the distal nerve stump by formation of the 'bands of Büngner' (Büngner, 1891). Together with the fibroblasts which proliferate at the injury site, they accumulate and extend towards the central stump, thereby bridging the gap between the central and peripheral stumps in addition to providing conducting protoplasmic bands inside the empty neurilemmal sheaths (Ramón y Cajal, 1928; Abercrombie and Johnson, 1946). Schwann cells therefore help to guide the regenerating axons from the central stump, across the injury site and through the distal stump to peripheral target organs. They finally myelinate the regenerating axons as described above.

The Schwann cells in the peripheral stump of a severed nerve provide an excellent environment for axonal regeneration; no other cell type, tissue or artificial conduit has been shown to support the regeneration of large numbers of axons over distances of many centimetres. However, a pathway of Schwann cells may not be obligatory for peripheral nerve regeneration over short distances. Axons can regenerate into segments of freeze-killed nerves which initally lack Schwann cells (Anderson *et al.*, 1983; Hall, 1986a, b; Ide *et al.*, 1983; Bresjanac and Sketelj, 1989; Sketelj *et al.*, 1989) but if there is no continuity with the peripheral stump the distances covered are relatively short (Zalewski and Gulati, 1982; Nadim *et al.*, 1990). Similarly, axons can regenerate across small gaps of 1–2cm in severed peripheral nerves (Mackinnon *et al.*, 1991). Schwann cells which migrate into the gap of freeze killed nerve segment from both the central and peripheral stumps (Thomas, 1966) may play an essential part in this type of regeneration (Anderson *et al.*, 1983; Hall, 1986b; but see also Ide *et al.*, 1983; Bresjanac and Sketelj, 1989). Although there is little evidence that basal laminae can support prolonged axonal regeneration in the absence of Schwann cells, the inner aspect of basal lamina tubes derived from killed peripheral nerve or skeletal muscle does constitute a preferred substrate for both axonal elongation and Schwann cell migration (Anderson *et al.*, 1983; Ide *et al.*, 1983; Glasby *et al.*, 1986). This may partly explain why regeneration through acellular tissues containing basal lamina tubes, although inferior to regeneration through living nerve (Anderson *et al.*, 1983; Gulati, 1988) is successful over longer distances than that bridged by artificial conduits such as silastic tubes (Lundberg *et al.*, 1982; Rich *et al.*, 1989).

The interactions of Schwann cells with degenerating and regenerating axons are complex and have been studied intensively in recent years.

In principle, Schwann cells have been shown to produce a variety of neurotrophic factors and cytokines including nerve growth factor (**NGF**), brain derived neurotrophic factor (BDNF) and cilliary neurotrophic factor (**CNTF**), and express important cell surface and extracellular matrix molecules including the neural cell adhesion molecules L1/NILE, N-CAM, N-Cadherin, laminin and tenascin (Bunge *et al.*, 1990; Richardson, 1991). All of these molecules are able to exert a profound influence over neurite outgrowth from various types of neurones in culture, but the identity of the factors which are critical for the unique ability of Schwann cells to support prolonged axonal regeneration *in vivo* remains unclear. A discussion of the activities of these molecules is beyond the scope of this chapter (for review see Bixby and Harris, 1991; Carbonetto, 1991).

Other cells including macrophages, perineurial cells, fibroblasts and endothelial cells may also play important roles in regeneration but cannot replace Schwann cells. It is characteristic of other repair mechanisms, including the co-ordination of blood coagulation factors and platelets in the repair of injured blood vessels, that there is considerable co-operation between the many different components which repair damage and support cellular growth. As a result there may be some redundancy, yet the absence of one particular factor may be critical. Furthermore, the number and function of extracellular matrix proteins may be underestimated at this time when many different genes encoding new isoforms of extra-cellular matrix proteins are being identified (Reichardt and Tomaselli, 1991; Bixby, 1992).

9.1.3 DIRECTION OF REGENERATION: SPECIFICITY?

Generally, the results of nerve regeneration, muscle reinnervation and functional recovery are best after nerve crush injuries in which the normal alignment of basement membranes remains intact (Sunderland, 1978; Haftek and Thomas, 1968). After the disconnected axons undergo Waller-ian degeneration, the regenerating nerve sprouts enter the same 'Schwann cell tubes' as the removed axons previously occupied. Under these conditions, regenerating nerves are 'led' to their original targets. The normal numbers and diameters of myelinated fibres are often fully reconstituted (Gutmann and Sanders, 1943), the normal mosaic distri-bution of muscle fibre types in the reinnervated muscles is often recov-ered (Warszawski *et al.*, 1975) and function is fully restored (reviewed by Sunderland, 1978). In contrast, the number of regenerating fibres in the peripheral stump and fibres which are successful in remaking peripheral contacts is far more variable after lesions which disrupt the Schwann cell tubes. The nerve fibres which succeed in remaking nerve–muscle connections show full recovery (Gordon and Stein, 1982a). The capacity of these fibres to reinnervate more muscle fibres than normal compensates for the loss of nerves which fail to regenerate (Rafuse, 1992). Rein- nervation, therefore, can lead to good recovery of muscle fibres from denervation atrophy and the reinnervated muscles can regain their former weight and develop normal muscle tensions (Gordon and Stein, 1982a, b). However, functional recovery may still be very disappointing. Regeneration of axons into distal pathways which guide the nerves to different and often inappropriate targets has long been recognized as a possible mechanism for incomplete or absent functional recovery (Langley and Hashimoto, 1917; Sunderland, 1978). This poor functional recovery is the consequence of a relative inability of axons to find and selectively grow within their original Schwann cell tube.

Experiments in which severed nerves were re-anastomosed illustrate the lack of specific guidance. Longitudinal sections through the suture line show extensive criss-crossing of fibres (Gutmann and Sanders, 1943) so that the chances of axons finding their original Schwann sheaths is remote. Most axons are likely to enter tubes which were formerly occu- pied by others and they can readily be directed to different and often inappropriate targets. Normally, axons are confined to a single Schwann cell tube throughout their distal course (Holmes and Young, 1942; Brown et al., 1981; Brown and Hardman, 1987). The loss of fibres to inappropriate target organs is particularly high in large mixed nerves whose sensory, motor and autonomic axons are normally destined for quite different targets. Results of recent experiments in which cut femoral nerves were given the choice to regenerate into nerve branches to the skin and to muscle suggested that some specific motor nerve markers remained in the motor branch to favour the regeneration of motor nerves, particularly in neonatal animals (Brushart, 1988, 1994). Brushart reported that there were significantly more motor fibres which regenerated into the 'motor' branch than into a sensory branch, irrespective of whether the nerves were permitted to grow to the denervated target. When parent sensory axons sent daughter axons into both channels, the motor nerves in the 'sensory' pathway appeared to be preferentially withdrawn (Brushart, 1990). However, these results are difficult to interpret in view of the massive branching that persists in target-deprived axons. The target connections also have a retrograde influence: more motor nerves rein- nervated the muscle than sensory targets even when the nerves regener- ated through a 'sensory' graft. The 'L2 carbohydrate' identified with L2 and HNK–1 antibodies to the carbohydrate epitope has been identified as a strong candidate for this 'motor' specificity because this epitope of several N-CAMs is selectively expressed on Schwann cells and basement membranes of motor axons but not on sensory axons and it persists after Wallerian degeneration (Martini et al., 1992; Brushart, 1994).

However, this specificity is limited and may apply to motor versus sensory nerves, but regenerating motor nerves do not appear to be able to 'select' the pathways to their original or appropriate muscles when their endoneuronal sheaths have been disrupted. Early experimental studies in which original and foreign nerves competed for reinnervation of mature muscle provided conflicting data. Some suggested preferential reinnervation (Hoh, 1975; Politis *et al.*, 1982) while others did not (Weiss and Hoag, 1946; Miledi and Stefani, 1969; Bernstein and Guth, 1961; Gerding *et al.*, 1977). Suggestions were also made that there may be segmental factors which persist in the mature peripheral nervous system and which determine specificity in a cranio-caudal direction (Sanes, 1993). However, the experiments are often complicated by the surgical manœuvres used to set up the competition between regenerating nerves, making it possible that intervening factors other than specificity of regenerating nerves could account for the findings. A more natural competitive condition is that of section and resuture of the common LGS nerve which supplies two separate and synergistic muscles, the slow soleus and fast lateral gastrocnemius muscles. If the nerves show any preference for their former muscles, the 'fast' nerves should preferentially reinnervate the 'fast' LG muscle and the 'slow' nerves should continue to grow through the LG muscle to the more distant 'slow' soleus muscle, which they did not: slow and fast nerves reinnervated each of the muscles randomly (Gillespie *et al.*, 1986, 1987). Reinnervation of intrinsic muscles of the hand after surgical repair of either the median or ulnar nerves also occurred in a random fashion (Thomas *et al.*, 1987) consistent with the random process of reinnervation demonstrated using retrograde marking of motoneurones (Brushart and Mesulam, 1980). The functional consequences of misdirection may be devastating, as in the case of regenerated facial motor nerves where voluntary activation of muscles results in inappropriate movements which may be extremely distressing and maladaptive (reviewed by Gordon, 1994). In primates and humans, these maladaptive functions can be to some extent corrected by training (Sperry, 1945).

Results of studies on reinnervation in neonatal animals suggests that some guidance cues that control axonal growth in early development may persist when nerves regenerate in neonatal animals. Thus there is some degree of specificity of reinnervation to facial musculature after facial nerve transection (Aldskogius and Thomander, 1986), and of intercostal muscles by segmental nerves (Hardman and Brown, 1987; DeSantis *et al.*, 1992). Thus some degree of specificity may exist in the immature neuromuscular system, but this seems to be absent later (e.g. Aldskogius and Thomander, 1986; see also Farel and Bemelmans, 1986; Wigston, 1986; Wigston and Kennedy, 1987; Wigston and Donahue, 1988).

9.1.4 RATE OF REGENERATION

The rate at which motor nerves regenerate varies and is thought to depend mainly on the speed at which the cell body can resynthesize and supply the required components for the reconstitution of the missing axon. Since regeneration rates correspond well with the rate of the SCb component of slow axonal transport, it is thought that the SCb transport is rate-limiting (Hoffman and Lassek, 1980). Kučerová (1949) found that myelinated axons with larger diameters regenerate faster but there is little information that compares rates of growth of axons from clearly defined motoneurones. An accurate estimate of the rate of growth of motor axons was first obtained by Gutmann (1942). The common peroneal nerve of the rabbit supplies muscles that spread the toes when the rabbit is lifted off the ground. On crushing the nerve, this reflex naturally disappears and its reappearance can be used as an indicator of reinnervation. The motor nerve fibres were crushed at different distances from the peroneal muscles and the times of the reappearance of the spreading reflex were noted. This allowed an assessment of the rate of growth. Figure 9.2 illustrates results from such experiments. It was found that there was a marked delay in the reappearance of the spreading reflex when the injury was further from the muscle. From these experiments it was calculated that the rate of functional motor nerve regeneration was 2.77 ± 0.09 mm/day (Gutmann et al., 1942).

The growth rate of an already regenerating axon can be significantly accelerated by a second crush lesion. This effect has come to be known as the **conditioning lesion**, following the original observations of Gutmann (1942) and McQuarrie and Grafstein (1973) that axons reach their destination sooner when the test crush is preceded by the 'conditioning' crush. It is as if the cell body is already primed to replace its missing axon, the basis being the up-regulation of growth-associated proteins and their transport into the axon prior to the outgrowth of nerve sprouts (McQuarrie and Jacob, 1991). The priming effect of the conditioned lesion appears to be generalized for all nerve processes, because a peripheral axonal transection of sensory nerves accelerated axonal regeneration of the central process of dorsal root ganglion cells (Richardson and Verge, 1987). In addition to the effect on the axotomized neurone, the first lesion primes the growth environment by initiating the proliferation of the Schwann cells in the distal nerve stump (Sjoberg and Kanje, 1990). This effect is of considerable interest in nerve repair, particularly with reference to the preparation and storage of nerve grafts where these are necessary to bridge gaps between severed nerves. For example, early regeneration of nerve sprouts proceeded more rapidly with more myelinated nerve fibres than normal when the nerve grew through a distal nerve stump which had been denervated a month previously (Sorenson

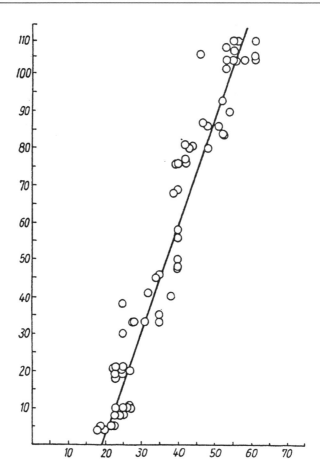

Figure 9.2 Recovery of motor function (toe-spreading reflex) in days (abscissa) after a crush injury to the peroneal nerve at various distances from the muscle (ordinate). (Data based on Gutmann, 1942, *J. Neurol. and Psychiat.* 5, 81.)

and Windebank, 1993). The more successful regeneration was attributed to increased neurotrophic activity in the denervated nerve stump but no other factors were considered by these authors. It has also been found that a nerve stump becomes a less effective conduit with time (Fu and Gordon, 1995 a, b; You *et al.*, 1994).

It is not known whether activity of the damaged neurone influences its rate of growth, despite considerable interest. Experiments of Beránek and Gutmann (1958) showed that, following section of the spinal cord, regenerating motor nerve fibres re-established functional connections much later than when the cord was intact, while the regeneration of sensory fibres in the same nerve trunk was unaffected. Spinal cord section was assumed to have reduced the activity of the motoneurone and thus reduced the rate of growth. It was therefore surprising to find that extensive exercise induced by swimming not only did not speed up the recovery of the function from a nerve crush but also considerably delayed it (Gutmann and Jakoubek, 1963). Since in these experiments it was the recovery of muscle function that was studied, it was impossible to say whether these procedures affected the growth of the axons or the establishment of connections between nerve and muscle. It is possible that the apparent discrepancy is due to the fact that the growth of axons may be increased by activity of the neurone, while establishment of connections is impaired by muscle activity (Chapters 3 and 6). Magnetic sensory pathway stimulation, possibly mediated by Ca^{2+} mobilization, has a significant accelerating effect on the rate of sensory regeneration (Kerns and Lucchinetti, 1992; Rusovan et al., 1992).

It is arguable whether the rate of regeneration declines with the distance from the cell body. While Sunderland (1950) claims this to be the case in humans, others found that the distance of the lesion from cell body does not alter the speed of regeneration (Gutmann et al., 1942). It may be that, since most of the experiments have been performed on relatively small laboratory animals, the nerve may not have been long enough to display such effects.

From the point of view of functional recovery, the critical issues are the number of regenerating nerves that succeed in traversing the long distances to reach their denervated targets and their ability to form neuromuscular contacts. The long distances necessitate long periods in which the axotomized neurones must maintain their ability to regenerate their axons, the distal stumps maintain their supportive growth environment and the denervated targets maintain their capacity to accept reinnervation. Even when functional connections are made, recovery will depend on the capacity to reverse changes of axotomy and denervation on the nerves and muscles, respectively, and on the maturation and myelination of the regenerated axons.

The ability of the neurone to replace its missing axon is apparently maintained for long periods since nerves can regenerate after long delays incurred by either repeated crush injuries (Holmes and Young, 1942) or by first suturing a cut nerve to a nearby innervated muscle to prevent regeneration and later detaching the nerve for cross-anastomosis to a freshly denervated nerve stump to encourage nerve regeneration (Fu and Gordon, 1995a). Nonetheless, regenerative capacity declines significantly the longer the nerves remain axotomized before nerve repair, even when the delayed regenerating nerves grow in an optimal growth environment (Fu and Gordon, 1995a). Once they reach the muscle, they demonstrate a normal capacity to branch and reinnervate muscle fibres and thereby compensate for the reduced numbers of regenerating nerves that reinnervate denervated muscle fibres.

The reverse condition of a freshly cut nerve regenerating into an 'old' denervated nerve stump, on the other hand, was much more detrimental for functional recovery. A freshly cut nerve which was forced to regenerate in a long-term denervated nerve stump showed a marked drop in the number of nerves which successfully traversed the distal nerve stump (Fu and Gordon, 1995b). It has been suggested that a decline in the number of Schwann cells with time may contribute to poor regeneration after delayed nerve repair (Hudson, 1984; Pellegrino and Spencer, 1985). Other suggestions include the possibility that the channels between Schwann cells and basement membrane become restrictive to growth (Sunderland, 1978), that the basement membrane fragments (Giannini and Dyck, 1990) and that expression of growth-associated gene products declines with time (You *et al.*, 1994). Once the nerves reach the muscle and grow within intramuscular nerve sheaths, the regenerating nerves may stray from the degenerating endoneurium and grow over the denervated muscle surface (Gutmann and Young, 1944). Finally, the capacity of long-term denervated muscle to receive innervation and recover from denervation atrophy declines with time (Gutmann and Young, 1944; Anzel and Wernig, 1989).

9.2 ESTABLISHMENT OF NERVE–MUSCLE CONNECTIONS BY THE REGENERATING NERVE FIBRE

9.2.1 FORMATION OF FUNCTIONAL CHOLINERGIC NERVE–MUSCLE CONTACTS

When the regenerating motor axon reaches the skeletal muscle, it branches and these branches continue to grow inside the muscle, following guidelines formed by the Schwann cells. Figure 9.2b shows such regenerating axons enclosed in the 'bands of Büngner' formed by the Schwann cells in the extensor digitorum longus muscle of a rat. Figure 9.3a shows denervated endplates that are stained for cholinesterase and the rows of Schwann cells leading to them. Figure 9.3c illustrates that the regenerating fibre is stopped at the old endplate site.

In their systematic and thorough study of the interaction between the regenerating motor nerve fibres and the denervated muscle fibre, Gutmann and Young (1944) noted that growing motor axons are usually seen to terminate at the original motor endplate and re-establish contact at this site, confirming the first observations of Tello (1917). Moreover, within the terminal, new active zones of transmitter release assemble precisely opposite post-synaptic junctional folds, where the density of ACh receptors is highest, thereby reconstituting the original geometry

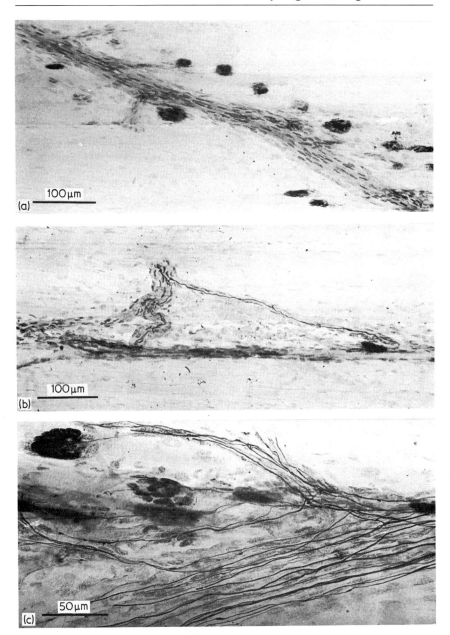

Figure 9.3 Changes occurring in a rat EDL muscle on reinnervation by its motor nerve, in preparations stained for cholinesterase and impregnated by silver; (a) vacated endplates can be seen with empty bands of Büngner leading to them; (b) thin regenerating fibres invading the guiding bands of Büngner and contacting endplates; (c) profusely growing nerve fibres contacting the original endplate sites.

of the synapse (Rich and Lichtman, 1989). Almost as soon as the regenerating motor nerve terminals arrive at the endplate, transmission from nerve to muscle is possible (Tonge, 1974; Miledi, 1960b). Such rapid recovery of neuromuscular transmission can take place only when the muscle was denervated for a short period by crushing the nerve close to the entry of the muscle. After longer periods of denervation this is not the case; the nerve fibres sometimes fail to find the original endplate region, often forming new 'ectopic' endplates on these very atrophic muscle fibres (Gutmann and Young, 1944). Figure 9.4 illustrates this point.

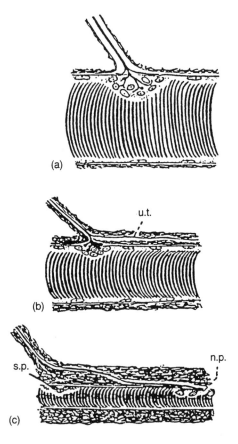

Figure 9.4 Rabbit nerves and muscles after various delays before reinnervation: (a) reinnervation of the muscle occurred soon after injury and the muscle fibres are not yet severely wasted; (b) reinnervation is slightly delayed and muscle fibres are atrophic – note the ultraterminal (u.t.) sprout at the endplate; (c) reinnervation after a long delay: the nerve ending grows past the old endplate (s.p.) and elaborates a new endplate (n.p.) some distance away from the old one. (After Gutmann (1992) *J. Neurol., Neurosurg. and Psychiat.*, **5**, 81–85.)

At the beginning of the century, Langley and Anderson (1904) performed a series of experiments to show that, while any cholinergic nerve can replace a skeletomotor nerve and establish functional connections with the muscle fibres, adrenergic nerves cannot assume the function of a skeletomotor nerve and are unable to establish functional connections with skeletal muscle fibres. Consistent with this was the observation of Gutmann and Young (1944) that a cutaneous nerve, although able to grow and sprout in a denervated muscle, did not form functional contact with it nor was it able to prevent the muscle from becoming reinnervated by a motor nerve. These results clearly show that functional contact between nerve and muscle can only be established if the nerve releases acetylcholine (ACh), the transmitter normally released from skeletomotor nerve endings.

As described in Chapter 8, following denervation, the muscle membrane becomes more sensitive to the transmitter outside the endplate region, but the ACh receptors (AChRs) at the old endplate are remarkably stable so that this region remains the most sensitive area of the muscle fibre (Miledi, 1960a; McArdle and Albuquerque, 1973). There is evidence from *in vitro* studies of growth cones that the growing nerve terminals release transmitter instantaneously upon contact with the muscle fibre, and even while growing in search of contact (Hume *et al.*, 1983; Young and Poo, 1983). Once the nerve makes contact with muscle, spontaneous and evoked release increases rapidly (Xie and Poo, 1986), possibly mediated by retrograde signals from the muscle which trigger a rapid and transient rise in intracellular levels of Ca^{2+} in the growth cone (Hall and Sanes, 1992). The special features of transmitter release of immature nerve endings after early reinnervation *in vivo* have been described for the frog neuromuscular junction by Miledi (1960b) and for the mammalian and avian neuromuscular junction by a number of authors (McArdle and Albuquerque, 1973; Bennett *et al.*, 1973; Bennett and Florin, 1974; Tonge, 1974). Initially the endings of regenerating axons release less transmitter as indicated by the reduced frequency of miniature endplate potentials and by the low quantal content. Moreover, the axons are unable to follow high-frequency activity so that not every impulse will invade the presynaptic terminal (Miledi, 1960b; Dennis and Miledi, 1971, 1974; Bennett *et al.*, 1973). The low quantal content is probably related to the small size of the nerve terminal since it is known that quantal content and nerve terminal size and length are directly correlated (Kuno *et al.*, 1971) and there is an inverse relationship between synaptic efficacy (transmitter release per unit terminal length) and terminal length (reviewed by Wernig and Herrara, 1986).

The structure of such immature contacts is different from that of normal ones. Gutmann and Young (1944) noticed in silver-stained preparations that terminal branches of regenerated axons appear thicker, and more recently it was found that individual endings contacting endplates branch less (Ip and Vrbová, 1973). On the ultrastructural level, several small axon profiles are seen to contact the muscle fibre, not unlike those seen during early development (Chapters 3 and 6). Many of the reinnervated endplates are initially supplied by nerve branches from two or more different axons, and only gradually does this 'polyneuronal' innervation subside (McArdle, 1975; Jansen and Van Essen, 1975; Letinsky et al., 1976; Rothshenker and McMahan, 1976; Gorio et al., 1983).

In a recent series of experiments, Rich and Lichtman (1989) used a vital nerve-terminal dye and fluorescently tagged α-bungarotoxin to label AChRs to study reinnervation of a mouse muscle. This elegant method allows repeated observation of the nerve terminal and AChRs in the post-synaptic membrane. The results of this study show that, on reinnervation, axon terminals converge on to the original endplate. A transient period of polyneuronal innervation is followed by elimination of excess terminals at individual neuromuscular junctions. A loss of AChR under the terminal that retracts seems to precede the process of terminal withdrawal, indicating that there is a link between the loss of AChR and elimination of the terminal.

These interesting results again indicate that the muscle participates in the reorganization of its synaptic inputs (Chapters 6 and section 9.4).

Thus the re-establishment of transmission from nerve to muscle at the original endplate sites, and the redistribution of the AChR at the endplate, is one of the earliest events on reinnervation (Miledi, 1960b). This is followed by the maturation of the synapse as well as the recovery of the reinnervated muscle and of the regenerating axon. The maturation of the synapse involves both the redevelopment of the presynaptic and post-synaptic parts of the synapse, which seems to occur relatively quickly (section 9.4), while the recovery of the nerve and muscles takes a remarkably long time. All these stages of reinnervation depend on the intimate interaction between the nerve and muscle.

9.2.2 GROWTH ENVIRONMENT IN THE DENERVATED MUSCLE

Why do regenerating nerves reinnervate denervated muscle fibres at the original endplate site? Attraction of the cholinergic nerve to the site of highest ACh sensitivity was a possible explanation. If so, regenerating nerves should not be attracted back to the endplate region when the muscle membrane is destroyed and AChR receptors are removed. In experiments in which the muscle fibres were cut and irradiated to prevent muscle regeneration, leaving only the empty basal lamina sheath 'ghosts', McMahan and his colleagues showed that regenerating nerves still grew to the original endplate sites (Letinsky et al., 1976; Marshall et al., 1977). Furthermore, once the nerves reached these sites, they ceased to grow and developed secreting terminals despite the absence of a receptive muscle (Glicksman and Sanes, 1983; Sanes, 1989). These striking findings showed that components of the basal lamina at the endplate site accounted, at least in part, for the attraction of regenerating nerves to that site and their formation of the neuromuscular synapse. Generally the nerves regenerated inside the naked basal lamina sheaths of the original nerve tubes and were accompanied by Schwann cells (see Kuffler, 1986), suggesting that the guidance could be attributed, at least in part, to the naked sheaths leading the regenerating nerves to the endplate sites. Nonetheless, the number of synaptic sites occupied was relatively low and increased significantly only if muscle fibres regenerated from satellite cells (Kuffler, 1986). The regenerated nerves were myelinated by Schwann cells which accompanied the nerves and, similar to the situation in normal junctions, became capped by a Schwann cell at the re-formed neuromuscular junction. If agarose implants containing antibodies against N-CAM were applied to the damaged muscles during the reinnervation of naked basal lamina sheaths, fewer synaptic contacts were made at the original endplate sites; most contacts were made at ectopic sites where Schwann cells were mostly absent from the nerve terminal (Rieger et al., 1988). Myelination of the regenerating nerves was also poor.

As illustrated in Figures 9.3. and 9.4, regenerating nerves normally traverse the relatively long distances required to reach the old endplate sites in the denervated muscles by following the guidelines formed by the Schwann cells in the intramuscular nerve sheaths. Presumably the supportive role of Schwann cells and the basement membrane within the intramuscular sheaths is similar to that in the distal nerve sheaths which guides the regenerating nerves to the denervated muscles (section 9.1.2). Experiments in which regenerating nerves were forced to grow outside the intramuscular sheaths, by implanting the nerves far from the nerve entry zone, showed that the number of nerve fibres which successfully regenerated into the muscle and made functional connections was reduced (Gordon, 1994) but the contacts were still made preferentially at the endplate regions.

9.2.3 TROPHIC AND SUBSTRATE SUPPORT INDUCED BY NON-MUSCLE CELLS IN DENERVATED MUSCLES

Infiltration of denervated muscles by numerous non-muscle cells and their link to fibrosis in long-term denervated muscles has long been recognized (Gutmann and Young, 1944; Žak et al., 1969; Murray and Robbins, 1982a, b). The cells proliferate and fill the interstitial spaces of denervated muscles (Chapter 8). Of the several non-muscle cells found at the denervated endplate site, fibroblasts are the most numerous (Connor and McMahan, 1987) and have been identified as the cellular source of the enhanced N-CAM expression in denervated muscles (Gatchalian *et al.*, 1989). These cells appear to synthesize the extracellular matrix macromolecules, tenascin, fibronectin and heparan sulphate glycoprotein. Perhaps the blood-borne cells that infiltrate the denervated muscles release cytokines and FGF, which can induce proliferation of intramuscular fibroblasts and Schwann cells and thereby indirectly contribute to the guidance of regenerating axons to the endplate region. The Schwann cells migrate some distance from the endplate site and extend processes which might facilitate guidance of regenerating nerves toward the original endplate site (Gutmann and Young, 1944; Reynolds and Woolf, 1992).

The products of Wallerian degeneration and the non-neuronal cells remaining in the vacated intramuscular nerve sheaths have a powerful inducing effect on nerve growth (e.g. Brown and Hopkins, 1981). A nerve stump placed a few millimetres from the nerves supplying a normally innervated muscle stimulated profuse terminal sprouting of the nerves (Jones and Tuffery, 1973) and induced the sprouts to grow along the muscle fibres and to enter the implanted nerve sheath (Diaz and Pécot-Déchavassine, 1990). In muscles such as the medial gastrocnemius with widely disparate nerve branch points that delineate the different muscle compartments, section of one branch induced sprouting from distant intact nerve branches (Torigoe, 1985). The capacity of regenerating axons to be attracted across gaps of several millimetres gave rise to the original idea of Ramón y Cajal (1928) that cells of the denervated nerve tube release trophic factors that attract the regenerating nerves (see also Politis *et al.*, 1982; Lundberg *et al.*, 1982). The ability of intramuscular sheaths and denervated muscle to attract regenerating nerves over a long distance (Kuffler, 1986, 1989) provides evidence for the release of diffusible factors which attract the regenerating nerve fibres. The identity of these factors and their source remains to be determined. Several candidate molecules have been suggested because they have been located in denervated muscle and have been shown to promote nerve sprouting. They include insulin-like growth factor 2 (**IGF–2**), basic fibroblast growth factor (**FGF**) and ciliary neurotrophic factor (**CNTF**) (Coroni and Grandes, 1990; Gurney *et al.*, 1992). Several of these are produced by many cell types and are known to stimulate DNA synthesis in quiescent fibroblasts *in vitro* (Deuel, 1987). It is therefore possible that they act indirectly by inducing proliferation of resident non-muscle cells such as fibroblasts and Schwann cells, which in turn have a more direct influence on regeneration or sprouting. In addition, the inflammatory component of the response of the denervated intramuscular nerve stump and muscle must be considered. After denervation, muscle fibres express the MHCI class antigen (Maehlen *et al.*, 1989) and this may atttract blood-borne cells which could release cytokines in a cascade in which one or several factors could induce nerve growth and attract regenerating nerve fibres. As described in Chapter 8, inflammation and subsequent infiltration by mononucleated cells (Jones and Lane, 1975) elicited by implantation of a segment of a nerve or a foreign body, has the profound effect of inducing expression of acetylcholine receptors (Jones and Vrbová, 1974; Jones and Vyskočil, 1975). Perhaps a simultaneous expression of N-CAM and other adhesive macromolecules attracts nerve fibres to form contacts at these ectopic endplate sites.

9.2.4 MATURATION OF THE REGENERATED AXON AND RECOVERY OF THE MUSCLE

Many of the changes that take place during denervation are reversed on reinnervation. The maturation of the regenerating sprouts resembles the peripheral nerve maturation in developing animals. Once new motor terminals become functional, axotomized neurones begin to recover their former properties. Rate of recovery of axon size after reinnervation is very similar to the rate of atrophic change in the axons after axotomy and presumably reflects the reversal of the altered gene expression in the neurones from the growth to the transmitting state (Watson, 1970; reviewed by Gordon, 1983).

Even when the axon regenerates and establishes contact with muscle fibres, the process of regeneration is not completed, for the regenerating axons are initially split into many small nerve fibres. In rabbits, even after a simple crush injury, it takes at least 250 days for the normal diameter and number of nerve fibres to become re-established (Gutmann and Sanders, 1943). The size of the regenerating axons is proportional to their parent axons but the distance between the successive nodes of Ranvier in regenerating nerves remains equal along fibres of different diameters (Young, 1945; Vizoso and Young, 1948). However, the axons are an order of magnitude smaller; therefore, they need to grow considerably more in diameter to regain normal fibre size, equal to their parent axon diameter. The increase in fibre diameter of the regenerating nerve fibres goes hand in hand with the reduction of the number of nerve fibres in the peripheral stump (Gutmann and Sanders, 1943; Sanders and Young, 1946; Aitken et al., 1947) and can be considered as part of the same process.

This maturation of the nerve fibres is critically dependent upon whether connections are established between nerve and end organ. If the regenerating nerve fibres are prevented from establishing connections with the muscle fibres, their diameter remains small and their number in the peripheral stump large (Aitken et al., 1947). Figure 9.1 illustrates this point. Thus the muscle in some way induces the axon to resume its normal shape. The importance of the peripheral connections for the axon can also be demonstrated on the parent axons of the regenerated fibres. As described in Chapter 8, when the nerve fibres are disconnected from their end organs, their diameter and conduction velocity decrease and do not recover unless peripheral connections are reformed (Gutmann and Sanders, 1943; Cragg and Thomas, 1964; Gordon *et al.*, 1991; Gordon and Stein, 1982a). Recovery of size and properties of motoneurones and their axons is slow (Gordon and Stein, 1982a; Foehring *et al.*, 1986a, b) consistent with the considerable time needed for nerve fibres to recover their normal size (Gutmann and Saunders, 1943). Thus it appears that the reformation of functional target connections is a prerequisite for the reversal of the effects of axotomy and for the maturation of the regenerated nerves.

Young had concluded in 1950 that the target exercises an important influence on the newly formed nerve fibres that reinnervate it, but the means by which the effect is produced are not clear. The evidence shows that regenerating fibres mature more completely if contact with the target is encouraged, and that such peripheral connections can completely reverse the effects of axotomy. Activity of new transmitting nerve terminals may signal to the neurone cell body to up-regulate proteins associated with transmission (see Watson, 1970). Moreover, motoneurones recover their properties and appropriate axon size, irrespective of the type of muscle they reinnervate (Gordon and Stein, 1982a; Gordon *et al.*, 1986). Yet elaboration of transmitting terminals cannot promote maturation of regenerating axons without intact muscle fibres. Regenerating axons acquire synaptic vesicles and thickened presynaptic membrane patches when they make contact with the extracellular basal lamina muscle sheath at the original endplate sites (section 9.2.1). Although these terminals develop even if the muscle is irradiated and killed and only the basal lamina remains, their axons do not mature. Therefore transmitting terminals and muscle are required for axonal maturation.

9.3 RESTORATION OF THE DISTRIBUTION OF MOTOR UNITS AFTER REINNERVATION

Even though several daughter regenerating axons emanate from the proximal nerve stump and become myelinated during their growth toward their peripheral targets, all but one of these are withdrawn after peripheral connections are made and one axon remains which branches within the intramuscular nerve sheaths to form the reinnervated motor unit. The nerve to most muscles normally divides into two or more discrete branches which supply different regions of muscle known as muscle compartments (English and Letbetter, 1982; English and Weeks, 1984, 1987; Hammond *et al.*, 1989). Particularly in large pinnate muscles, the nerve branching is essential to distribute the nerve supply to muscle fibres which may be separated by considerable distances (English and Weeks, 1984, 1987; Weeks and English, 1987). For complete reinnervation of the muscles, it is essential that regenerating nerves follow these main intramuscular trunks, otherwise whole muscle compartments remain denervated. This is because the branching of each motor axon occurs distal to the nerve branch points which delineate the muscle compartments, and branches do not appear to cross the compartmental boundaries (Totosy de Zepetnek *et al.*, 1991). The branching of single motor nerves cannot be visualized directly but has been inferred from the spatial distribution of the muscle fibres in single motor units. Muscle fibres in individual motor units are identified in muscle cross-sections by their absence of glycogen after repetitive exhaustive stimulation of single axons or motoneurons (Kugelberg *et al.*, 1970; Burke and Tsairis, 1974; Nemeth *et al.*, 1981).

Comparison of the spatial distribution of muscle fibres in normal and reinnervated motor units shows clearly that the regenerating nerves supply different muscle fibres from normal, except after crush injuries, when normal patterns are often restored. Muscle fibres of single motor units are normally distributed in a discrete area of territory and they are intermingled with fibres belonging to several different motor units. This intermingling is the basis for the typical mosaic of histochemical types seen in normal muscles (Kugelberg *et al.*, 1970). After reinnervation, the motor unit territories are again confined to compartments but the distribution of fibres within the compartments are obviously different, as illustrated in Figure 9.5 (Kugelberg *et al.*, 1970; Peyronnard and Charron, 1980; Totosy de Zepetnek *et al.*, 1992a; Bodine-Fowler *et al.*, 1993) and the fibres within the territory are often segregated in groups. The extent of grouping varies according to:

- the muscle, being far more obvious in smaller rather than larger muscles where the absolute size of the unit territories are small and large, respectively (e.g. Kugelberg *et al.*, 1970; Gordon, 1987; Totosy de Zepetnek *et al.*, 1992a); and
- the number of reinnervated motor units.

As the number of reinnervated motor units declines, the extent of grouping of fibres increases significantly (Rafuse *et al.*, 1992).

Since the muscle fibres reinnervated by a new axon formerly belonged to different motor units, it is clear that the regenerating nerves do not find and reinnervate their original muscle fibres. The finding that the fibres in reinnervated motor units are of the same histochemical type demonstrates the capacity of the nerve to alter muscle properties (Kugelberg *et al.*, 1970; Chapter 7) and also explains the frequent finding of grouping of fibres of the same histochemical type after reinnervation (Karpati and Engel, 1968). The observation that type grouping occurs

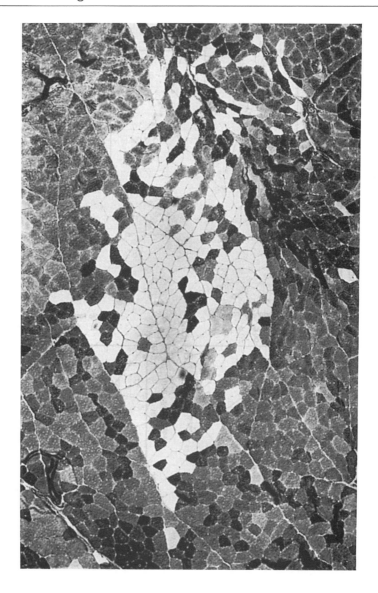

Figure 9.5 Glycogen-depleted fibres in a reinnervated motor unit observed in a periodic acid-Schiff stained cross-section of the rat TA muscle. The magnification is 84 ×. Note the adjacencies of muscle fibres within the depleted motor unit.

more readily after cutting and repairing peripheral nerves than after crush injuries suggests that the regenerating nerves can find their original muscle fibres when they are led to them via their intact Schwann cell tubes while the regenerating axons which enter tubes in the distal stump at random do not (Warszawski *et al.*, 1975; Totosy de Zepetnek *et al.*, 1992a). Since the spatial distribution of reinnervated motor units is more or less normal after crush injuries but not after lesions which sever the nerve, it appears that the regenerating axons which fail to grow within their original pathways also reinnervate denervated muscle fibres that are segregated around the ingrowing axon.

Findings that the number of muscle fibres reinnervated by each motor nerve (i.e. the innervation ratio) and unit tetanic force are directly correlated to the size of the parent axon (Totosy de Zepetnek *et al.*, 1992a) indicate that the regenerated axons of large motoneurones are able to supply a larger number of muscle fibres than smaller ones and thus restore the normal size relationships between IR, tetanic force and nerve size (Totosy de Zepetnek *et al.*, 1992a). Interestingly, when motor units increase their size by sprouting to reinnervate denervated muscle fibres after partial nerve injuries, the large motoneurones enlarge their motor units more than smaller motoneurones (Rafuse *et al.*, 1992). Consistent with this is the finding that, when regenerating axons emanate as sprouts from the proximal nerve stump of a crushed or severed nerve at a distance from the denervated muscles, the number as well as the size of the sprouts is directly proportional to the size of the parent fibres (Devor and Govrin-Lippman, 1979a, b). It is therefore possible that the extent of branching of regenerating or sprouting nerves depends on the size of the parent axon, and this may help to restore the normal range of motor unit tensions in the reinnervated muscles (but see section 9.4). The restoration of the size relationships between motoneurone, IR and motor unit force is important: on reinnervation, gradation of force during movement is still obtained by orderly recruitment of motor units which develop progressively larger forces as described in Chapter 5.

The number of muscle fibres innervated will ultimately be limited. A 5–8 fold increase in motor unit size in partially denervated muscles has been attributed to vigorous terminal and collateral sprouting within the muscle (reviewed by Brown *et al.*, 1981). However, in the rat soleus the ratio of collateral and terminal sprouts changes with time after partial denervation. Soon after partial denervation this ratio is 2.3:1, whereas at later times it becomes 18:1. Thus terminal sprouts seem to decrease, whereas collateral ones increase (Fisher *et al.*, 1989). This suggests that the anatomy of the motor unit in partially denervated or reinnervated muscles continues to change over long periods.

Like undamaged nerves, regenerating nerve fibres have the same capacity to increase the size of their peripheral field when few nerves reinnervate denervated muscles. Factors that determine the increase of motor unit territory include the local growth environment, the interaction between the ingrowing nerve and the denervated muscle fibres, and the intrinsic capacity of axons to branch and reinnervate many muscle fibres.

9.4 EFFECTS OF MUSCLE ON NORMAL AND REGENERATING MOTOR NERVE TERMINALS

The effects of the muscle on the maintenance of normal and regenerating axons is much less understood than the reverse, i.e. the effects of nerve on muscle (Chapters 3, 4, 5 and 7). In a normal skeletal muscle of an adult vertebrate, the neuromuscular junction appears to be a more or less stable structure, yet in his Ferrier lecture Young (1951) suggested that a synapse is in a state of continuous degeneration and regeneration, and that such events may be modified by activity. Continuous degeneration and regeneration of normal peripheral nerves was noticed in the last century by Meyer (1896) and considered to be a normal feature of peripheral nerves.

The concept of a continuous degeneration and regeneration of axon terminals at the neuromuscular junction was taken up by Barker and Ip (1966), who proposed that, like many structures in the body, the neuromuscular junction is being continually destroyed and replaced. Their evidence for the continuous replacement was the presence of numerous ultra-terminal sprouts in apparently normal muscles. The dynamic situation of the presynaptic elements of the neuromuscular junction is further highlighted by changes that take place at this site during the lifespan of an individual (Woolfe and Coërs, 1974). In the newborn vertebrate, each endplate is supplied by more than one axon but this type of innervation is eliminated during postnatal development (for review, see Navarrete and Vrbová, 1993, and Chapter 6). For some time thereafter, no dramatic changes – apart from rather subtle ones, noted by Barker and Ip (1966) and Tuffery (1971) – can be seen. Recently it has been reported that it is possible to induce rapid morphological changes of the branching pattern of nerve terminals at adult neuromuscular junctions by varying the concentrations of extracellular Ca^{2+} or by inhibiting the CANP (Swanson and Vrbová, 1987). These findings indicate that mechanisms similar to those that control the NMJ during development (Chapter 6) also modify the adult NMJ.

During old age the structure of the endplates becomes more complex, with many more terminal and ultraterminal branches trying to make contact with the muscle fibre (Tuffery, 1971; Woolf and Coërs, 1974). This increased complexity of the terminal axons occurs at the time when contact between many nerve terminals and muscle fibres disappears, as it was noticed that it is not uncommon for whole motor units to disappear in old individuals (Campbell and McComas, 1970). The decrease of the number of motor units is not due to a decrease of motor axons, since the number of axons remains unaltered in old age (Gutmann and Hanzlíková, 1976). There is, however, a decrease in choline acetyl-transferase, the enzyme that synthesizes ACh (Tuček and Gutmann, 1973) and a reduction of the frequency of MEPPS (Gutmann *et al.*, 1971). It may be that the neurone in old animals is no longer able to support its axonal expansion. These findings indicate that the maintenance of normal contact between nerve and muscle depends on continuous growth and renewal.

Figure 9.6 illustrates the changes in the connections between nerve and muscle during postnatal development of mammals. In immature animals, more than one axon is in contact with individual muscle fibres (a); during adult life only one axon contacts an individual muscle fibre (b); in old age many muscle fibres lose contact with the nerve terminal (c).

Similar to muscle fibres in neonates, reinnervated neuromuscular junctions of mammals are transiently supplied by several nerve endings (McArdle, 1975; Gorio *et al.*, 1983; Jansen and Van Essen, 1975; Rich and Lichtman, 1989).

It is possible that, as in neonates, regenerating nerve terminals are affected by neuromuscular activity (Chapter 6). When rats were subjected to exercise during nerve regeneration, the establishment of functional connections between the motor nerves and muscle fibres was delayed (Gutmann and Jakoubek, 1963). Results that show a greater expansion

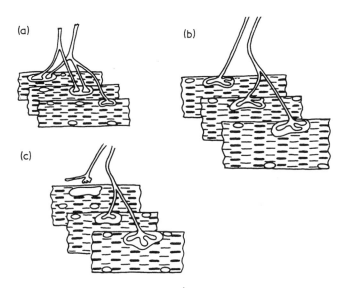

Figure 9.6 Changes in pattern of innervation at endplates of striated muscles during life: (a) during early stages of development, individual muscle fibres are supplied by more than one axon; (b) during adult life, each muscle fibre is supplied by only one axon; (c) in old animals, contact between motor nerve and muscle fibre is occasionally lost so that the size of the individual motor units decreases.

of motor unit territory in partially denervated adult soleus muscles treated with an inhibitor of the CANP, an enzyme involved in nerve terminal elimination (Vrbová and Fisher, 1989) (Chapter 6), further suggest that the same mechanisms that regulate synapse development in neonates operate during reinnervation. If neuromuscular activity leads to a reduction of contacts between the motor nerve endings and muscle fibre, on reinnervation nerve terminals that usually belong to more active motor units may be eliminated faster, and therefore fewer muscle fibres will maintain contact with the motoneurone. In this way the small motor uits will remain small on reinnervation, and the inactive large ones will regain their large territory. It is interesting that, when activity of regenerating axons is reduced by treatment with TTX, their motor units enlarge (Ribchester, 1993). The enlargement of the motor unit territory in these experiments was attributed in fact to more extensive branching of axons in inactive muscles. This, however, is unlikely to be the cause of the expansion, since branching alone is unable to secure an expansion of motor unit territory. Establishment and maintenance of neuromuscular contacts is a prerequisite for this to take place (Rich and Lichtman, 1989).

Taken together, these findings show that the accurate restoration of the size of motor units within a mixed muscle as observed by Gordon and her colleagues is likely to be regulated by the different activity patterns of the various types of motor units, which in turn controls their innervation ratio by mechanisms described in Chapter 6.

CONCLUSIONS

Regeneration of the motor nerve depends on the ability of the motoneurone to replace the severed portion of its axon, i.e. to regenerate. This process of restoration of the axon depends not only on the motoneurone's growth potential but also on the interaction of the growing axon with Schwann cells. Many extracellular matrix molecules actively contribute to the process of regeneration of axons and their guidance to their appropriate targets. The contribution of these different factors that influence regeneration of the motor axon is discussed.

After reaching the target muscle, neuromuscular contacts are established between the ingrowing axons and the denervated muscle fibres. The new neuromuscular junctions formed have some features of immature synapses and some time is required before they acquire adult characteristics. The re-establishment of these contacts allows the reversal of changes that occurred in the motoneurone during its separation from the target, as well as those that occurred in the muscle during denervation. Nevertheless, the distribution of motor unit territory is often altered on reinnervation since the regenerated axons do not always return to the muscle fibres they originally supplied. The number of muscle fibres occupied by a particular motoneurone seems to be related to the size of the motoneurone in that large cells will have a bigger innervation ratio than small ones. This means that the size principle is re-established after regeneration. The possibility that this re-establishment of motor unit size is determined by activity-dependent interactions with the target muscle fibres is discussed.

REFERENCES

Abercrombie, M. and Johnson, M. L. (1946) Quantitative histology of Wallerian degeneration. 1. Nuclear population in rabbit sciatic nerve. *J. Anat.* **80**, 37–50.

Aitken, J. T., Sharman, M. and Young, J. Z. (1947) Maturation of peripheral nerve fibres with various peripheral connections. *J. Anat.* **81**, 1–22.

Aguayo, A. J., Epps, J., Charron, L. and Bray, G. M. (1976) Multipotentiality of Schwann cells in cross anastomosed and grafted myelinated and unmyelinated nerves: Quantitative microscopy and radioautography. *Brain Res.* **104**, 1–20.

Aguayo, A. J., Attiwell, M., Trecarten, J., Perkins, S. and Bray, G. (1977) Abnormal myelination in transplanted Trembler mouse Schwann cells. *Nature,* **265**, 73–75.

Aldskogius, H. and Thomander, L. (1986) Selective reinnervation of somatotopically appropriate muscles after facial nerve transection and regeneration in the neonatal rat. *Brain Res.* **375**, 126–134.

Allt, G. (1976) Pathology of the peripheral nerve, in *The Peripheral Nerve* (ed. D. N. Landon), Chapman & Hall, London, pp. 666–739.

Anderson, P. N., Mitchell, J., Major, D. and Stauber, V. V. (1983) An ultrastructural study of the early stages of axonal regeneration through rat nerve grafts. *Neuropathol. Appl. Neurobiol.* **9**, 455–466.

Anzel, A. and Wernig, A. (1989) Muscle fibre loss and reinnervation after long-term denervation. *J. Neurocytol.* **19**, 833–845.

Baichwal, R. R., Bigbee, J. W. and DeVries, G. H. (1988) Macrophage-mediated myelin-related mitogenic factor for cultured Schwann cells. *Proc. Natl. Acad. Sci. USA* **85**, 1701–1705.

Barker, D. and Ip, M. C. (1966) Sprouting and degeneration of mammalian motor axons in normal and deafferented skeletal muscle. *Proc. R. Soc. B* **163**, 538–554.

Bennett, M. R. and Florin, T. J. (1974) A statistical analysis of the release of acetylcholine at newly formed synapses in striated muscle. *J. Physiol.* **238**, 93–107.

Bennett, M. R., Pettigrew, A. G. and Taylor, R. S. (1973) The formation of synapses in reinnervated mammalian striated muscle. *J. Physiol.* **288**, 481–500.

Beránek, R. and Gutmann, E. (1953) Influence of nociceptive stimulation on regeneration of the peripheral nerve. *CS Fysiol,* **2**, 140–147.

Bernstein, J. J. and Guth, L. (1961) Nonselectivity in establishment of neuromuscular connections following nerve regeneration in the rat. *Exp. Neurol.* **4**, 262–275.

Bethe, A. (1901) Über die Regeneration der Nervenfasern. *Arch. f. Psych.* **34**, 234–280.

Beuche, W. and Friede, R. L. (1984) The role of non-resident cells in Wallerian degeneration. *J. Neurocytol.* **13**, 767–796.

Bixby, J. L. (1992) Diversity of axonal growth-promoting receptors and regulation of their function. *Curr. Opin. Neurobiol.* **2**, 66–69.

Bixby, J. L. and Harris, W. A. (1991) Molecular mechanisms of axon growth and guidance. *Ann. Rev. Cell Biol.* **7**, 117–159.

Black, M. M. and Lasek, R. J. (1979) Slowing of the rate of axonal regeneration during growth and regeneration. *Exp. Neurol.* **63**, 108–119.

Bodine-Fowler, S. C., Unguez, G. A., Roy, R. R. *et al.* (1993) Innervation patterns in the cat tibialis anterior six months after self-reinnervation. *Muscle & Nerve,* **16**, 379–391.

Bray, G. M. and Aguayo, A. J. (1974) Regeneration of peripheral unmyelinated nerves. Fate of the axonal sprouts which develop after injury. *J. Anat.* **117**, 517–529.

Bray, D., Thomas, C. and Shaw, G. (1978) Growth cone formation in cultures of sensory neurones. *Proc. Natl. Acad. Sci. USA* **75**, 5226–5229.

Bresjanec, M. and Sketelj, J. (1989) Neurite promoting influences of proliferating Schwann cells and target tissues are not prerequisite for rapid axonal elongation after nerve crush. *J. Neurosci. Res.* **24**, 501–507.

Brown, M. C. and Hardman, V. J. (1987) A reassessment of the accuracy of reinnervation by motoneurons following crushing or freezing of the sciatic or lumbar spinal nerves of rats. *Brain* **110**, 695–705.

Brown, M. C., Holland, R. L. and Hopkins, W. G. (1981) Motor nerve sprouting. *Ann. Rev. Neurosci.* **4**, 17–42.

Brown, M. C. and Hopkins, W. G. (1981) Role of degenerating axon pathways in regeneration of mouse soleus motor axons. *J. Physiol.* **318**, 365–373.

Brushart, T. M. E. (1988) Preferential reinnervation of motor nerves by regenerating motor axons. *J. Neurosci.* **8**, 1026–1031.

Brushart, T. M. E. (1990) Preferential motor reinnervation: a sequential double-labeling study. *Restor. Neurol., Neurosci.,* **1**, 281–287.

Brushart, T. M. (1994) Motor axons preferentially reinnervate motor pathways. *J. Neurosci.* (in press).

Brushart, T. M. and Mesulam, M. M. (1980) Alteration in connections between muscle and anterior horn neurons. *Science* **208**, 603–605.

Bunge, R. P. and Bunge, M. B. (1983) Interrelationship between Schwann cell function and extracellular matrix production. *TINS* **6**, 499–505.

Bunge, R. P., Bunge, M. B. and Eldridge, C. F. (1986) Linkage between axonal ensheathment and basal lamina production by Schwann cells. *Ann. Rev. Neurosci.* **9**, 305–328.

Bunge, M. B., Bunge, R. P., Kleitman, N. and Dean, A. C. (1989) Role of peripheral nerve extracellular matrix in Schwann cell function and in neurite regeneration. *Dev. Neurosci.* **11**, 348–360.

Bunge, M. B., Clark, M. B., Dean, A. C. *et al.* (1990) Schwann cell function depends on axonal signals and basal lamina components. *Ann. NY Acad. Sci.* **580**, 281–287.

Bunge, M. B., Williams, A. K. and Wood, P. M. (1982) Neuron–Schwann cell interaction in basal lamina formation. *Dev. Biol.* **92**, 449–460.

von Büngner, O. (1891) Über die Degenerations – und Regenerations – vorgänge an Nerven nach Verletzungen. *Zieglers Beitr. z. Path. u. path. Anat.* **10**, 321.

Burke, R. E. and Tsairis, P. (1974) Anatomy and innervation ratios in motor units in cat gastrocnemius. *J. Physiol. (Lond.)* **234**, 749–765.

Campbell, M. J. and McComas, A. J. (1970) The effect of aging on muscle function, in, *5th Symp. on Current Research on Muscular Dystrophy and Related Disease. London Abst. 6*, MDGB, London.

Carbonetto, S. (1991) Glial cells and extracellular matrix in axonal regrowth. *Curr. Opin. Neurobiol.* **1**, 407–413.

Caroni, P. and Grandes, P. (1990) Nerve sprouting in innervated adult skeletal muscle induced by exposure to elevated levels of insulin-like growth factors. *J. Cell Biol.* **110**, 1307–1317.

Clark, M. B. and Bunge, M. B. (1989) Cultured Schwann cells assemble normal-appearing basal lamina only when they ensheath axons. *Dev. Biol.* **133**, 393–404.

Clemence, A., Mirsky, R. and Jessen, K. R. (1989) Non-myelin-forming Schwann cells proliferate rapidly during Wallerian degeneration in the rat sciatic nerve. *J. Neurocytol.* **18**, 185–192.

Connor, E. A. and McMahan, U. J. (1987) Cell accumulation in the junctional region of denervated muscle. *J. Cell Biol.* **104**, 109–120.

Cragg, B. G. (1970) What is the signal for chromatolysis? *Brain Res.* **23**, 1–21.

Cragg, B. G. and Thomas, P. K. (1964) The conduction velocity of regenerated peripheral nerve fibres. *J. Physiol.* **171**, 164–175.

Dennis, M. J. and Miledi, R. (1971) Lack of correspondence between the amplitudes of spontaneous potentials and unit potentials evoked by nerve impulses at regenerating neuromuscular junctions. *Nature* **232**, 126–128.

Dennis, M. J. and Miledi, R. (1974) Non-transmitting neuromuscular junctions during an early stage of endplate reinnervation. *J. Physiol.* **239**, 553–570.

DeSantis, M., Berger, P. K., Laskowski, M. B. and Norton, A. S. (1992) Regeneration by skeletomotor axons in neonatal rats is topographically selective at an early stage of regeneration. *Exp. Neurol.* **116**, 229–239.

Deuel, T. F. (1987) Polypeptide growth factors: roles in normal and abnormal cell growth. *Annu. Rev. Cell Biol.* **3**, 443–492.

Devor, M. and Govrin-Lippman, R. (1979a) Maturation of axonal sprouts after nerve crush. *Exp. Neurol.* **64**, 260–270.

Devor, M. and Govrin-Lippman, R. (1979b) Selective regeneration of sensory fibers following nerve crush injury. *Exp. Neurol.* **65**, 243–254.

Diaz, J. and Pécot-Déchavassine, M. (1990) Nerve sprouting induced by a peripheral nerve placed over a normally innervated frog muscle. *J. Physiol.* **421**, 123–133.

English, A. W. and Letbetter, W. D. (1982) Anatomy and innervation patterns of cat lateral gastrocnemius and plantaris muscles. *Am. J. Anat.* **164**, 67–77.

English, A. W. and Weeks, O. I. (1984) Compartmentalization of single motor units in cat lateral gastrocnemius. *Exp. Brain Res.* **56**, 361–368.

English, A. W. and Weeks, O. I. (1987) An anatomical and functional analysis of cat biceps femoris and semitendinosus muscles. *J. Morphol.* **191**, 161–175.

Evans, D. H. L. (1947) Endings produced by somatic nerve fibres growing into adrenal gland. *J. Anat.* **81**, 225–232.

Farel, P. B. and Bemelmans, S. E. (1986) Restoration of neuromuscular specificity following ventral rhizotomy in the bullfrog tadpole, *Rana catebeiana. J. Comp. Neurol.* **254**, 125–132.

Fawcett, J. W. and Keynes, R. J. (1986) Muscle basal lamina: A new graft material for periphal nerve repair. *J. Neurosurg.* **65**, 354–363.

Fawcett, J. W. and Keynes, R. J. (1990) Peripheral nerve regeneration. *Annu. Rev. Neurosci.* **13**, 43–60.

Fisher, T. J., Vrbová, G. and Wijetunge, A. (1989) Partial denervation of the rat soleus muscle at two different developmental stages. *Neurosci.* **28**, 755–763.

Foehring, R. C., Sypert, G. W. and Munson, J. B. (1986a) Properties of self-reinnervated motor units of medial gastrocnemius of the cat. II. Long-term reinnervation. *J. Neurophysiol.* **56**, 931–946.

Foehring, R. C., Sypert, G. W. and Munson, J. B. (1986b) Properties of self-reinnervated motor units of medial gastrocnemius of the cat. II Axotomised neurones and time course of recovery. *J. Neurophysiol.* **55**, 947–965.

Fontana, F. (1781) *Traité sur le venin de la vipère, sur le paissins americains. Sur la structure primitive du corps animal*, Vols. I and II, Florence.

Friede, R. L. and Bischhausen, R. (1980) The fine structure of stumps of transected nerve fibers in subserial sections. *J. Neurol. Sci.* **44**, 181–203.

Fu, S. and Gordon, T. (1995a) Contributing factors to poor functional recovery after delayed nerve repair: Prolonged escotomy. *J. Neurosci.* in press.

Fu, S. and Gordon, T. (1995b) Contributing factors to poor functional recovery after delayed nerve repair. *J. Neurosci.* in press.

Gatchalian, C. L., Schachner, M. and Sanes, J. R. (1989) Fibroblasts that proliferate near denervated synaptic sites in skeletal muscle synthesize the adhesive molecules tenascin (J1), N-CAM, fibronectin, and a heparin sulfate proteoglycan. *J. Cell Biol.* **108**, 1873–1890.

Gerding, R., Robbins, N. and Antosiak, J. (1977) Efficiency of reinnervation of neonatal rat muscle by original and foreign nerves. *Dev. Biol.* **61**, 177–183.

Giannini, C., Lais, A. C. and Dyck, P. J. (1989) Number, size and class of peripheral nerves regenerating after crush, multiple crush, and graft. *Brain Res.* **500**, 131–138.

Giannini, C. and Dyck, P. J. (1990) The fate of Schwann cell basement membranes in permanently transected nerves. *J. Neuropathol. Exp. Neurol.* **49**, 550–563.

Gillespie, M. J., Gordon, T. and Murphy, P. A. (1986) Reinnervation of the lateral gastrocnemius and soleus muscles in the rat by their common nerve. *J. Physiol.* **372**, 485–500.

Gillespie, M. J., Gordon, T. and Murphy, P. A. (1987) Motor units and histochemistry in the rat lateral gastrocnemius and soleus muscles: evidence for dissociation of physiological and histochemical properties after reinnervation. *J. Neurophysiol.* **57**, 921–937.

Glasby, M. A., Geschmeissner, S. G., Hitchcock, R. J. I. and Huang, C. L. -H. (1986) The dependence of nerve regeneration through muscle grafts in the rat on the availability and orientation of basement membrane. *J. Neurocytol.* **15**, 497–510.

Glicksman, M. and Sanes, J. R. (1983) Differentiation of motor nerve terminals formed in the absence of muscle fibers. *J. Neurocytol.* **12**, 661–671.

Gordon, T. (1983) Dependence of peripheral nerves on their target organs, in *Somatic and Autonomic Nerve–Muscle Interactions* (eds G. Burnstock, G. Vrbová and R. A. O'Brien), Elsevier, New York, pp. 289–323.

Gordon, T. (1987) Muscle plasticity during sprouting and reinnervation. *Am. Zool.* **27**, 1055–1066.

Gordon, T. (1994) Mechanisms for functional recovery of the larynx after surgical repair of injured nerves, *J. Voice*, **8**, 70–78.

Gordon, T., Gillespie, J., Orozco, R. and Davis, L. (1991) Axotomy-induced changes in rabbit hindlimb nerves and the effects of chronic electrical stimulation. *J. Neurosci.* **11**, 2157–2169.

Gordon, T. and Stein, R. B. (1982a) Time course and extent of recovery in reinnerv-ated motor units of cat triceps surae muscles. *J. Physiol.* **323**, 307–323.

Gordon, T. and Stein, R. B. (1982b) Reorganization of motor-unit properties in reinnervated muscles of the cat. *J. Neurophysiol.* **48**, 1175–1190.

Gordon, T., Stein, R. B. and Thomas, C. (1986) Motor unit organisation in extensor muscles cross-reinnervated by flexor motoneurones. *J. Physiol.* **374**, 443–456.

Gordon, T., Yang, J. F., Stein, R. B. and Tyreman, N. (1993) Recovery potential of the neuromuscular system after nerve injuries. *Brain Res. Bull.* **30**, 477–482.

Gorio, A., Campignoto, G., Finesso, P. *et al.* (1983) Muscle reinnervation. II. Sprouting, synapse formation and repression. *Neurosci.* **8**, 403–416.

Greenman, M. J. (1913) Studies on the regeneration of the peroneal nerve of the albino rat: number and sectional area of fibers: area relation of axis to sheath. *J. Comp. Neurol.* **23**, 479–513.

Gulati, A. K. (1988) Evaluation of acellular and cellular nerve grafts in repair of rat peripheral nerve. *J. Neurosurg.* **68**, 117–123.

Gurney, M. E., Yamamoto, H. and Kwon, Y. (1992) Induction of motor neuron sprouting in vivo by ciliary neurotrophic factor and basic fibroblast growth factor. *J. Neurosci.* **12**, 3241–3247.

Gutmann, E. (1942) Fators affecting recovery of motor function after nerve lesions. *J. Neurol. Neurosurg. and Psychiat.* **5**, 81–95.

Gutmann, E. (1958) *Die functionelle Regeneration der peripheren Nerven*, Akademie-Verlag, Berlin.

Gutmann, E., Guttmann, L., Medawar, P. B. and Young, J. Z. (1942) The rate of regeneration of nerve. *J. Exp. Biol.* **19**, 14–44.

Gutmann, E. and Hanzlíková, V. (1976) Fast and slow motor units in ageing. *Gerontology* **22**, 280–300.

Gutmann, E., Hanzlíková, V. and Vyskočil, F. (1971) Age changes in cross-striated muscle of the rat. *J. Physiol.* **216**, 331–342.

Gutmann, E. and Jakoubek, B. (1963) Effect of increased motor activity on regen-eration of the peripheral nerve in young rats. *Physiol. Bohem.* **12**, 463–468.

Gutmann, E. and Sanders, F. K. (1943) Recovery of fibre numbers and diameters in the regeneration of peripheral nerves. *J. Physiol.* **101**, 489–518.

Gutmann, E. and Young, J. Z. (1944) The reinnervation of muscle after various periods of atrophy. *J. Anat.* **78**, 15–43.

Haftek, J. and Thomas, P. K. (1968) Electron microscope observations on the effects of localized crush injuries on the connective tissues of peripheral nerves. *J. Anat.* **103**, 233–243.

Hall, S. M. (1986a) Regeneration in cellular and acellular autografts in the periph-eral nervous system. *Neuropathol. Appl. Neurobiol.* **12**, 27–46.

Hall, S. M. (1986b) The effect of inhibiting Schwann cell mitosis on the reinner-vation of acellular autografts in the peripheral nervous system of the mouse. *Neuropathol. Appl. Neurobiol.* **12**, 401–414.

Hall, Z. W. and Sanes, J. R. (1992) Synaptic structure and development: The neuromuscular junction. *Cell* **72**/*Neuron* **10** (suppl.) 99–121.

Hammond, C. G. M., Gordon, D. C., Fisher, J. T. and Richmond, F. J. R. (1989) Motor unit territories supplied by primary branches of the phrenic nerve. *J. Appl. Physiol.* **66**, 61–71.

Hardman, V. J. and Brown, M. C. (1987) Accuracy of reinnervation of rat intercos-tal muscles by their own segmental nerves. *J. Neurosci.* **7**, 1031–1036.

Havton, J. O. and Kellerth, (1987) Regeneration by supernumerary axons with synaptic terminals in spinal motoneurones of cats. *Nature* **325**, 711–714.

Heslop, J. P. (1975) Axonal flow and fast transport in nerves. *Adv. in Comp. Physiol. and Biochem.* **6**, 75–165.

Hoffman, P. L. and Lasek, R. J. (1980) Axonal transport of the cytoskeleton in regenerating motoneurones: Constancy and change. *Brain Res.* **202**, 317–333.

Hoh, J. F. Y. (1975) Selective and non-selective reinnervation of fast-twitch and slow-twitch rat skeletal muscle. *J. Physiol.* **251**, 791–801.

Holmes, W. and Young, J. Z. (1942) Nerve regeneration after immediate and delayed suture. *J. Anat.* **77**, 63–96.

Hudson, A. R. (1984) Peripheral nerve surgery, in *Peripheral Neuropathy*, Vol. 1 (eds P. J. Dyck, P. K. Thomas, E. H. Lambert and R. Bunge), Saunders, Philadelphia, pp. 420–438.

Hume, R. I., Role, L. W. and Fischbach, G. D. (1983) Acetylcholine release from growth cones detected with patches of acetylcholine receptor-rich membranes. *Nature* **305**, 632–634.

Ide. C., Tohyama, K., Yokota, R. *et al.* (1983) Schwann cell basal lamina and nerve regeneration. *Brain Res.* **288**, 61–75.

Ip, M. C. and Vrbová, G. (1973) Motor and sensory reinnervation of fast and slow mammalian muscles. *Z. Zellforsch. mikrosk. Anat.* **146**, 261–279.

Jansen, J. K. S. and Van Essen, D. C. (1975) Reinnervation of rat skeletal muscle in the presence of α-bungarotoxin. *J. Physiol.* **250**, 651–667.

Jenq, C. -B. and Coggeshall, R. E. (1985) Numbers of regenerating axons in parent and tributary peripheral nerves in the rat. *Brain Res.* **326**, 27–40.

Jones, R. and Lane, J. T. (1975) Proliferation of mononuclear cells in skeletal muscle after denervation or muscle injury. *J. Physiol.* **246**, 60–61P.

Jones, R. and Tuffery, A. R. (1973) Relationship of endplate morphology to junctional chemosensitivity. *J. Physiol.* **232**, 13–15.

Jones, R. and Vrbová, G. (1974) Two factors responsible for the development of denervation hypersensitivity. *J. Physiol.* **236**, 517–538.

Jones, R. and Vyskočil, F. (1975) An electrophysiological examination of the changes in skeletal muscle fibres in reponse to degenerating nerve tissue. *Brain Res.* **88**, 309–317.

Karpati, G. and Engel, W. K. (1968) 'Type grouping' in skeletal muscles after experimental reinnervation. *Neurology* **18**, 447–455.

Kerns. J. M. and Lucchinetti, C. (1992) Electrical field effects on crushed nerve regeneration. *Exp. Neurol.* **117**, 71–80.

Kučerová, M. (1949) Cited by E. Gutmann, (1958) in *The Functional Regeneration of Peripheral Nerves*, Verlag, Berlin.

Kuffler, D. P. (1986) Accurate reinnervation of motor end plates after disruption of sheath cells and muscle fibres. *J. Comp. Neurol.* **250**, 228–235.

Kuffler, D. P. (1989) Regeneration of muscle axons in the frog is directed by diffusible factors from denervated muscle. *J. Comp. Neurol.* **281**, 416–425.

Kugelberg, E., Edström, L. and Abbruzzese, M. (1970) Mapping of motor units in experimentally reinnervated rat muscle. Interpretation of histochemical and atrophic fibre patterns in neurogenic lesions. *J. Neurol. Neurosurg. Psychiat.* **33**, 319–329.

Kuno, M., Tukanis, S. A. and Weakly, J. N. (1971) Correlation between nerve terminal size and transmitter release at the neuromuscular junction of the frog. *J. Physiol.* **213**, 545–556.

Langley, J. N. and Anderson, H. K. (1904) The union of different kinds of nerve fibres. *J. Physiol.* **31**, 365–391.

Langley, J. N. and Hashimoto, M. (1917) On the suture of separate nerve bundles in a nerve trunk and on internal nerve plexuses. *J. Physiol. (Lond.)* **51**, 318–345.

LeBlanc, A. C. and Poduslo, J. F. (1990) Axonal modulation of myelin gene expression in the peripheral nerve. *J. Neurosci. Res.* **26**, 317–326.

Lemke, G. (1988) Unwrapping the genes of myelin. *Neuron* **1**, 535–543.

Letinsky, M. S., Fischbach, D. G. and McMahan, U. J. (1976) Precision of reinnervation of original post synaptic sites in frog muscle after nerve crush. *J. Neurocytol.* **5**, 691–718.

Linda, H., Risling, M. and Culheim, S. (1985) Dendraxons in regenerating motoneurones in the rat: do dendrites generate new axons after central axotomy? *Brain Res.* **358**, 329–333.

Ljubinska, L. and Niemierko, S. (1971) Velocity and intensity of bidirectional migration of acetylcholinesterase in transected nerves. *Brain Res.* **27**, 329–342.

Lundberg, G., Dahlin, L. B., Danielsen, N. *et al.* (1982) Nerve regeneration in silicon chambers: Influence of gap length and of distal stump components. *Exp. Neurol.* **76**, 361–375.

McArdle, J. J. (1975) Complex endplate potentials at the regenerating neuromuscular junction of the rat. *Exp. Neurol.* **49**, 629–638.

McArdle, J. J. and Albuquerque, E.X. (1973) A study of the reinnervation of fast and slow mammalian muscles. *J. Gen. Physiol.* **61**, 1–23.

McQuarrie, J. C. and Grafstein, B. (1973) Axon outgrowth enhanced by previous nerve injury. *Arch. Neurol.* **29**, 53–55.

McQuarrie, I. G. (1985) Effect of a conditioning lesion on axonal sprout formation at nodes of Ranvier. *J. Comp. Neurol.* **231**, 239–249.

McQuarrie, I. G. and Jacob, J. M. (1991) Conditioning nerve crush accelerates cytoskeletal protein transport in sprouts that form after a subsequent crush. *J. Comp. Neurol.* **305**, 139–147.

McQuarrie I. G. and Lasek, R. J. (1989) Transport of cytoskeletal elements from parent axons into regenerating daughter axons. *J. Neurosci.* **9**, 436–446.

Mackinnon, S., Dellon, L. and O'Brien, J. (1991) Changes in nerve fibre numbers distal to a nerve repair in the rat sciatic nerve model. *Muscle and Nerve* **14**, 1116–1122.

Maehlen, J., Nennesmo, J., Olson, A. B., Olson, T., Daa-Schröder, H. D. and Kristenson, K. (1989) Peripheral nerve injury causes transient expression of MHC class 1 antigens in rat neurones and skeletal muscles. *Brain Res.* **481**, 368–372.

Marshall, L. M., Sanes, J. R. and McMahan, U. J. (1977) Reinnervation of original synaptic sites on muscle fiber basement membrane after disruption of the muscle cells. *Proc. Natl. Acad. Sci. USA* **74**, 3073–3077.

Martini, R., Xin, Y., Schmitz, B. and Schachner, M. (1992) The L2/HNK–1 carbohydrate epitope is involved in the preferential outgrowth of motor neurons on ventral roots and motor nerves. *Eur. J. Neurosci.* **4**, 628–639.

Meyer, S. (1896) Über eine Verbindungsweise de Neurone. *Arch. für mikr. Anat.* **47**, 120–132.

Miledi, R. (1960a) The acetylcholine sensitivity of frog muscle fibres after complete or partial denervation. *J. Physiol.* **151**, 1–23.

Miledi, R. (1960b) Properties of regenerating neuromuscular synapses in the frog. *J. Physiol.* **154**, 190–205.

Miledi, R. and Stefani, E. (1969) Non-selective reinnervation of slow and fast muscle fibres in the rat. *Nature* **222**, 569–571.

Miller, R. J., Lasek, R. J. and Katz, M. J. (1987) Preferred microtubules for vesicle transport in lobster axons. *Science* **235**, 220–222.

Mirsky, R. and Jessen, K. R. (1988) Axon control of Schwann cell differentiation, in *The Current Status of Peripheral Nerve Regeneration* (eds T. Gordon, R. B. Stein and P. A. Smith), Alan R. Liss, Inc., pp. 91–97.

Morris, J. H., Hudson, A. R. and Weddell, G. (1972) A study of degeneration and regeneration in the divided rat sciatic nerve based on electron microscopy. II. The development of the 'regenerated unit'. *Z. Zellforsch Mikrosk. Anat.* **24**, 103–130.

Murray, M. A. and Robbins, N. (1982a) Cell proliferation in denervated muscle: time course, distribution, and relation to disuse. *Neurosci.* **7**, 1817–1822.

Murray, M. A. and Robbins, N. (1982b) Cell proliferation in denervated muscle: identity and origin of the dividing cells. *Neurosci.* **7**, 1823–1834.

Nadim, W., Anderson, P. N. and Turmaine, M. (1990) The role of Schwann cells and basal lamina tubes in the regeneration of axons through long length of freeze killed nerve grafts. *Neuropathol. Appl. Neurobiol.* **16**, 411–421.

Nathaniel, E. J. H. and Pease, D. C. (1963a) Degenerative changes in rat dorsal roots during Wallerian degeneration. *J. Ultrastruct. Res.* **9**, 511–532.

Nathaniel, E. J. H. and Pease, D. C. (1963b) Regenerative changes in rat dorsal roots following Wallerian degeneration. *J. Ultrastruct. Res.* **9**, 533–549.

Navarrete, R. and Vrbová, G. (1993) Activity dependent interactions between motoneurones and muscles: Their role in the development of the motor unit. *Progr. Neurobiol.* **41**, 93–124.

Nemeth, P. M., Pette, D. and Vrbová, G. (1981) Comparison of enzyme activities among single muscle fibres within defined motor units. *J. Physiol.* **311**, 489–495.

Peroncito, A. (1907) Die Regeneration der Nerven. *Zieglers Beitr. Path. und Path. Anat.* **42**, 354.

Perry, V. H. and Brown, M. C. (1992) Role of macrophages in peripheral nerve degeneration and repair. *Bioessays* **14**, 401–406.

Pestronk, A., Drachman, D. B. and Griffen, J. W. (1980) Effects of aging on nerve sprouting and regeneration. *Exp. Neurol.* **70**, 65–80.

Peyronnard, J.-M. and Charron, L. (1980) Muscle reorganization after partial denervation and reinnervation. *Muscle and Nerve* **3**, 509–518.

Politis, M. J., Ederle, K. and Spencer, P. S. (1982) Trophism in nerve regeneration in vivo. Attraction of regenerating axons by diffusible factors derived from cells in distal nerve stumps of transected peripheral nerves. *Brain Res.* **53**, 1–12.

Rafuse, V. F., Gordon, T. and Orozco, R. (1992) Proportional enlargement of motor units after partial denervation of cat triceps surae muscles. *J. Neurophysiol.* **68**, 1261–1276.

Ranson, S. W. (1912) Degeneration and regeneration of nerve fibres. *J. Comp. Neurol.* **22**, 487–546.

Ranvier, L. (1874) De quelques faits relatifs à l'histologie et à la physiologie des muscles striés. *Arch. Physiol. Norm. Path.* **6**, 1–15.

Reichardt, L. F. and Tomaselli, K. (1991) Extracellular matrix molecules and their receptors: Functions in neural development. *Ann. Rev. Neurosci.* **14**, 531–570.

Reynolds, M. L. and Woolf, C. J. (1992) Terminal Schwann cells elaborate extensive processes following denervation of the motor endplate. *J. Neurocytol.* **21**, 50–66.

Ribchester, R. R. (1993) Co-existence and elimination of convergent motor nerve terminals in reinnerated and paralysed adult rat skeletal muscle. *J. Physiol.* **466**, 421–441.

Rich, K. M., Alexander, T. D., Pryor, J. C. and Hollowell, J. P. (1989) Nerve growth factor enhances regeneration through silicone chambers. *Exp. Neurol.* **105**, 162–170.

Rich, M. M. and Lichtmann, J. W. (1989) In vivo visualisation of pre- and postsynaptic changes during synapse elimination in reinnervated mouse muscle. *J. Neurosci.* **9**, 1781–1805.

Richardson, P. M. (1991) Neurotrophic factors in regeneration. *Curr. Opin. Neurobiol.* **1**, 401–406.

Richardson, P. M. and Verge, V. M. K. (1987) Axonal regeneration in dorsal spinal roots is accelerated by peripheral axonal transection. *Brain Res.* **411**, 406–408.

Rieger, F., Nicolet, M., Pincon-Raymond, M. *et al.* (1988) Distribution and role of N-CAM in the basal laminae of muscle and Schwann cells. *J. Cell Biol.* **107**, 707–719.

Rothshenker, S. and McMahan, V. J. (1976) Altered patterns of innervation in frog muscle after denervation. *J. Neurocytol.* **5**, 719–730.

Rusovan, A., Kanje, M. and Mild, K. H. (1992) The stimulatory effect of magnetic fields on regeneration of the rat sciatic nerve is frequency dependent. *Exp. Neurol.* **117**, 81–84.

Salonen, V., Aho, H., Roytta, M. and Peltonen, J. (1988) Quantitation of Schwann cells and endoneurial fibroblast-like cells after experimental nerve trauma. *Acta Neuropathol. (Berl.).* **75**, 331–336.

Salzer, J. L. and Bunge, R. P. (1980) Studies of Schwann cell proliferation. I. An analysis in tissue culture of proliferation during development, Wallerian degeneration, and direct injury. *J. Cell Biol.* **84**, 739–752.

Salzer, J. L., Williams, A. K., Glaser, L. and Bunge, R. P. (1980a) Studies of Schwann cell proliferation. II. Characterization of the stimulation and specificity of the response to a neurite membrane fraction. *J. Cell Biol.* **84**, 753–766.

Salzer, J. L., Bunge, R. P. and Glaser, L. (1980b) Studies of Schwann cell proliferation. III. Evidence for the surface localization of the neurite mitogen. *J. Cell Biol.* **84**, 767–778.

Sanders, F. K. and Young, J. Z. (1946) The influence of peripheral connections on the diameter of regenerating nerve fibres. *J. Exp. Biol.* **22**, 203–212.

Sanes, J. R. (1989) Extracellular matrix molecules that influence neural development. *Am. Rev. Neurosci.* **12**, 521–546.

Sanes, J. R. (1993) Topographic maps and molecular gradients. *Curr. Opin. Neurobiol.* **3**, 67–74.

Scaravilli, F. (1984) Regeneration of the perineurium across a surigically induced gap in a nerve encased in a plastic tube. *J. Anat.* **139**, 411–424.

Shaw, G. and Bray, D. (1977) Movement and extension of isolated growth cones. *Exp. Cell Res.* **104**, 55–62.

Simpson, S. A. and Young, J. Z. (1945) Regeneration of fibre diameter after cross-union of visceral and somatic nerves. *J. Anat.* **79**, 48–64.

Sjoberg, J. and Kanje, M. (1990) The initial period of peripheral nerve regeneration and the importance of the local environment for the conditioning lesion effect. *Brain Res.* **529**, 79–84.

Sketelj, J., Bresjanac, M. and Popovic, M. (1989) Rapid growth of regenerating axons across the segments of sciatic nerve devoid of Schwann cells. *J. Neurosci. Res.* **24**, 153–162.

Sorenson, E. J. and Windebank, A. J. (1993) Relative importance of basement membrane and soluble growth factors in delayed and immediate regeneration of rat sciatic nerve. *J. Neuropath. Exp. Neurol.* **52**, 216–222.

Thomas, P. K. (1964) The deposition of collagen in relation to Schwann cell basement membrane during peripheral nerve regeneration. *J. Cell Biol.* **23**, 375–382.

Thomas, P. K. (1966) The cellular response to nerve injury. I. The celluar outgrowth from the distal stump of transected nerve. *J. Anat.* **100**, 287–303.

Toft, P. B., Fugleholm, K. and Schmalbruch, H. (1988) Axonal branching following crush lesions of peripheral nerves of rats. *Muscle and Nerve* **11**, 880–889.

Tonge, D. A. (1974) Physiological characteristics of re-innervation of skeletal muscle in the mouse. *J. Physiol.* **241**, 141–153.

Torigoe, K. (1985) Distribution of motor nerve sprouting in the mouse gastrocnemius muscle after partial denervation. *Brain Res.* **330**, 273–282.

Totosy de Zepetnek, J. E., Gordon, T., Stein, R. B. and Zung, H. V. (1991) Comparison of force and EMG measures in normal and reinnervated tibialis anterior muscles of the rat. *Can. J. Physiol. Pharmacol.* **69**, 1774–1783.

Totosy de Zepetnek, J. E., Zung, H. V., Erdebil, S. and Gordon, T. (1992a) Innervation ratio is an important determinant of force in normal and reinnervated rat Tibialis Anterior muscles. *J. Neurophysiol.* **67**, 1385–1403.

Totosy de Zepetnek, J. E., Zung, H. V., Erdebil, S. and Gordon, T. (1992b) Motor unit categorisation on the basis of contractile and histochemical properties: a glycogen depletion analysis of normal and reinnervated rat tibialis anterior muscle. *J. Neurophysiol.* **67**, 1404–1415.

Tuček, S. and Gutmann, E. L. (1973) Choline acetyltransferase activity in muscle of old rats. *Expl. Neurol.* **38**, 349–360.

Tuffery, A. R. (1971) Growth and degeneration of motor endplates in normal rat hind-limb muscles. *J. Anat.* **110**, 221–247.

Vanlair, C. (1882) *Arch. Biol. Paris* **3**, 379. Cited by E. Gutmann (1958) in *The Functional Regeneration of Peripheral Nerves*, Verlag, Berlin.

Vizoso, A. D. and Young, J. Z. (1948) Internode length and fibre diameter in developing and regenerating nerves. *J. Anat.* **82**, 110–134.

Vrbová, G. and Fisher, T. (1989) The effect of inhibiting the calcium activated neutral protease, on motor unit size after partial denervation of the rat soleus muscle. *Euro. J. Neurosci.* **1**, 616–625.

Warszawski, M., Telerman-Toppet, N., Durdu, J. *et al.* (1975) The early stages of neuromuscular regeneration after crushing the sciatic nerve in the rat. Electrophysiological and histochemical study. *J. Neurol. Sci.* **24**, 21–32.

Watson, W. E. (1970) Some metabolic responses of axotomized neurons to contact between their axons and denervated muscles. *J. Physiol.* **210**, 321-343.

Webster, H. deF. (1962) Transient focal accumulation of axonal mitochondria during the early stages of Wallerian degeneration. *J. Cell Biol.* **12**, 361–383.

Weeks, O. I. and English, A. W. (1987) Cat triceps surae motor nuclei are organized topologically. *Exp. Neurol.* **96**, 163–177.

Weiss, P. (1943) Nerve regeneration in the rat following tubular splinting of severed nerves. *Arch. Surg.* **46**, 525–547.

Weiss, P. and Hiscoe, H. B. (1948) Experiments on the mechanism of nerve growth. *J. Exp. Zool.* **107**, 315–396.

Weiss, P. and Hoag, H. (1946) Competitive reinnervation of rat muscles by their own and foreign nerves. *J. Neurophysiol.* **9**, 413–418.

Weiss, P. and Taylor, A. C. (1944) Further experimental evidence against 'neurotropism' in nerve regeneration. *J. Exp. Zool.* **95**, 233–257.

Wernig, A. and Herrara, A. A. (1986) Sprouting and remodelling at the nerve–muscle junction. *Prog. Neurobiol.* **27**, 251–291.

Wessells, N. K., Johnson, S. R. and Nuttall, R. P. (1978) Axon initiation and growth cone regeneration in cultured motoneurons. *Exp. Cell. Res.* **117**, 335–345.

Wigston, D. J. (1986) Selective innervation of transplanted limb muscles regenerating motor axons in the axolotl. *J. Neurosci.* **6**, 2757–2763.

Wigston, D. J. and Kennedy, P. R. (1987) Selective reinnervation of transplanted muscles by their original motoneurons in the axolotl. *J. Neurosci.* **6**, 1857–1865.

Wigston, D. J. and Donahue, S. P. (1988) The location of cues promoting selective reinnervation of axolotl muscles. *J. Neurosci.* **8**, 3451–3458.

Woolf, A. L. and Coërs, C. (1974) *Pathological Anatomy of Intermuscular Nerve Endings. I. Disorders of voluntary muscle* (ed. J. N. Walton) 3rd edn, Churchill Livingstone, Edinburgh, pp. 274–309.

Wyrwicka, W. (1950) On the rate of regeneration of the rat sciatic nerve in white mouse. *Acta Biol. Exp.* **15**, 147–153.

Xie, Z.-p. and Poo, M.-m. (1986) Initial events in the formation of neuromuscular synapse: rapid induction of acetylcholine release from embryonic neuron. *Proc. Natl. Acad. Sci. USA* **83**, 7069–7073.

You, S., Raji, M. and Gordon, T. (1994) Decline in the expression of low affinity NGF receptors correlates with decline in number of regenerating nerves after delayed nerve repair. *Can. J. Physiol. Pharmacol.* (in press).

Young, J. Z. (1945) The history of the shape of a nerve fibre, in *Essays on Growth and Form* (ed. d'Arcey Thompson), Oxford University Press, Oxford.

Young, J. Z. (1948) Growth and differentiation of nerve fibres. *Symp. Soc. Exp. Biol. Number 2: Growth*, pp. 57–74.

Young, J. Z. (1950) The determination of the specific characteristics of nerve fibers, in *Int. Conf. on the Development, Growth and Regeneration of the Nervous System: Genetic Neurology* (ed. P. Weiss), University of Chicago Press, Chicago, pp. 92–104.

Young, J. Z. (1951) Growth and plasticity in the nervous system. The Ferrier Lecture. *Proc. R. Soc. B* **139**, 18–37.

Young, J. Z. and Medawar, P. B. (1940) Fibre suture of peripheral nerves. Measurement of the rate of generation. *Lancet* **ii**, 126–128.

Young, S. H. and Poo, M.-m. (1983) Spontaneous release of transmitter from growth cones of embryonic neurones. *Nature* **305**, 634–637.

Žak, R., Grove, D. and Rabinowitz, M. (1969) DNA synthesis in the rat diaphragm as an early response to denervation. *Am. J. Physiol.* **216**, 647–654.

Zalewski, A. A. and Gulati, A. K. (1982) Evaluation of histocompatibility as a factor in the repair of nerve with a frozen nerve allograph. *J. Neurosurg.* **56**, 550–554.

Zelená, J., Ljubinska L. and Gutmann, E. (1968) Accumulation of organelles at the ends of interrupted axons. *Z. Zellforsch. Mikrosk. Anat.* **91**, 200–219.

Disturbances of nerve–muscle interactions and their consequences

10

A disturbance of the interactions between nerve and muscle may have several causes:

- It may result from a change in the function of the motoneurone due to injury or disease within the central nervous system which, in turn, may lead to altered activity patterns in motoneurones with subsequent changes in muscle fibre properties.
- It can be due to disease of the motoneurone (and its possible death) or damage to the motor axon.

In the latter case, all interactions between the nerve and the muscle fibres it innervates cease and can only be restored if and when the damaged cell recovers, the axon regenerates or when other neurones take over the territory of the denervated muscle fibres. When motoneurones die, or when axons are unable to return to the muscle that they originally innervated, there may be sprouting from nearby intact axons of surviving motoneurones. If this is the case, then the degree of recovery and the characteristics of the reinnervated muscle fibres will depend on the origin of these new connections. Thus not only will the degree of recovery be dependent on the ability of intact axons within the affected muscle to sprout and form new connections, but also the properties of the recovered muscle fibres will depend primarily on the characteristics of the reinnervating nerve fibres, their ability to sustain these new connections and their activity patterns.

As well as changes in innervation or nerve activity patterns, other disturbances such as muscle immobilization in shortened or lengthened conditions may interfere with normal neuromuscular relationships.

Selected features of the disturbance of nerve-muscle interaction and some of their implications for the understanding and treatment of the neuromuscular aspects of neurological disease will be discussed in this chapter.

10.1 DISTURBANCES CAUSED BY A CHANGE IN ACTIVITY WITHIN THE CNS

In previous chapters, evidence has been presented to show that mammalian skeletal muscles can adjust their properties to the functional requirements of the motoneurones that supply them. It is not surprising that when these requirements are altered, especially with respect to patterns of activity and use, the muscle fibres undergo corresponding changes.

Gross patterns of movement are inevitably altered in the majority of diseased and injured states of the CNS, and the inability to perform **co-ordinated** movement is often the most disabling feature of many neurological conditions. A detailed discussion of the clinical features relating to movement disorder is far beyond the scope of this book and is well covered in a number of other texts. However, relatively little consideration is given to the possible effects of altered activity patterns on muscle in neurological conditions and little is known regarding the effects of particular disturbances of CNS function on the muscles or the motoneurones that are affected by such disturbances. This section is therefore restricted to a few examples of evidence for such disturbances.

The best explored example of the disturbance of motoneurone function as a result of loss of normal connections with higher centres is that which follows spinal cord transection. In this case, the influences that normally reach the spinal motoneurones from the brain are disrupted, and much of the fine control of the neural networks that initiate and regulate motor activity are lost (see discussion by Dimitrijevic and Faganel, 1985). However, the motoneurones are still attached to the muscles, and muscle contractions can be elicited by electrical stimulation of motor axons, or by activation of segmental afferents.

Following experimental spinal cord section, slow muscles are more affected than fast ones. If dorsal root section (de-afferentation) is added to spinal cord section, the slow soleus muscle atrophies as fast as a denervated soleus (Hník, 1966). It is possible that spinal, de-afferented soleus muscles become completely inactive, leading to greater atrophy, whereas fast muscles supplied by motoneurones that have more interneuronal inputs, do not atrophy as much since they may still have some residual activity.

In humans with spinal cord injury, muscles have been found to be more fatiguable (Lenman *et al.*, 1989) and this is in accord with the experimentally observed increased fatigue and change in muscle fibre type composition of muscles that have been rendered inactive (Chapter 7). However, the outcome of accidental spinal cord injury in humans may be very different from that induced experimentally, since in humans such injury is likely to cause incomplete spinal transection and may also include peripheral nerve damage, making it difficult to attribute any observed changes to specific causes. Nevertheless, stimulation of such paralysed muscles does reduce their fatiguability (see Stein *et al.*, 1992)

Damage to the CNS can also lead to an uncontrolled increase in activity – for example, spasticity. Studies that have been carried out on the properties of muscles of people who have suffered a stroke and showed signs of spasticity demonstrate that, with time, a gradual change in fibre composition towards increased numbers of slow fibres develops (see Young and Shahani, 1980; Dietz, 1992). This finding, together with changes of motoneurone firing patterns, might explain the gradual increase of spasticity, since slow muscle fibres develop larger forces at lower rates of firing (Chapter 5).

It has been reported recently that the speed of muscle contraction and rate of motoneurone firing patterns of limb muscles of MS patients decrease (Rice *et al.*, 1992). This is consistent with the findings of Jones *et al.* (1994) that the median frequency of the power spectrum of the EMG is also reduced in MS patients, indicating a greater than normal proportion of muscle fibres with a slowed conduction velocity. This change occurs early and may even be evident when there is little or no change in motor unit voluntary recruitment or force generation (Jones *et al.*, 1994). Whether this is due to actual muscle fibre conversion or to some other cause is not yet established, but it indicates that very early changes in muscle properties can be observed in a chronic condition such as multiple sclerosis, and that these changes may be linked to altered motoneurone firing patterns (Rice *et al.*, 1992). Dietz (1992) suggests that spasticity following damage to the CNS may be viewed as a mechanism to preserve gait and to support the weight of the body against gravity at the expense of fast reflex activity.

These brief notes on the possible activity-related changes in muscle in a few neurological disorders will not be discussed further, but examples indicating where the interactive processes between the nerve and muscle are more directly affected by disease follow.

10.2 NEUROMUSCULAR DISEASES VIEWED AS A DISTURBANCE OF NERVE–MUSCLE INTERACTIONS

The conventional distinction of neuromuscular diseases as neurogenic or myogenic indicates the prevailing attitude which considers these dis-orders as primary disease of either the nervous system or muscle. This division has been challenged by several workers (McComas, 1977; Haus-manowa-Petrusewicz, 1989) but the question is unresolved. It is increasingly obvious that the motoneurone and muscle are so closely linked with each other, particularly during development, that it is often difficult to decide whether the primary disturbance in a disease occurs in the muscle or in the nervous system. Here, three conditions will be discussed from the point of view of disturbed interactions between nerves and muscles:

1. A particular form of motoneurone disease: spinal muscular atrophy (SMA).
2. A disease of the neuromuscular junction, i.e. myasthenia gravis.
3. A primary muscle disease: Duchenne muscular dystrophy.

10.2.1 SPINAL MUSCULAR ATROPHY VIEWED AS A CONSEQUENCE OF DISTURBED NERVE–MUSCLE INTERACTION DURING DEVELOPMENT

SMA is an inherited disease and the mode of inheritance is autosomal recessive. The disease presents several different clinical pictures, ranging from the severe Werdnig Hoffman form to the milder Kugelberg–Welander type. It is beyond the scope of this chapter to give details of classifications or descriptions of clinical features; there are excellent reviews dealing with these aspects of the disease (for classifications of the various forms, see Hausmanowa-Petrusewicz, 1989). The site of the genetic mutation for SMA has been identified on chromosome 5 (5q11–q 13) and this appears to be the same for both the Werdnig Hoffmann and the Kugelberg-Welander forms (Gilliam *et al.*, 1990; Melki *et al.*, 1990a, b., Brzustowicz *et al.*, 1990). A characteristic feature of this disease is a substantial loss of motoneurones and this results in considerable muscle weakness. Unique to this condition, compared with various adult forms of motoneurone disease, is the fact that the only nerve cell affected in SMA is the motoneurone. The selective involvement of this cell, together with the evidence that shows the critical dependence of the motoneurone on contact and interaction with the muscle during development (as discussed in Chapters 2 and 6), suggests that the death of motoneurones might be caused by some deficiency of this interaction.

Some of the new developments in our understanding of the dependence of motoneurones on their target muscles during a critical stage of development, the mechanisms that regulate synaptic contacts between motor nerve endings and muscle fibres, and the dependence of developing skeletal muscle fibres on contact with their motoneurones for their maturation and survival may help to develop a scheme that could explain the underlying mechanism of this condition where a single neuronal type, i.e. the motoneurone, is uniquely affected.

The dependence of embryonic motoneurones on their target muscles has been known for many years, and the mechanisms by which the target might bring about the survival of motoneurones in the developing vertebrates has been discussed (Chapters 2 and 6). Not only is it necessary for the survival of the motoneurone to maintain contact with the target, but it is also important that the motoneurone should be able to interact with the muscle, since preventing functional interactions also causes motoneurones to die (Greensmith and Vrbová, 1992).

There are several similarities between the experimentally target-deprived motoneurones that survive the period of target deprivation and motoneurones in SMA patients. In rats, when motoneurones are experimentally disconnected from their targets during a critical developmental period the surviving motoneurones are small and more active than motoneurones of a similar age in normal rats (Lowrie *et al.*, 1987; Navarrete and Vrbová, 1984). Similar changes are observed in SMA patients. The size of motoneurones in these patients is smaller than normal (Banker, 1986) and their motor units are continuously active (Buchthal and Olsen, 1970; Hausmanowa-Petrusewicz, 1989). These changes are consistent with findings on axotomized motoneurones where a large increase in specific membrane resistance has been found (Chapter 8). Such high membrane resistance will amplify any excitatory inputs that impinge upon the motoneurones. During mammalian development there is an early and prodigious increase in these inputs, which are probably responsible for many features of foetal locomotor activity, and this process continues during postnatal development (for rat, see Navarrete and Vrbová, 1993). During this developmental period the motoneurone has to acquire properties that will enable it to process and accommodate this increase in excitary inputs. It could be that, to acquire such properties, continued interaction of the motoneurone and muscle is necessary. It is possible that, in the absence of nerve–muscle interactions, the motoneurone is unable to develop the characteristics that enable it to survive the consequences of these increasing synaptic inputs. If this is the case, the mechanisms that regulate neuromuscular contacts during development will be of utmost importance for the survival of motoneurones (for review see Vrbová and Lowrie, 1989; Lowrie and Vrbová, 1992).

The interaction of the motoneurone with the muscle and successful maintenance of connections are important considerations, particularly in view of recent results which indicate that application of some growth factors, such as BDNF or CNTF, seem to encourage neuronal survival *in vitro* and for short intervals *in vivo* (Sendtner *et al.*, 1990, 1992; Yan *et al.*, 1992; Henderson *et al.*, 1993). It must be borne in mind, however, that none of these studies has, as yet, provided evidence that:

- the rescue of motoneurones destined to die is permanent;
- rescued motoneurones are able to reinnervate and maintain contact with skeletal muscle fibres and reverse their deterioration;
- rescued motoneurones become appropriately integrated into the spinal cord circuitry and are used by the CNS as required during normal locomotor activity.

The solution of these questions remains open to experimental examination.

Another approach is based on the data that target-deprived motoneurones do not mature and acquire the properties necessary to cope with the increased excitory input. Motoneurones that remain in contact with their target are not destroyed by treatment with agonists of excitatory amino acids such as NMDA. However, motoneurones disconnected from the target die after a single exposure to NMDA (Greensmith et al., 1993, 1994). In addition, a proportion of motoneurones destined to die after neonatal axotony can be rescued by treatment with a non-competitive blocker of NMDA receptors MK801 (Mentis et al., 1993). It could be that the various growth factors achieve their effects by down-regulating the expression of glutamate receptors, and that the two approaches are not incompatible. Nevertheless, when considering treatment of patients, it is necessary to show experimentally not only that motoneurones can survive but also that they are able to establish functional connection with the peripheral musculature and appropriate interaction in the CNS.

Whether motoneurone disease, the most common condition affecting adults that involves motoneurone death, can be attributed to similar processes as those proposed in developing systems is still in debate. As during development, some motoneurones may die in old age (Gutmann and Hanzlíková, 1976) and the extensive alterations in motor unit territory observed in older adults further indicates that changes in motoneurones can occur beyond the 5th decade in humans (Campbell et al., 1973). The cause of this late-onset change in motoneurone function is not understood. It has been suggested that adult motoneurone disease may be an uncontrolled and exaggerated form of late motoneurone death since it is most common in humans in the later decades of life, although factors that may lead to toxic interference with neuromuscular interactions have also been proposed. (For a review of these issues see Williams, 1994.)

10.2.2 DISTURBANCES DUE TO DISEASES OF THE NEUROMUSCULAR JUNCTION

A distinct clinical condition that affects the neuromuscular junction is myasthenia gravis. The most characteristic symptom of this condition is 'fatiguability' of the muscle (Jolly, 1895; see De Smedt, 1973) with spectacular improvement of muscle function following the administration of anticholinesterases (Walker, 1934). The similarity of the symptoms of this disease to curare poisoning led Mary Walker to propose that it is the neuromuscular junction that is affected. Her finding that the symptoms of the disease abate after treatment with anticholinesterase drugs confirmed this suggestion.

In spite of the well-established fact that curare is a drug that acts on the post-synaptic membrane of skeletal muscle fibres and the well-recognized similarities between symptoms of curare poisoning and those of myasthenia (Walker 1934), there was a surprisingly long-lasting discussion as to whether this disease affects the pre- or post-synaptic part of the neuromuscular junction. The debate and subsequent further discoveries illustrate how difficult it may be to determine exactly which part of a synapse is at fault.

In the 1960s, when it was established that acetylcholine (ACh) was stored in presynaptic vesicles and released from them when an action potential invades the nerve terminal (Wittaker and Gray, 1962; Katz, 1966), emphasis centred on transmitter release. It was suggested that the failure of transmission in myasthenia gravis is due to reduced synthesis or release of ACh from nerve terminals (De Smedt, 1966; Dahlbäck *et al.*, 1961). At the time of this study, transmitter release was assessed by measuring potential changes produced by ACh release from the presynaptic terminal on the post-synaptic membrane. Nevertheless, changes so recorded can sometimes be accounted for by altered sensitivity of the post-synaptic membrane. These results were interpreted to show a reduction of ACh release and a strong case was made for myasthenia being due to a reduction of ACh synthesis or release from motor nerve terminals.

The hypothesis that the disease is caused by a deficient synthesis or release of transmitter seemed unlikely on several grounds. Firstly, the cholinergic terminal appeared little affected in patients with myasthenia, and contained all the usual organelles and a nearly normal complement of presynaptic terminals (Johnson and Woolfe, 1965; Santa *et al.*, 1972a, b); secondly, only the cholinergic neuromuscular junctions between motor nerves and striated muscles are affected. Since all cholinergic terminals use similar mechanisms for transmitter synthesis and release, it would be surprising if the synthesis or release mechanisms of all similar terminals were affected.

Another proposal, defended vigorously by Simpson (1960, 1966), was that myasthenia is an autoimmune disease and that antibodies to ACh receptors affect the sensitivity of the post-synaptic membrane. Simpson noted that a number of myasthenic patients in his care suffered from disorders which may be autoimmune in nature, such as diabetes, rheumatoid arthritis, systemic lupus erythematosus and sarcoidosis. This, taken together with the finding that thymectomy often has beneficial effects on myasthenia gravis, made Simpson's hypothesis an attractive one. His theory that an antibody to the acetylcholine receptor binds to the post-synaptic membrane, blocking the response to the transmitter, was later confirmed by evidence obtained by Goldstein and his colleagues (Goldstein and Whittingham, 1966; Goldstein and Hoffman, 1968). They found that muscles of animals injected with saline extracts from the thymus gland or skeletal muscle, in conjunction with Freund's adjuvant, developed symptoms and muscle abnormalities that resembled those in myasthenic patients. The autoimmune nature of the disease was substantiated by the finding that, in those animals in which the disease had been induced, the thymus showed an autoimmune inflammatory response together with physiological evidence of post-junctional blockade in skeletal muscles. Later a number of investigators used the ACh receptors purified from the electric organ of *Torpedo* to induce antibody production and symptoms similar to those noticed in myasthenic patients (Patrick and Lindström, 1973). When the autoimmune response to this disease was analysed, it was found that thymus-derived cells became sensitized to ACh receptor molecules. As a result, inflammatory cells accumulate in the synaptic cleft and destroy part of the post-synaptic membrane (Lennon *et al.*, 1975, 1976; Lindström *et al.*, 1988). Although there are some differences between the animal model of the disease and human myasthenia gravis, the similarities between the autoimmune model and the human disease are so close that it is difficult to argue against Simpsons' original hypothesis (Simpson, 1960, 1966).

It is astonishing that the morphological changes of the neuromuscular junction of patients suffering from myasthenia are not more conspicuous. The endplates are elongated, the terminal axon expansion slightly reduced, the synaptic cleft widened and the secondary folds increased in number (Zachs *et al.*, 1961; Johnson and Woolfe, 1965; Santa *et al.*, 1972a, b). The appearance of axon profiles changes little, indicating that the nerve terminals are healthy and functioning well. A possible means of controlling the disease may therefore be to induce the production of new neuromuscular junctions when the antibody titre is low.

More recently a myasthenia-like syndrome, Lambert-Eaton Myasthenic syndrome (LEMS), has been described; it often occurs in association with small cell carcinomas of the lung (reviewed by Vincent *et al.*, 1989). In this case it is the presynaptic element of the neuromuscular junction that is affected. In particular, voltage-sensitive calcium channels at the active sites of presynaptic nerve terminals appear to be blocked or disrupted as a result of antibody reactivity. As in myasthenia gravis, there is little gross morphological evidence of nerve terminal degeneration, but the normally well-organized arrays of calcium gate associated protein are disorganized and reduced in number.

It is possible that early confusion over the nature of myasthenia gravis resulted from failure to recognize Lambert-Eaton syndrome as a different disease with an autoimmune component (Vincent *et al.*, 1989).

10.2.3 DISEASES OF THE MUSCLE: MISMATCH BETWEEN FUNCTION AND STRUCTURE

In the last century the Scottish surgeon Charles Bell (1830, see McComas, 1977) and later the French physician Duchenne (1872) described the first cases of a muscle disease that was later to carry Duchenne's name. The hereditary nature of the disease was recognized by Meryon (1892; see McComas, 1977), who also noticed that it affected predominantly males. Duchenne (1872) and others after him described several forms of these hereditary wasting diseases. It was Erb (1884) who, while grouping these conditions as primary diseases of the muscle or, as he called them, 'dystrophia muscularis progressiva', defined and classified the then newly described syndromes. He postulated that they were due to a 'complex nutritional disturbance of the muscle fibre', which they may still turn out to be. A much used classification of muscular dystrophies is that by Walton and Gardner Medwin (1974).

The study of Duchenne muscular dystrophy (DMD) represents a real triumph of molecular genetics. The dramatic advances of recent years have revealed the location of the gene, and its structure as well as its gene product (dystrophin) have been identified (Monaco *et al.*, 1986; Burghes *et al.*, 1987; Koenig *et al.*, 1987; Hoffman *et al.*, 1987).

Attempts to remedy the disease by introducing the missing gene directly into muscles of animal models where the gene product is also missing have been only moderately successful (Ascadi *et al.*, 1991). A much publicized way of achieving the expression of the missing protein into diseased muscles has been that of myoblast transfer (Partridge *et al.*, 1989). The partial success of this method is evaluated by Dubowitz (1993).

However, remarkably little is known about the function of the missing protein. This is particularly puzzling, since its absence does not seem to cause great discomfort to the mouse mutant (mdx) where this protein is also absent (Dangain and Vrbová, 1984). The only muscle affected is the diaphragm (Stedman *et al.*, 1991) though at least the laboratory mdx mouse manages extremely well with this disability. In humans and dogs the absence of dystrophin appears to be responsible for relentlessly pro-gressive muscle weakness, but even in these cases there is a selective involvement of some, but not other, muscle groups.

Thus dystrophin, the product of the missing gene, might be specifically required for particular muscle functions. It is interesting to mention in this context that the only time when the mdx mouse shows distinct symptoms of locomotor disturbance is when the animal is beginning to use its limbs during postnatal development. It is also at this point that its muscles undergo massive, rapid degeneration followed subsequently by almost complete recovery (Dangain and Vrbová, 1984; Cornworth and Shotton, 1987). In the human disease, however, muscle deterioration is progressive.

Several proposals have been made as the role that dystrophin may have for normal muscle function. It has been suggested that it is important for maintaining the structural integrity of the muscle fibre, or that it is associated with surface glycoproteins with which it forms cross-links. However, the findings in affected mice and the clinical picture of the disease in humans, where the weight-bearing muscles are most affected, indicates that the need for dystrophin is related to the contractile activity of the muscle together with certain mechanical conditions when this activity is carried out. Thus it appears that in this condition there is a mismatch between the demands imposed upon the muscle during movement and its structural equipment to sustain activity in the long term. This is supported by results that show that, while mdx mice can cope with acute exercise, chronic overload leads to a progressive deterioration of their muscle fibres (Sacco *et al.*, 1992; Dick and Vrbová, 1993). An important finding related to the explanation of some of the symptoms is that muscles of young animals, and probably children, respond unfavourably to overload: instead of becoming stronger, as in adults, they become weaker (Dangain and Vrbová, 1986, 1988). In Duchenne muscular dystrophy, the function of the destroyed muscles is taken over by regenerating muscle fibres and this, in turn, may lead to their overload, deterioration and further weakness. This continuous overuse of muscles that are still functional may then contribute to the relentless progress of the disease.

CONCLUSIONS

In this chapter the impact of the information on nerve-muscle interaction on some neurological and muscle disorders is considered. Since the motoneurone and its activity have such profound effects on the muscles it supplies, it is not surprising that in conditions where motoneurone activity is altered the muscle is also going to be changed. These changes in muscle fibre properties are not always advantageous to the patient, and the possible malfunction due to such alterations is discussed.

Neuromuscular diseases highlight the importance of nerve–muscle interactions for the development and progress of the disease. Three examples of neuromuscular diseases are given, and for each the possible role played by disturbed neuromuscular interactions is considered. In the case of Duchenne and Becker muscular dystrophy, it is proposed that the lack of dystrophin leads to muscle degeneration mainly as a result of a particular physiological function of the muscle for which this protein is essential. In diseases of the neuromuscular junction, the interaction between the nerve terminal and post-synaptic membrane is also very marked, and it appears that the nerve terminal does show some adaptive change to the less responsive muscle membrane. Finally, perhaps the most interesting example where disturbed neuromuscular interaction (particularly during early development) could lead to dire consequences, such as motoneurone death seen in patients with SMA, is discussed with the view in mind that these may be causing motoneurones to die. Although only a few examples are given, it is hoped that they highlight the significance of neuromuscular interactions and their disturbance in the disease state.

REFERENCES

Ascadi, G., Dickson, G., Love, D. R. *et al.* (1991), Human dystrophin expression in mdx mice after intramuscular injection of DNA constructs. *Nature* **352**, 815–818.

Banker, B. Q. (1986) The pathology of motoneurone disorders, in *Myology*, (eds A. G. Engel and B. Q. Banker), McGraw Hill, New York, pp. 2037–2068.

Brzustowicz, L., Lehner, T., Castilla, L. *et al.* (1990) Genetic mapping of chronic childhood-onset spinal muscular atrophy to chromosome 5q 11.2–13.3. *Nature* **344**, 540–541.

Buchtal, F. and Olsen, C. B. (1970) Electromyography and muscle biopsy in infantile spinal muscular atrophy. *Brain* **93**, 15–30.

Burghes, A. H. M., Logan, C., Hu, K. *et al.* (1987) A cDNA clone from the Duchenne/Becker muscular dystrophy gene. *Nature* **328**, 434–437.

Campbell, M. J., McComas, A. J. and Pettito, F. (1973) Physiological changes in ageing muscle. *J. Neurol. Neurosurg. Psychiat.* **36**, 174–182.

Cornworth, J. W. and Shotton, D. M. (1987) Muscular dystrophy in the mdx mouse. Histopathology of the soleus and extensor digitorum longus muscle. *J. Neurol. Sci.* **80**, 39–54.

Dahlbäck, O., Elmquist, D., Johns, T. R. and Thesleff, S. (1961) An electrophysiological study of the neuromuscular junction in myasthenia gravis. *J. Physiol. (Lond.)* **156**, 336–343.

Dangain, J. and Vrbová, G. (1984) Muscle development in *mdx* mutant mice. *Muscle and Nerve* **7**, 700–704.

Dangain, J. and Vrbová, G. (1986) Response of normal and dystrophic muscle to increased functional demand. *Exp. Neurol.* **94**, 796–801.

Dangain, J. and Vrbová, G. (1988) Response of dystrophic muscle to reduced load. *J. Neurol. Sci.* **88**, 277–288.

DeSmedt, J. E. (1966) Presynaptic mechanisms in myasthenia gravis, in *Myasthenia Gravis* (ed. K. E. Osserman), *Ann. NY Acad. Sci.* **135**, 209–246.

DeSmedt, J. E. (1973) The neuromuscular disorder in myasthenia gravis, in *New*

Developments in Electromyography and Clinical Neurophysiology, Vol. 1 (ed. J. E. DeSmedt), Karger, Basel, pp. 242–304.

Dick, J. and Vrbová, G. (1993) Progressive deterioration of muscles in *mdx* mice induced by overload. *Clin. Sci.* **84**, 145–150.

Dietz, V. (1992) Spasticity: Exaggerated reflexes or movement disorder?, in *Movement Disorders in Children* (eds H. Forssberg and H. Hirschfeld), Karger, Basel, pp. 225–233.

Dimitrijević, M. R. and Faganel, J. (1985) Motor control in the spinal cord, in *Upper Motoneurone Functions and Dysfunctions* (eds J. C. Eccles and M. J. Dimitirijevic), Karger, Basel, pp. 150–163.

Dubowitz, V. (1993) Myoblast transfer in muscular dystrophy: Panacea or pie in the sky? *Neuromusc. Disorders* **2**, 305–310.

Duchenne, G. B. (1872) *De l'électricisation localisée et son application à la pathologie et à la therapie,* 2nd and 3rd edns, Baillière et Fils, Paris.

Erb, W.H. (1884) Über die 'Juvenile Form' der progressiven Muskelatrophie; ihre Beziehungen zur sogennanten Pseudohypertrophie der Muskeln. *D. Arch. Klin. Med.* **34**, 466–519.

Gilliam, T. Brzustowicz, L., Castilla, L. *et al.* (1990) Genetic homogeneity between acute and chronic forms of spinal muscular atrophy. *Nature* **345**, 823–825.

Goldstein, G. and Hoffmann, W. W. (1968) Electrophysiological changes similar to those of myasthenia gravis in rats with experimental autoimmune thymitis. *J. Neurol. Neurosurg. Psychiat.* **31**, 453–459.

Goldstein, G. and Whittingham, S. (1966) Experimental autoimmune thymitis. An experimental model of myasthenia gravis. *Lancet* **ii**, 315–318.

Greensmith, L., Hassan, H and Vrbová, G. (1994) Nerve injury increases the susceptibility of motoneurones to MNDA induced neurotoxicity. *Neurosci.* **58**, 727–733.

Greensmith, L. and Vrbová, G. (1992) Alterations of nerve–muscle interactions during postnatal development influence motoneurone survival in rats. *Dev. Brain Res.* **69**, 125–131.

Greensmith, L., Sieradzan, K and Vrbová, G. (1993) Possible consequences of disruption of neuromuscular contacts in early development for motoneurone survival. *Acta Neurobiol. Exp.* **53**, 319–324.

Gutmann, E. and Hanzlíková, V. (1976) Fast and slow motor units in aging. *Gerontology* **22**, 280–300.

Hausmanowa-Petrusewicz, I. (1989) A research strategy for the resolution of childhood spinal muscular atrophy, in *Current Concepts in Childhood Spinal Muscular Atrophy* (eds L. Merlini, C. Grouche and V. Dubowitz), Springer-Verlag, pp. 21–33.

Henderson, Ch. E., Camu, W., Mettling, C. *et al.* (1993) Neurotrophins promote motoneurone survival and are present in embryonic limb bud. *Nature* **363**, 266–270.

Hoffman, E. P., Brown, R. H. and Kunkel, L. M. (1987) Dystrophin: the protein product of the Duchenne muscular dystrophy locus. *Cell* **51**, 919–928.

Hník, P. (1966) *Muscle Atrophy,* Czechoslovak Academy of Science Publishing House, Prague.

Johnson, A. G. and Woolfe, A. L. (1965) Replacement at the neuromuscular junction of the terminal axonic expansion by the Schwann cell. *Acta Neuropath.* **4**, 436–441.

Jolly, F. (1895) Über Myasthenia gravis pseudo-paralytica. *Berliner Klinische Wochenschrift* **32**, 1–7.

Jones, R., Rees, D. P. and Campbell, M. J. (1994) Tibialis anterior surface EMG parameters change before force output in multiple sclerosis patients. *Clinical Rehab.* **8** (2) 10–16.

Katz, B. (1966) *Nerve Muscle and Synapse*, McGraw Hill, New York.

Koenig, M., Hoffman, E., Bertelson, C. *et al.* (1987) Complete cloning of the Duchenne muscular dystrophy (DMD) cDNA and preliminary genomic organisation of the DMD gene in normal and affected individuals. *Cell* **50**, 509–517.

Lenman, A. J. R., Tulley, F. M., Vrbová, G. *et al.* (1989) Muscle fatigue in some neurological disorders. *Muscle and Nerve* **12**, 938–942.

Lennon, V. A., Lindström, M. J. and Seybold, M. E. (1975) Experimental autoimmune myasthenia: A model of myasthenia gravis in rats and guinea pigs. *J. Exp. Med.* **141**, 1365–1375.

Lennon, V. A., Lindström, M. J. and Seybold, M. E. (1976) Experimental autoimmune myasthenia gravis: Cellular and hormonal immune responses. *Ann. New York Acad. Sci.* **274**, 283–299.

Lindström, J., Shelton, J. and Fuigi, Y. (1988) Myasthenia Gravis. *Adv. Immunol.* **42**, 233–284.

Lowrie, M. B., Subramaniam, K. and Vrbová, G. (1987) Permanent changes in muscle and motoneurones induced by nerve injury during a critical period of development of the rat. *Devel. Brain Res.* **31**, 91–101.

Lowrie, M. B. and Vrbová, G. (1992) Dependence of postnatal motoneurones on their targets: a review and hypothesis. *TINS* **15**, 75–80.

McComas, A. J. (1977) *Neuromuscular Function and Disorders*, Butterworth, London.

Mentis, G. Z., Greensmith, L. and Vrbová, G. (1993) Motoneurones destined to die are rescued by blocking N-Methyl-D-Aspartate receptors by MK–801. *Neurosci.* **54**, 283–285.

Melki, J., Seth, P., Abdelhak, S. *et al.* (1990a) Mapping of acute (Type 1) spinal muscular atrophy to chromosome 5q12-q14. *Lancet* **336**, 271–273.

Melki, J., Abdelhak, S., Seth, P. *et al.* (1990b) Gene for chronic proximal spinal muscular atrophies maps to chromosome 5q. *Nature* **344**, 767–768.

Monaco, A. P., Neve, R., Colelhi-Feener, R. *et al.* (1986) Isolation of candidate cDNAs for portions of the Duchenne muscular dystrophy gene. *Nature* **232**, 646–650.

Navarrete, R. and Vrbová, G. (1984) Differential effect of nerve injury at birth on the activity pattern of reinnervated slow and fast muscles of the rat. *J. Physiol. (Lond.)* **351**, 675–685.

Navarrete, R. and Vrbová, G. (1993) Activity dependent interactions between motoneurones and muscles: Their role in the development of the motor unit. *Prog. Neurobiol.* **41**, 93–124.

Partridge, T. A., Morgan, J. E., Coulton, G. R. *et al.* (1989) Conversion of mdx myofibres from dystrophin negative to positive by injection of normal myoblasts. *Nature* **337**, 176–179.

Patrick, J. and Lindström, M.J. (1973) Autoimmune reponse to acetylcholine receptor. *Science* **180**, 871–872.

Rice, C. L., Vollmer, T. L. and Bigland-Ritchie, B. (1992) Neuromuscular reponses of patients with multiple sclerosis. *Muscle and Nerve* **15**, 1123–1132.

Sacco, P., Jones, D. A., Dick, J. T. R. and Vrbová, G. (1992) The contractile properties and susceptibility to exercise induced damage of normal and *mdx* mouse tibialis anterior muscle. *Clin. Sci.* **82**, 227–236.

Santa, T., Engel, A. G. and Lambert, E. H. (1972a) Histometric study of neuro-muscular junction ultrastructure. I. Myasthenia gravis. *Neurology* **22**, 370–376.

Santa, T., Engel, A. G. and Lambert, E. H. (1972b) Histometric study of neuro-muscular ultrastructure. II. Myasthenic syndrome. *Neurology* **22**, 370–376.

Sendtner, M., Kreutzberg, G. W. and Thoenen, H. (1990) Ciliary neurotrophic factor prevents the degeneration of motoneurones by axotomy. *Nature* **345**, 440–441.

Sendtner, M., Holtman, B., Kolbeck, R. *et al.* (1992) Brain derived neurotrophic factor prevents the death of motoneurones in newborn rats after nerve section. *Nature* **360**, 757–759.

Simpson, J. A. (1960) Myasthenia Gravis: a new hypothesis. *Scottish Med. J.* **5**, 419–436.

Simpson, J. A. (1966) Myasthenia Gravis as an autoimmune disease: clinical aspects, in *Myasthenia Gravis* (ed. K.E. Ossaman), *Ann. NY Acad. Sci.* **135**, 506–516.

Stedman, H. H., Sweeney, H. L., Steuger, J. B. *et al.* (1991) The *mdx* mouse diaphragm reproduces the degenerative changes of Duchenne muscular dys-trophy. *Nature* **352**, 536–539.

Stein, R. B., Gordon, T., Jefferson, J. *et al.* (1992) Optimal stimulation of paralysed muscle after human spinal cord injury. *J. Appl. Physiol.* **72(4)**, 1393–1400.

Vincent, A., Lang, B. and Newsom-Davis, J. (1989) Autoimmunity to the voltage-gated calcium channel underlies the Lambert-Eaton myasthenic syndrome, a para-neoplastic disorder. *TINS* **12**, 496–502.

Vrbová, G. and Lowrie, M. B. (1989) The role of nerve–muscle interactions in the pathogenesis of SMA, in *Current Concepts in Childhood Spinal Muscular Atrophy* (eds L. Merlini, C. Grouche and V. Dubowitz), Springer-Verlag, pp. 33–43.

Walker, M. (1934) Treatment of myasthenia gravis with prostigmine. *Lancet* **226**, 1200–1201.

Walton, J. N. and Gardner Medwin, D. (1974) Progressive muscular dystrophy and the myotonic disorders, in *Disorders of Voluntary Muscle* (ed. J. N. Walton), Churchill Livingstone, Edinburgh.

Williams, A. C. (1994) *Motoneurone Disease*, Chapman & Hall, London.

Wittaker, V. P. and Gray, E. G. (1962) The synapse – Biology and morphology. *Brit. Med. Bull.* **18**, 223–228.

Yan, Q., Elliot, J. and Snider, W. D. (1992) Brain derived neurotrophic factor rescues neurones from axotomy induced cell death. *Nature* **360**, 753–755.

Young, R. R. and Shahani, B. T. (1980), in *Spasticity Disordered Motor Control* (eds R. G. Feldman, R. R. Young and W. P. Koella), Year Book Medical Publishers, Chicago, pp. 206–219.

Zachs, S. I., Bauer, W. C. and Blumberg, J. J. (1961) Abnormalities in the fine structure of the neuromuscular junction in patients with myasthenia gravis. *Nature* **190**, 280–281.

Index

Page numbers in **bold** type refer to figures